From Bench to Bedside: Motor-Cognitive Interactions

Contents

brain sciences

MDPI

Editorial

From Bench to Bedside: Motor–Cognitive Interactions

Daniele Corbo

Department of Medical and Surgical Specialties, Radiological Sciences and Public Health, University of Brescia, 25123 Brescia, Italy; daniele.corbo@unibs.it

1. Introduction

The capacity of humans to learn new motor abilities is known as motor learning, which is often understood as increasing movement precision over time and space through repetition [1–3]. Theories suggest that motor learning involves cognitive processes such as working memory and is not only a physical function [4–6]. Although motor and cognitive deficits are often studied separately, a connection between the two is emphasized in our growing understanding of the task-dependent interaction between the motor and cognitive systems, each of which has different neuroanatomic substrates [7–10]. The importance of motor–cognitive interactions in neurodegenerative illnesses and other clinical groups, including dementia, stroke, Parkinson's disease, and multiple sclerosis, is becoming increasingly evident. As such, considerable effort has been devoted to creating rehabilitative procedures that focus on motor–cognitive connections to address the circumstances associated with these disorders.

The ten papers and six reviews in this Special Issue of *Brain Sciences* provide an intriguing and well-matched mixture of all these research areas in terms of understanding the fundamental processes of motor–cognitive connections and novel therapies. The first group of papers focuses on experimental and clinical investigations dealing with relevant aspects of the motor–cognitive interactions, and a second group of articles consists of review papers, which collate the existing literature on various aspects concerning the connections between motor skills and neural effects and their possible clinical applications. The articles in this collection, based on the topics covered, can be classified into three sugroups: mechanisms of cognitive–motor interactions, diagnostic tools, and intervention strategies.

Citation: Corbo, D. From Bench to Bedside: Motor–Cognitive Interactions. *Brain Sci.* **2024**, *14*, 886. https://doi.org/10.3390/brainsci14090886

Received: 19 August 2024
Revised: 22 August 2024
Accepted: 23 August 2024
Published: 30 August 2024

2. Cognitive–Motor Interaction Mechanisms

Xiao et al. (contribution 1) used a mouse-tracking technique to analyze the hand motions of participants to investigate the role of attention in subliminal semantic processing. Their findings suggest that the temporal–spatial features extracted from cursor motion trajectories can reliably reveal subliminal semantic processing and attentional status, proving that, for a wide range of topics, the mouse-tracking approach is a suitable tool for uncovering implicit dynamic cognition in future studies.

In an investigation into the brain mechanisms underpinning the perception of others' activities, Urgen et al. (contribution 2) provided a series of videos featuring 100 human behaviors captured in real environments. The study observation of the 100 events triggered a well-established action observation network, and they used fMRI to validate the dataset. This extensive collection of videos is a valuable tool for studying the brain and perception.

Dahm et al. (contribution 3) investigated whether leg vs. arm left–right judgments are harder and if limb type affects these judgments. A combined score for accuracy and speed was investigated to further avoid any trade-offs and accurately assess each subject's unique ability. They concluded that realistic stimulus material enhances the effects of perspective and facilitates the understanding of tasks. The linear speed–accuracy score was found to be a reliable indicator of performance in mental body rotations by repeating earlier research findings.

The flexibility of the peripersonal area was examined by Ferroni et al. (contribution 4) both before and after actual or virtual motor training using a tool. Their findings demonstrate that the peripersonal area only expands in response to the use of real-world tools, not in response to virtual ones, underscoring the possibility that the two forms of training depend on distinct mechanisms. The state of the art regarding the malleability of the peripersonal area in both real and virtual environments is enhanced by this study. The authors discuss about their data's applicability to the creation of training, learning, and rehabilitation immersive environments.

Van Hove et al. (contribution 6) aimed to examine how speech production and cognitive load levels affect timed up and go (TUG) and static equilibrium tasks. They assessed the impact of speech production (SP), cognitive load (CL), and dual-task cost (DTC) on these variables. The center of pressure oscillation velocity during static balancing was substantially higher when both tasks were completed orally than in the control scenario. The cognitive load was linked to an increase in TUG, but the oral or mental component did not appear to have any impact. SP more strongly impacted mobility in complicated cognitive tasks. This might be crucial for test selection and comprehending abnormalities in postural control.

3. Diagnostic Tools

The discriminating value of the trail-walking test was assessed by Klotzbier et al. (contribution 5) as a prospective diagnostic tool to enhance the prediction ability during clinical evaluation regarding the severity of Parkinson's disease and documenting the many walking-related characteristics. Patients with Parkinson's disease who exhibit postural instability or problems with walking must be diagnosed using reliable, thoroughly assessed clinical tests that provide an accurate evaluation of each patient's unique fall risk, illness severity, and disease prognosis. The trail-walking test allows for the distinction of motor phenotypes in Parkinson's disease and covers a variety of mobility-related topics, including the link between walking and cognitive functioning.

Chen et al. (contribution 7) created a prediction model of cognitive degradation in patients with Parkinson's disease using a machine learning technique. The clinical information, plasma biomarkers, and cognitive test results of people with Parkinson's disease were gathered as model predictors. Machine learning techniques such as principal component analysis and support vector machines were used to create a cognitive categorization model. The classifier achieved an accuracy of 92.3% and an area under the receiver operating characteristic curve (AUC) of 0.929 employing 32 comprehensive predictive criteria. Furthermore, with 13 well-selected features, the accuracy of the classifier was increased to 100% and the AUC to 1.0. The priorities for future perspectives include expanding the sample size and conducted a longitudinal investigation.

Beauchet et al. (contribution 8) investigated the relationship between incident major neurocognitive disorders (MNCDs) in older community-dwelling individuals and the inability to name the date (a sign of cognitive impairment), the use of a walking assistance, and/or a history of falls (a sign of motor impairment). The incidences of MNCDs was found to be higher when the inability to recall the date and the use of a walking assistance and/or a history of falls were combined. This suggests that the combination of items may be used for screening for the risk of MNCD in the older population, particularly for the incidence of AD. This study opens new avenues for identifying MNCD risk and managing its modifiable risk factors due to the ease with which older populations may obtain data on both of these factors.

Corbo et al. (contribution 9) trained and evaluated a unique automated and image-derived scoring system to improve the capability to discriminate the sensorimotor impairments that are predictive of sensorimotor dysfunction with the Luria–Nebraska Neuropsychological Battery (LNNB) for neuro-motor tasks. The conventional scores, which were evaluated and verified by numerous administrators to reduce subjectivity, showed a strong association with the image-derived LNNB task scores (Pearson's correlation > 0.70).

The innovative image-based scoring method distinguished between individuals with poor motility (<mean population values) with 70–83% specificity and 70% sensitivity. The new image-derived LNNB task scores have potential for use in telemedicine and in the timely evaluation of sensorimotor skills and delays.

Chen et al. (contribution 10) investigated the variables linked to the fear of falling (FOF) in people with moderate cognitive impairment (MCI) owing to Parkinson's disease (PD-MCI) and minor cognitive impairment (MCI) owing to Alzheimer's disease (AD-MCI). In the AD-MCI group, the FOF was strongly linked with gait speed, stride length, Tinetti assessment scale score, executive function, attention and working memory, and global cognitive function. Furthermore, the primary causes of the FOF were working memory and attention. The FOF strongly correlated with both gait speed and timed up and go subtask performance in the PD-MCI group. Moreover, the primary cause of the FOF was turn-to-walk behavior. Therapies targeting attention and working memory and turn-to-walk, respectively, may be used to reduce the FOF in people with AD-MCI and PD-MCI.

4. Intervention Strategies

The review of Saviola et al. (contribution 11) summarizes the research on the neuroimaging outcomes of physical therapy in cohorts of patients with psychosis. The twenty-one studies included in this narrative review were all research publications. Saviola et al. suggest that physical intervention is now considered the standard for helping patients with psychosis experience brain alterations. This means that physical intervention is beneficial not only when the disease first manifests but also for enhancing the illness's course and functional result. However, additional data are required to further the understanding of the long-term plastic reorganization of the psychotic brain, particularly in areas of the brain that have not been thoroughly studied, including motor circuits.

Xiao et al. (contribution 12) summarized the available data on the impact of dual-task training on motor and cognitive skills in patients with Parkinson's disease to support the therapeutic practice of dual-task training. The present views on the mechanism underlying the interplay between motor and cognitive training were also covered. In summary, dual-task training can help people with PD with varying lengths of illness to enhance their motor performance. Dual-task training can help with balance, single-task steep length, single-task gait speed, objective experience of gait freezing in Parkinson's disease, and motor symptoms. This review has several restrictions as well: Because the control intervention and dual-task training design differed among the studies, studies that were not written in English were excluded. Additionally, study quality varied because both RCT and non-RCT studies were included.

Deste et al. (contribution 13) conducted an overview of the literature on physical exercise as a treatment for cognitive impairment in schizophrenia and of the studies that combined physical exercise and cognitive remediation as an integrated rehabilitation intervention. More research is currently required to better understand how to incorporate physical exercise and cognitive remediation in psychiatric rehabilitation practice, even though these interventions seem to be effective treatments for cognitive impairment in people with schizophrenia.

Pertichetti et al. (contribution 14) systematically reviewed the scientific literature on both neuropsychological tests and fMRI tasks for preoperative planning. Changes in functions during the neuropsychological evaluation may assist in identifying patients who can benefit from fMRI and, potentially, functions that should be examined, according to the correlation between the findings of the two tests. fMRI and neuropsychological testing play complimentary roles in the preoperative evaluation. The small number of studies that satisfied the inclusion criteria is the main constraint of this study. This dearth of information is a reflection of the diversity of the literature in terms of behavioral experimental design, neuropsychological testing, fMRI investigations, and particular objectives examined.

Kamińska et al. (contribution 15) assessed the efficacy of different treadmill training outcomes in individuals with Down syndrome (DS), including adults and children. With a total of 687 people, they chose 25 trials for analysis and found 25 distinct results, which are then narratively presented. They found favorable benefits in every case, with the treadmill training being the most effective. People with DS see improvements in their physical and emotional health when they incorporate treadmill exercise into their regular physiotherapy regimen.

Jylänki et al. (contribution 16) conducted a systematic review to better understand the methodological quality and the impact of physical exercise and fundamental motor skill therapies on academic and cognitive skills in 3- to 7-year-old children with special educational needs. The effects of the intervention seemed to vary depending on the severity of the learning difficulty. Regarding language and cognitive skills, children who were at risk because of their family background benefited the most from the intervention, whereas children with learning disabilities benefited most in terms of executive functions. However, providing a broadly applicable summary of the results is difficult because of the wide variation in the included studies and the relatively low methodological quality. Therefore, more thorough studies are needed to evaluate the efficacy of these therapies.

Taken together, the papers gathered in this Special Issue of *Brain Sciences* dealing with motor–cognitive interactions should therefore be of considerable interest for neuroscientists interested in understanding of key mechanisms of motor–cognitive interactions and innovative treatments.

Funding: This study received no external funding.

Conflicts of Interest: The author declares no conflicts of interest.

List of Contributions:

1. Xiao, K.; Zhang, A.; Qu, J.; Deng, F.; Guo, C.; Yamauchi, T. Hand Motions Reveal Attentional Status and Subliminal Semantic Processing: A Mouse-Tracking Technique. *Brain Sci.* **2023**, *13*, 1267. https://doi.org/10.3390/brainsci13091267.
2. Urgen, B.A.; Nizamoğlu, H.; Eroğlu, A.; Orban, G.A. A Large Video Set of Natural Human Actions for Visual and Cognitive Neuroscience Studies and Its Validation with fMRI. *Brain Sci.* **2023**, *13*, 61. https://doi.org/10.3390/brainsci13010061.
3. Dahm, S.F.; Muraki, E.J.; Pexman, P.M. Hand and Foot Selection in Mental Body Rotations Involves Motor-Cognitive Interactions. *Brain Sci.* **2022**, *12*, 1500. https://doi.org/10.3390/brainsci12111500.
4. Ferroni, F.; Gallese, V.; Soccini, A.M.; Langiulli, N.; Rastelli, F.; Ferri, D.; Bianchi, F.; Ardizzi, M. The Remapping of Peripersonal Space in a Real but Not in a Virtual Environment. *Brain Sci.* **2022**, *12*, 1125. https://doi.org/10.3390/brainsci12091125
5. Klotzbier, T.J.; Schott, N.; Almeida, Q.J. Profiles of Motor-Cognitive Interference in Parkinson's Disease—The Trail-Walking-Test to Discriminate between Motor Phenotypes. *Brain Sci.* **2022**, *12*, 1217. https://doi.org/10.3390/brainsci12091217.
6. Van Hove, O.; Pichon, R.; Pallanca, P.; Cebolla, A.M.; Noel, S.; Feipel, V.; Deboeck, G.; Bonnechère, B. Influence of Speech and Cognitive Load on Balance and Timed up and Go. *Brain Sci.* **2022**, *12*, 1018. https://doi.org/10.3390/brainsci12081018.
7. Chen, P.-H.; Hou, T.-Y.; Cheng, F.-Y.; Shaw, J.-S. Prediction of Cognitive Degeneration in Parkinson's Disease Patients Using a Machine Learning Method. *Brain Sci.* **2022**, *12*, 1048. https://doi.org/10.3390/brainsci12081048.
8. Beauchet, O.; Matskiv, J.; Rolland, Y.; Schott, A.-M.; Allali, G. Using Interaction between Cognitive and Motor Impairment for Risk Screening of Major Neurocognitive Disorders: Results of the EPIDOS Observational Cohort Study. *Brain Sci.* **2022**, *12*, 1021. https://doi.org/10.3390/brainsci12081021.
9. Corbo, D.; Placidi, D.; Gasparotti, R.; Wright, R.; Smith, D.R.; Lucchini, R.G.; Horton, M.K.; Colicino, E. The Luria-Nebraska Neuropsychological Battery Neuromotor Tasks: From Conventional to Image-Derived Measures. *Brain Sci.* **2022**, *12*, 757. https://doi.org/10.3390/brainsci12060757.
10. Chen, P.-H.; Yang, Y.-Y.; Liao, Y.-Y.; Cheng, S.-J.; Wang, P.-N.; Cheng, F.-Y. Factors Associated with Fear of Falling in Individuals with Different Types of Mild Cognitive Impairment. *Brain Sci.* **2022**, *12*, 990. https://doi.org/10.3390/brainsci12080990.

11. Saviola, F.; Deste, G.; Barlati, S.; Vita, A.; Gasparotti, R.; Corbo, D. The Effect of Physical Exercise on People with Psychosis: A Qualitative Critical Review of Neuroimaging Findings. *Brain Sci.* **2023**, *13*, 923. https://doi.org/10.3390/brainsci13060923.
12. Xiao, Y.; Yang, T.; Shang, H. The Impact of Motor-Cognitive Dual-Task Training on Physical and Cognitive Functions in Parkinson's Disease. *Brain Sci.* **2023**, *13*, 437. https://doi.org/10.3390/brainsci13030437.
13. Deste, G.; Corbo, D.; Nibbio, G.; Italia, M.; Dell'Ovo, D.; Calzavara-Pinton, I.; Lisoni, J.; Barlati, S.; Gasparotti, R.; Vita, A. Impact of Physical Exercise Alone or in Combination with Cognitive Remediation on Cognitive Functions in People with Schizophrenia: A Qualitative Critical Review. *Brain Sci.* **2023**, *13*, 320. https://doi.org/10.3390/brainsci13020320.
14. Pertichetti, M.; Corbo, D.; Belotti, F.; Saviola, F.; Gasparotti, R.; Fontanella, M.M.; Panciani, P.P. Neuropsychological Evaluation and Functional Magnetic Resonance Imaging Tasks in the Preoperative Assessment of Patients with Brain Tumors: A Systematic Review. *Brain Sci.* **2023**, *13*, 1380. https://doi.org/10.3390/brainsci13101380.
15. Kamińska, K.; Ciołek, M.; Krysta, K.; Krzystanek, M. Benefits of Treadmill Training for Patients with Down Syndrome: A Systematic Review. *Brain Sci.* **2023**, *13*, 808. https://doi.org/10.3390/brainsci13050808.
16. Jylänki, P.; Mbay, T.; Byman, A.; Hakkarainen, A.; Sääkslahti, A.; Aunio, P. Cognitive and Academic Outcomes of Fundamental Motor Skill and Physical Activity Interventions Designed for Children with Special Educational Needs: A Systematic Review. *Brain Sci.* **2022**, *12*, 1001. https://doi.org/10.3390/brainsci12081001.

References

1. Hands, B.; McIntyre, F.; Parker, H. The General Motor Ability Hypothesis: An Old Idea Revisited. *Percept. Mot. Ski.* **2018**, *125*, 213–233. [CrossRef] [PubMed]
2. Corbo, D.; Orban, G.A. Observing Others Speak or Sing Activates Spt and Neighboring Parietal Cortex. *J. Cogn. Neurosci.* **2017**, *29*, 1002–1021. [CrossRef] [PubMed]
3. Mao, X.; Huang, S.; Ouyang, M.; Xie, Y.; Tan, X. Effect of skill proficiency on motor imagery ability between amateur dancers and non-dancers. *Front. Psychol.* **2022**, *13*, 899724. [CrossRef] [PubMed] [PubMed Central]
4. Stuhr, C.; Hughes, C.M.L.; Stöckel, T. Task-specific and variability-driven activation of cognitive control processes during motor performance. *Sci. Rep.* **2018**, *8*, 10811. [CrossRef] [PubMed] [PubMed Central]
5. Latino, F.; Tafuri, F. Physical Activity and Cognitive Functioning. *Medicina* **2024**, *60*, 216. [CrossRef] [PubMed] [PubMed Central]
6. Bläsing, B.; Calvo-Merino, B.; Cross, E.S.; Jola, C.; Honisch, J.; Stevens, C.J. Neurocognitive control in dance perception and performance. *Acta Psychol.* **2012**, *139*, 300–308. [CrossRef] [PubMed]
7. Chou, Y.-H.; Sundman, M.; That, V.T.; Green, J.; Trapani, C. Cortical excitability and plasticity in Alzheimer's disease and mild cognitive impairment: A systematic review and meta-analysis of transcranial magnetic stimulation studies. *Ageing Res. Rev.* **2022**, *79*, 101660. [CrossRef] [PubMed] [PubMed Central]
8. Ryman, S.G.; Poston, K.L. MRI biomarkers of motor and non-motor symptoms in Parkinson's disease. *Park. Relat. Disord.* **2020**, *73*, 85–93. [CrossRef] [PubMed]
9. Argento, O.; Piacentini, C.; Bossa, M.; Caltagirone, C.; Santamato, A.; Saraceni, V.; Nocentini, U. Motor, cognitive, and combined rehabilitation approaches on MS patients' cognitive impairment. *Neurol. Sci.* **2023**, *44*, 1109–1118; Erratum in *Neurol. Sci.* **2023**, *44*, 1835. https://doi.org/10.1007/s10072-023-06630-1. [CrossRef] [PubMed]
10. Wu, Q.; Chan, J.S.; Yan, J.H. Mild cognitive impairment affects motor control and skill learning. *Rev. Neurosci.* **2016**, *27*, 197–217. [CrossRef] [PubMed]

brain sciences

MDPI

Systematic Review

Neuropsychological Evaluation and Functional Magnetic Resonance Imaging Tasks in the Preoperative Assessment of Patients with Brain Tumors: A Systematic Review

Marta Pertichetti, Daniele Corbo, Francesco Belotti, Francesca Saviola, Roberto Gasparotti and Marco Maria Fontanella et al.

[1] Neurosurgery Unit, Department of Medical and Surgical Specialties, Radiological Sciences and Public Health, University of Brescia and ASST Spedali Civili Hospital, 25123 Brescia, Italy; marco.fontanella@unibs.it (M.M.F.); pierpaolo.panciani@unibs.it (P.P.P.)
[2] Department of Medical and Surgical Specialties, Radiological Sciences and Public Health, University of Brescia, 25123 Brescia, Italy; daniele.corbo@unibs.it (D.C.); francesca.saviola@unibs.it (F.S.); roberto.gasparotti@unibs.it (R.G.)
[3] Neuroradiology Unit, ASST Spedali Civili of Brescia, 25123 Brescia, Italy
* Correspondence: frbelotti90@gmail.com; Tel.: +39-030-3995585

Abstract: Background: Current surgical treatment of gliomas relies on a function-preserving, maximally safe resection approach. Functional Magnetic Resonance Imaging (fMRI) is a widely employed technology for this purpose. A preoperative neuropsychological evaluation should accompany this exam. However, only a few studies have reported both neuropsychological tests and fMRI tasks for preoperative planning—the current study aimed to systematically review the scientific literature on the topic. Methods: PRISMA guidelines were followed. We included studies that reported both neuropsychological tests and fMRI. Exclusion criteria were: no brain tumors, underage patients, no preoperative assessment, resting-state fMRI only, or healthy sample population/preclinical studies. Results: We identified 123 papers, but only 15 articles were included. Eight articles focused on language; three evaluated cognitive performance; single papers studied sensorimotor cortex, prefrontal functions, insular cortex, and cerebellar activation. Two qualitative studies focused on visuomotor function and language. According to some authors, there was a strong correlation between performance in presurgical neuropsychological tests and fMRI. Several papers suggested that selecting well-adjusted and individualized neuropsychological tasks may enable the development of personalized and more efficient protocols. The fMRI findings may also help identify plasticity phenomena to avoid unintentional damage during neurosurgery. Conclusions: Most studies have focused on language, the most commonly evaluated cognitive function. The correlation between neuropsychological and fMRI results suggests that altered functions during the neuropsychological assessment may help identify patients who could benefit from an fMRI and, possibly, functions that should be tested. Neuropsychological evaluation and fMRI have complementary roles in the preoperative assessment.

Keywords: glioma; brain tumor; surgery; fMRI; neuropsychological evaluation

check for updates

Citation: Marta Pertichetti, Daniele Corbo, Francesco Belotti, Francesca Saviola, Roberto Gasparotti and Marco Maria Fontanella et al. Neuropsychological Evaluation and Functional Magnetic Resonance Imaging Tasks in the Preoperative Assessment of Patients with Brain Tumors: A Systematic Review. *Brain Sci.* **2023**, *13*, 1380. https://doi.org/10.3390/brainsci13101380

Academic Editor: Xiaoming Jiang

Received: 24 August 2023
Revised: 19 September 2023
Accepted: 25 September 2023
Published: 28 September 2023

1. Introduction

Patients diagnosed with brain tumors may have variable prognoses influenced by histology and the molecular profile of the neoplastic formation, and by the degree of resection achieved in the tumor during surgery [1–3]. Indeed, among the several prognostic factors suggested in the literature, the extent of resection (EOR) based on objective tumoral volume analysis seems to be one of the main predictors of overall survival [4–7]. Surgical treatment, however, can only rarely be considered radical due to the infiltrating nature of gliomas per se [2]. Therefore, in recent years, treatment paradigms shifted from surgery focused on gross-total resection (GTR) to a maximally safe and function-preserving resection approach [8–11]. This transition was possible thanks to changes in the clinical approach by applying technological and conceptual innovations to improve safety during surgery, such as intraoperative ultrasound, cortical mapping, awake surgery, and tumor margin detection with fluorescence dye [4,5,12–14]. Parallel to this, the preoperative evaluation of patients and accurate presurgical planning also improved, including different noninvasive methods to identify the relationship between the brain tumor and eloquent areas, both at the cortical and subcortical levels [15].

Firstly, Functional Magnetic Resonance Imaging (fMRI) started to be widely employed for this purpose [16,17], allowing the creation of a functional map of the eloquent brain regions based on modifications in blood oxygenation levels by using the blood oxygen level-dependent (BOLD) contrast [18,19]. Specifically, fMRI can be employed at rest while the subject is lying still in the scanner and instructed to think about nothing in particular. In this case, fMRI measures the inter-regional dependencies across the brain and can be applied in presurgical functional mapping [20]. Otherwise, fMRI can be employed during the execution of a cognitive task (i.e., task-based fMRI). Task-based fMRI compares BOLD signal changes while performing specific tasks to baseline conditions, assuming that increased cerebral blood flow reflects neuronal activity [21]. Among several potential tasks suitable for fMRI experiments, a defined group of tasks are commonly applied for presurgical mapping in gliomas: sensorimotor, language-related, and executive function tasks [19,22,23]. In this context, the sensorimotor paradigm demonstrated high reliability [19,24,25], whereas mapping language and other higher cognitive functions is more debated regarding both the anatomical specificity and the paradigm's sensitivity [18,19,26]. Paradigms for identifying visual and somatosensory areas have also been proposed.

Secondly, the neuropsychological evaluation is another important aspect regarding the preoperative assessment of patients with brain tumors [27,28]. Patients can develop impairments in multiple cognitive domains before or after surgery [29]. Therefore, the evaluation of cognitive function is essential for informing management and monitoring the long-term effects of tumors [29]. However, the wide range of existing tests reflects the fractionation of the cognitive system, and an in-depth assessment can take several hours. Furthermore, the wide range and variety of available tests may lead to reduced overlap between those used from one center to another, making the comparison of outcomes complex [29]. The presurgical combination of fMRI and neuropsychological assessment should help define (i) the tumor's anatomical features, such as tumor site compared to fMRI-positive areas, plasticity phenomena, and prediction of EOR; (ii) the tumor's functional effects on cognition. These data not only inform the surgeon during surgical planning, but also suggest the possible cognitive outcome and potential recovery. Preoperative findings can also be compared postoperatively to assess surgical results and to monitor cognitive functions during the follow-up. Nevertheless, only a few studies reported the tests included in the neuropsychological assessment together with the fMRI tasks for the preoperative planning.

Previous reviews focused on the multimodal MRI assessment of healthy brain aging and neurodegenerative diseases [30,31], but the literature focusing on patients with gliomas is scarce. In this context, motivated by the need for defined functional treatment protocols in brain tumor surgery, this review specifically targeted this patient population. The current study aimed to systematically review scientific evidence to identify studies including both the preoperative neuropsychological assessment and the fMRI task protocol, contributing to improving the neurosurgical treatment of gliomas.

2. Materials and Methods

Preferred Reporting Items for Systematic Reviews and Metanalyses (PRISMA) guidelines were followed for the systematic review [32,33]. A systematic search of the PubMed electronic database was conducted in February 2023 by cross-matching the following keywords: functional, magnetic resonance imaging, fMRI, MRI, brain tumor, glioma, task, neuropsy*. We included English studies published before February 2023. After duplicate removal, two researchers (MP and FB) independently reviewed titles and abstracts to identify articles of interest. Disagreement was resolved with a discussion that involved a third researcher (DC). We included studies that reported both neuropsychological tests and functional neuroimaging studies. Exclusion criteria included tumors not affecting the brain, secondary brain tumors, patients under 18 years old (given that cognitive results may be affected by developmental brain plasticity mechanisms), no preoperative assessment, resting state fMRI only, or healthy sample population/preclinical studies.

The articles were then evaluated, looking for correlations between neuropsychological assessment and fMRI task results, defined as the quantitative outcome. Papers reporting tests and tasks but without specifying the results attained by the patients in them, or exclusively showing the results of the postoperative assessment, were only included qualitatively according to the PRISMA guidelines (i.e., not meeting the review criteria but reporting additional beneficial results). The current review has been registered in the Open Science Framework (OSF) registry (https://doi.org/10.17605/OSF.IO/8DCZG; accessed on 10 June 2023).

The "Risk Of Bias In Systematic reviews" (ROBIS) assessment tool was employed to check for bias in the review process [34].

3. Results

We identified 123 papers after duplicate removal. After title and abstract analysis, 62 articles were identified for full-text analysis. Eligibility evaluation led to the inclusion of 15 articles in the systematic review (Figure 1). According to the ROBIS assessment tool, we identified a low risk of bias in the "study eligibility criteria", "identification and selection of studies", and "data collection and study appraisal" domains. Based on the heterogeneity of results and the low number of eligible studies, a high risk of bias was identified in the "synthesis and findings" domain.

Nevertheless, we aimed to address all the concerns while interpreting the findings. We highlighted the relevance of each included study but avoided emphasizing results based on statistical significance. We also state the limitations of the current review in the paper's discussion section.

Detailed results about the included papers and the employed neuropsychological and fMRI tasks are shown in Table 1 (we report specific fMRI tasks and neuropsychological tests extensively in Supplementary Materials). The aims of those studies were rather heterogeneous, as were the methods and conclusions. Only one study was published in 2006, reporting a single patient [35]. All the remaining papers were published after 2010: one in 2010 [36], one in 2011 [37], one in 2014 [38], one in 2015 [39], two in 2017 [40,41], two in 2019 [42,43], three in 2020 [44–46], and three in 2022 [47–49].

Their main results are discussed in the following sections according to the cognitive function investigated in the attempt to draw general conclusions about the clinical relevance of integrating pre-surgical neuropsychological assessment and fMRI. As a qualitative result, we found two more studies reporting only the details of the neuropsychological evaluation administered postoperatively [50,51].

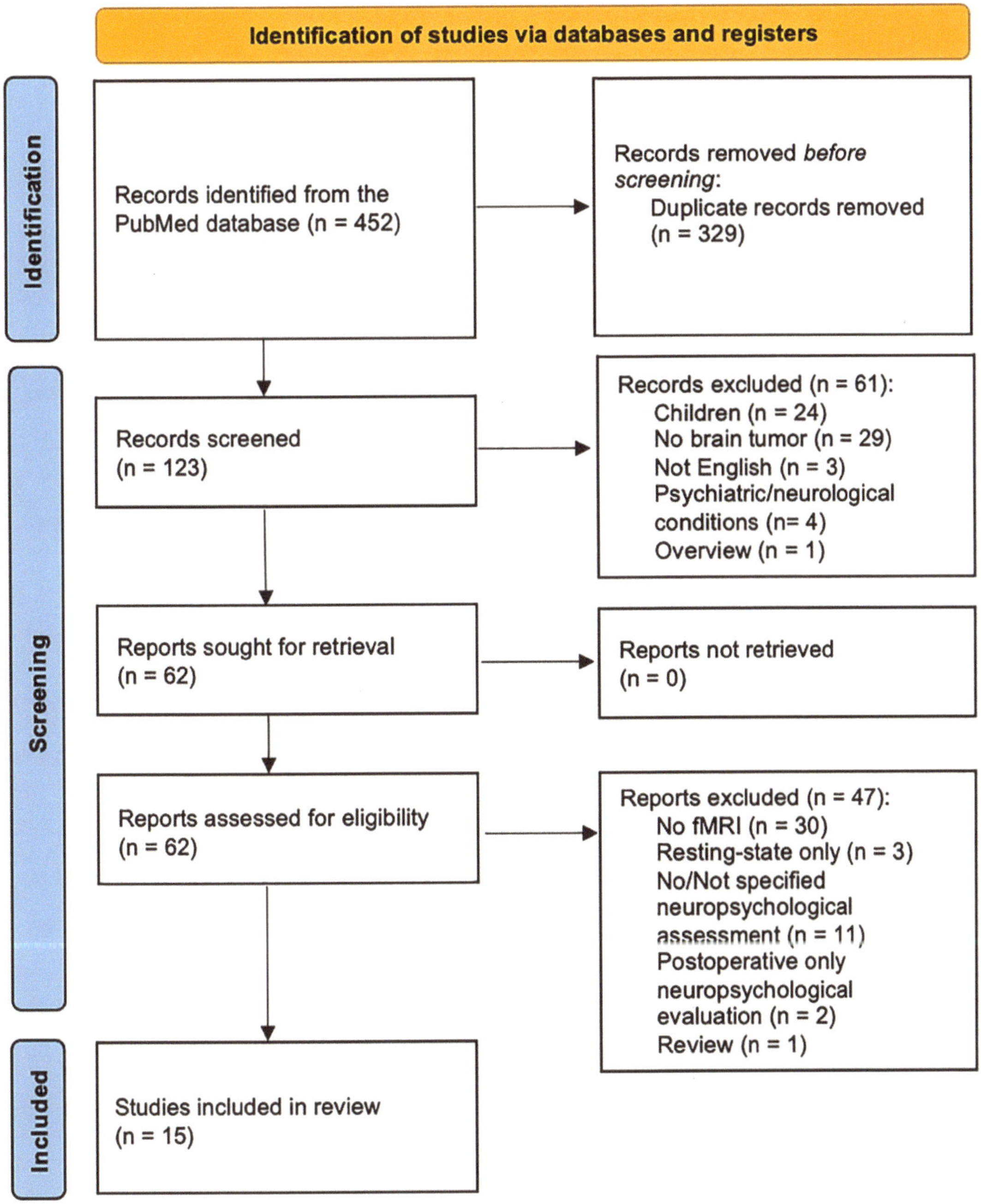

Figure 1. PRISMA flow diagram of the systematic review [33].

Table 1. Results of the systematic review.

Author	Patients (N)	WHO Grade	Hemisphere and Location			Surgery	Cognitive Function Domain	fMRI Tasks	Neuropsychological Tests	Awake Surgery Assessment
			L	R	NA/Both					
[40]	25	18 HGG 7 LGG	7 prefrontal HGG 1 prefrontal LGG	4 prefrontal HGG 3 prefrontal LGG	7 non-prefrontal HGG 3 non-prefrontal LGG	Y	Executive	Go/No-Go task	MMSE, IQ, Verbal short-term memory and working memory Selective and divided attention visuospatial short-term memory Phonemic fluency	N
[44]	20	13 HGG 7 LGG	2 premotor HGG 4 motor HGG 2 sensorimotor HGG 1 parasagittal HGG 2 premotor LGG 2 motor LGG 1 L sensorimotor LGG	1 premotor HGG 2 motor HGG 1 sensorimotor HGG 1 premotor LGG 1 motor LGG	-	N	Sensorimotor Language	Motor localizer tasks, general motor imagery ability, conceptual knowledge of actions, lexical grammar processing, verb naming	Nonverbal intelligence Visuospatial short-term memory Constructional apraxia Visuospatial/constructive ability and planning Attentional neglect Visuoconceptual and visuomotor tracking Verbal short-term memory Buccofacial and ideomotor apraxia Noun naming and phonemic fluency	N
[37]	19	13 HGG 6 LGG	ventrolateral frontal (anterior and posterior groups)	-	-	N	Language	Verb generation task, orofacial apraxia	Phonemic fluency Semantic fluency Orofacial apraxia	N
[42]	26	13 HGG 13 LGG	8 HGG 7 LGG	5 HGG 6 LGG	4 of the HGG were frontal 10 of the LGG were frontal (Side NA)	N	Executive	N-back task	Cognitive flexibility (shifting attention)	N
[48]	1	II	1 fronto-insular	-	-	Y	Executive	Stroop task	Executive functions and attention Working memory Inhibition Mental flexibility Set shifting abilities Verbal fluency (semantic and phonological) Language production and naming Verbal comprehension Insular-related functioning (Empathy scale and emotion recognition) Mood	Y

Table 1. *Cont.*

Author	Patients (N)	WHO Grade	Hemisphere and Location			Surgery	Cognitive Function Domain	fMRI Tasks	Neuropsychological Tests	Awake Surgery Assessment
			L	R	NA/Both					
[35]	1	II	-	1 insular	-	Y	Language	Verb generation task, abstract/concrete categorization	Language functions IQ Memory Visual retention	N
[38]	20	3 HGG 15 LGG	6 frontal 1 fronto-temporal 1 parietal 1 fronto-parietal 2 temporo-occipital 1 temporo-parietal 2 temporal	4 frontal 1 fronto-temporal 1 parietal	-	Y	Language	Verb generation task	IQ Abstract reasoning Cognitive processing speed Executive functioning Attention Working memory	Y
[41]	16	11 HGG 5 LGG	8 frontal 2 insular 1 temporal 1 frontoparietal	2 frontal 1 temporoparietal 1 temporal	-	Y	Executive Sensorimotor Language	N-back task, motor and language tasks	National Institutes of Health Cognitive Battery	N
[45]	18	11 HGG 7 LGG	10 frontal 5 temporal 1 insular 2 parietal	-	-	Y	Language Sensorimotor	Verb generation, semantic and syntactic decision tasks, motor tasks	Picture-naming Nonverbal visual semantic decision Verb-generation task	Y
[47]	15	10 HGG 5 LGG	10 frontal	5 frontal	-	Y	Language	Phonemic fluency task	MMSE, IQ Naming Phonemic verbal fluency Category fluency Short-term verbal memory and episodic memory Visuospatial short-term memory span and long-term visuospatial memory Visuoconstructive and planning abilities Attention and executive functions Depression and anxiety Cognitive reserve	Y

Table 1. *Cont.*

Author	Patients (N)	WHO Grade	Hemisphere and Location			Surgery	Cognitive Function Domain	fMRI Tasks	Neuropsychological Tests	Awake Surgery Assessment
			L	R	NA/Both					
[36]	44	19 HGG 25 LGG	frontal and temporal	-	-	Y	Language	Word generation, picture naming tasks	Non-verbal intelligence Verbal and visuospatial, short- and long-term memory Selective and divided attention Orofacial, ideomotor, and constructional apraxia Spatial cognition Phonemic and semantic fluency Naming tasks Sentence comprehension Repetition	Y
[46]	46	25 HGG 21 LGG	13 HGG 10 LGG (possible involvement of central executive network or default mode network)	12 HGG 10 LGG (possible involvement of central executive network or default mode network)	1 LGG both	N	Executive	N-back task	Cognitive performance	N
[39]	1	III	premotor	-	-	Y	Language	Verb generation task	SMA functions Picture description Semantic and phonological verbal fluency Motor execution Processing speed Working memory Verbs and nouns generation	Y
[49]	19	HGG and LGG	-	3 posterosuperior temporal lobe	16 supratentorial (controls, NA)	Y	Language	Speech perception, object recognition, auditory short-term memory holding	Behavioral testing with language-related and cognitive non-language tasks	N
[43]	23	NA	-	-	23	N	Sensorimotor Language Executive	Sensorimotor processing, language, working memory, executive function, visual function, auditory function	MoCa	N

Legend: HGG, High-grade glioma. IQ, Intelligence quotient. LGG, Low-grade glioma. MMSE, Mini mental state examination. N, No. NA, Not available. WHO, World Health Organization. Y, Yes. MoCa, Montreal Cognitive Assessment.

3.1. Sensorimotor Functions

In the neurological context, the sensorimotor domain aims to integrate the sensory/perceptual component for processing stimuli and the motor response. Within this domain, four of the studies selected in this review reported the fMRI mapping of motor cortices in glioma patients (Table 2).

Most of these studies focused on the general mapping of motor cortices. This result was usually obtained by employing standard and well-established experimental fMRI paradigms (e.g., finger tapping) [43], together with neuropsychological scores investigating motor skills and praxis.

Concerning the investigation of specific functions of the sensorimotor cortex, only two studies were found. The first one, by Argiris et al., suggested how the use of specific motor and sensory neuropsychological tests can be related to the tumor-affected hemisphere. This proposal underlined the concept of functional cerebral lateralization: gliomas located in the right hemisphere are more susceptible to impact visuospatial domains. Right-lateralized gliomas more frequently result in neglect conditions. Therefore, an accurate estimation of correlated neuropsychological visuospatial profile (i.e., visuospatial short-term memory, constructional apraxia, constructive ability, etc.) could help the pre-surgical planning by focusing first on the impairments and then on the related brain functional areas specific for this type of tumors. The same context applies to left-lateralized brain tumors, where the focus will be on linguistic functioning instead. By doing so, the authors showed a discrimination between tumor hemispheric localization in the performance of tasks previously considered unrelated to hemispheric lateralization, such as motor imagery processes [44]. Indeed, left tumor patients presenting a lesion near somatotopic hand representations performed significantly worse on the mental rotation hand fMRI task, correlating with motor-evoked potential (MEP) amplitudes in the upper limb motor region, and highlighting the involvement of the motor system in motor imagery processes [44].

The second study, by Zacharia et al., focused instead on investigating the role the cerebellum plays in cognitive, motor, and emotional functions, potentially acting during the development and refinement of internal models in motor and cognitive functions. Specifically regarding the motor domain, authors employed classic experimental designs to test motor activation (e.g., finger tapping, lip movement, etc.). They demonstrated that besides the presence of gliomas, the cerebellar activation patterns noted on functional MRI and cerebro-cerebellar connections remain intact [43].

Hence, regarding the sensorimotor functions, this novel review provides insights about (i) the importance of considering brain tumor lateralization for an accurate neuropsychological assessment, and shows that (ii) the cerebellar function and its associated cognitive performance seem to be preserved even in the presence of cortical gliomas.

Table 2. Selected studies investigating presurgical sensorimotor functions.

Author	Patients (N)	WHO Grade	Surgery	fMRI Tasks	fMRI Measures	Neuropsychological Tests (Related to Task)	Main Results	Awake Surgery	Task during Awake Surgery
[44]	20	13 HGG 7 LGG	N	Motor localizer tasks General motor imagery ability Conceptual knowledge of actions	Somatotopic cortical mapping (mouth, hand and feet) Imagery questions (joint movement, hands spatial position during action production) Mental rotation task Kissing and Dancing Test	Visuospatial short-term memory Constructional apraxia Visuospatial/constructive ability and planning Attentional neglect Visuoconceptual and visuomotor tracking Buccofacial and ideomotor apraxia	Involvement of the motor system in motor imagery processes	N	
[41]	16	11 HGG 5 LGG	Y	Motor tasks		NA	FPN functional connectivity is related to cognitive outcomes after surgery	Y	NA
[45]	18	11 HGG 7 LGG	Y	Motor tasks	NA	Picture-naming Nonverbal visual semantic decision task Verb-generation task	Navigated fMRI data did not influence DCS in practice	Y	Picture naming, nonverbal visual semantic decision task
[43]	23	NA	N	Sensorimotor processing	Finger tapping Toe movement Lip movement	MoCa (Montreal Cognitive Assessment)	Simultaneous cerebellar activation across different cognitive domains (except visual)	N	

Legend: HGG, High-grade glioma. IQ, Intelligence quotient. LGG, Low-grade glioma. MMSE, Mini-mental state examination. N, No. NA, Not available. WHO, World Health Organization. Y, Yes. DCS, direct cortical stimulation. FPN, Frontoparietal network.

3.2. Language

Language is the cognitive ability to communicate through a system of conventional and structured rules. Given its paramount importance, the linguistic function was the cognitive domain mainly investigated in the presurgical phases of treatment for glioma patients, with ten studies exploring the topic (Table 3).

The results of studies focusing on language were highly heterogeneous, with several experimental paradigms and neurocognitive assessments exploited. However, some main findings can be drawn.

Firstly, a large part of the studies focused on investigating the relationship between these three components: areas belonging to the neoplastic formation, the neuropsychological assessment of linguistic performance, and fMRI mapping of the language network. This mainly ensured the mapping of anatomo-functional language-related relationships within the brain. Preoperatively, the cognitive impairment of linguistic performance was associated with neoplastic formations involving the ventral precentral gyrus and the arcuate fasciculus [37]. Despite this, conserved landmarks of functional *pars opercularis* were observed with task-fMRI in these patients, making the region highly relevant in presurgical planning [37].

Furthermore, gliomas involving the uncinate fasciculus or right temporal lobe also significantly impacted linguistic performance in denomination tasks [36] and speech perception [49], respectively. In this context, fMRI effectively described task response in different sub-components of the language domain, with high specificity. For example, in right superior temporal lobe gliomas, speech perception was characterized by a lower activation within the tumor site and enhanced activation of the contralateral hemisphere, which was reversed during speech production [49]. Additional evidence concerns left-lateralized tumor response during phonemic fluency tasks, which not only exhibited classic activation in left temporal and parietal regions, but included increased activity in frontal regions strongly correlated with the behavioral executive components of this linguistic skill [47].

Additionally, other authors highlighted aspects related to intra-operative language assessment and its relationship with presurgical evaluation. Indeed, Leote et al. focused on the intraoperative consequences of impaired presurgical cognitive performance. They described how presurgical cognitive deficits led to a decreased DCS duration and consequently to lower reliability of the methodology [45], which was also reflected by an ineffective fMRI mapping of the relevant cognitive functions [45].

Moreover, the compensatory capability of the unaffected brain areas was also recently studied postoperatively in the context of the linguistic domain. The functional recovery of language functions seemed to rely on changes in activation near the surgical resection (not in the contralateral hemisphere) [38]. However, these changes in the activation pattern were unrelated to functional variations of the performance following surgery, as measured with neuropsychological testing [38]. Lastly, the histology of the tumors could also play a role in language function preservation, with grading being one of the most impacting factors on cognitive recovery. Mitolo et al. reported that low-grade tumors showed higher rightward frontal operculum fMRI activations and, therefore, better cognitive performance in tests measuring general cognitive abilities, semantic fluency, verbal short-term memory, and executive functions [47].

Therefore, some main findings can be drawn: (i) the anatomo-functional brain correspondence of linguistic performance is not straightforward; (ii) presurgical fMRI can help detect eloquent areas and regions specific for linguistic sub-components otherwise disregarded; (iii) linguistic cognitive impairments must be taken into account for an efficient intra-surgical cortical mapping; (iv) language plasticity processes depend on tumor features and are related to complex diaschisis changes. Nevertheless, results about the language domain across different studies were discrepant, and more robust evidence could be provided only by considering longitudinal studies.

Table 3. Selected studies investigating presurgical linguistic functions.

Author	Patients (N)	WHO Grade	Surgery	fMRI Tasks	fMRI Measures	Neuropsychological Tests (Related to Task)	Main Results	Awake Surgery	Task during Awake Surgery
[44]	20	13 HGG 7 LGG	N	Lexical grammar processing Verb naming	Verbs conjugation discrimination Verb oral naming task from BADA	Nonverbal intelligence Noun naming Phonemic fluency	Lexico-semantic processing of action not compromised by sensorimotor area lesion	N	
[37]	19	13 HGG 6 LGG	N	Verb generation task	Verb generation	Phonemic fluency Semantic fluency Orofacial apraxia	Functional activation of pars opercularis	N	
[35]	1	II	Y	Verb generation task Abstract/concrete categorization	Silent verb generation related to a noun Categorization of a word	Language functions IQ Memory Visual Retention	Activation of left frontal regions	N	
[38]	20	3 HGG 15 LGG	Y	Verb generation task	Covert articulation of a verb related to a noun	IQ	Perilesional functional reorganization of language areas	Y	Motor and language tasks
[45]	18	11 HGG 7 LGG	Y	Verb generation task Semantic and syntactic decision tasks	Silent verb generation related to a noun Judgment of the semantic correctness of phrases	Picture-naming Nonverbal visual semantic decision Verb-generation task	DCS duration is not reduced by the use of fMRI mapping	Y	Picture naming Nonverbal visual semantic decision task
[47]	15	10 HGG 5 LGG	Y	Phonemic fluency task	Covert generation of a noun starting with a given letter	MMSE, IQ Naming Phonemic verbal fluency Category fluency Short-term verbal memory and episodic memory	Left hemispheric dominance in temporal and parietal regions	Y	Specific language tests
[36]	44	19 HGG 25 LGG	Y	Word generation Picture naming tasks	Language dominance	Non-verbal intelligence Orofacial, ideomotor, and constructional apraxia Phonemic and semantic fluency Naming tasks Sentence comprehension Repetition	Role of the uncinate fasciculus in the retrieval of a word form for proper names	Y	Language with blocks of items (living, non-living, faces, verbs)

Table 3. *Cont.*

Author	Patients (N)	WHO Grade	Surgery	fMRI Tasks	fMRI Measures	Neuropsychological Tests (Related to Task)	Main Results	Awake Surgery	Task during Awake Surgery
[39]	1	III	Y	Verb generation task	Covert generation of a verb starting with a given noun	Picture description Semantic and phonological verbal fluency Verbs and nouns generation	No functional change post-surgically in the verb generation task	Y	Language tasks (ability to repeat words and non-words and to generate verbs)
[49]	19	HGG and LGG	Y	Speech perception Object recognition Auditory short-term memory holding	Recognition of the semantic relationship	Behavioral testing with language-related and cognitive non-language tasks	Importance of right temporal lobe for language processing	N	
[43]	23	NA	N	Language processing	Word generation Verb generation Sentence completion	MoCa	Simultaneous cerebellar activation across different cognitive domains (except visual)	N	

Legend: HGG, High-grade glioma. IQ, Intelligence quotient. LGG, Low-grade glioma. MMSE, Mini mental state examination. N, No. NA, Not available. WHO, World Health Organization. Y, Yes. MoCa, Montreal Cognitive Assessment.

3.3. Executive Functions

The last category of cognitive function investigated regards executive skills. These include many high-cognitive abilities, including sustained and selective attention, response and inhibition control, working memory, and processing speed. They comprise the capabilities necessary for monitoring and controlling our behavior to reach a chosen goal. In this review, we selected six studies investigating the above-cited cognitive processes (Table 4).

The majority of the selected studies focused their investigations on working memory capacity by explicitly looking at changes in large-scale functional networking. Lang et al. showed that the task-evoked reconfiguration of the frontoparietal network (FPN; i.e., the executive network) correlates with cognitive performance, suggesting that its reconfiguration may play a role in cognitive deficits in brain tumor patients [41]. Nevertheless, a higher average connectivity within the FPN or in the parietal region of the tumor-affected hemisphere was associated with lower cognitive scores, and a lower connectivity of the parietal region of the non-tumor hemisphere was associated with worse neuropsychologic outcomes [41]. Therefore, patients with less connectivity in the FPN in the tumor hemisphere had preserved cognition. Alternatively, the authors hypothesized that the presence of the glioma may result in inefficient processing in the FPN due to maladaptive brain reorganization [41]. Furthermore, the fMRI results reveal normal central executive network (CEN) activation in glioma patients but a reduced default mode network (DMN) deactivation. This reduced responsiveness of the DMN may suggest that cognitive deficits reflect a reduced capacity to achieve a brain state necessary for normal cognitive performance, rather than the abnormal functioning of executive brain regions [46].

Additionally, one paper underlined the role of histology, showing that high-grade gliomas were significantly associated with lower cognitive flexibility and working memory capacity [42]. Lastly, tumor site was also reported as a relevant variable to consider pre-surgically: frontal tumors and left hemisphere lesions led to lower working memory skills [42].

On the other hand, Arbula et al., studying sustained attention and response control, concluded that right prefrontal damage led to frequent target omissions, probably due to sustained attention lapses, and that left prefrontal patients showed both target omissions and high false alarm rates to warning stimuli, suggesting a decisional rather than inhibitory impairment [40]. The anatomo-functional correlation of gliomas showed that left ventrolateral and dorsolateral prefrontal lesions were associated with target discrimination failure. In contrast, right ventrolateral and medial prefrontal lesions correlated with target detection failure [40].

One study focused instead on the self-monitoring function, with a particular interest in the functional role of the insular cortex. It proposed a new multimodal protocol combining DCS, awake surgery, and fMRI to measure self-monitoring skills with a modified version of the Stroop task. They reported differences in metacognitive domains of glioma patients, showing (i) increased difficulties in detecting action–outcome mismatches during insular DCS, and (ii) significant insular BOLD activations during outcome incongruences for self-made actions [48]. This highlights the importance of considering the insula activation in executive processing, especially in the metacognitive domain, for patients undergoing surgery.

Globally, these preliminary findings may imply that (i) executive function performance, such as working memory, is highly susceptible to plastic connectivity changes in related brain networking; (ii) such changes can relate to both attentive (FPN) and general (DMN) functional networks; (iii) tumor features, such as grade and site, can be negative predictors of executive performance; (iv) there is an existing correspondence between tumor site and attentive performance; (v) executive processing together with its networking can also be related to metacognitive skills.

Table 4. Selected studies investigating presurgical executive functions.

Author	Patients (N)	WHO Grade	Surgery	fMRI Tasks	fMRI Measures	Neuropsychological Tests (Related to Task)	Main Results	Awake Surgery	Task during Awake Surgery
[40]	25	18 HGG 7 LGG	Y	Go/No-Go task	Omissions and false alarms	MMSE, IQ, Verbal short-term memory and working memory Selective and divided attention Visuospatial short-term memory	Prefrontal areas underlie broader cognitive control processes (response selection, target detection)	N	
[42]	26	13 HGG 13 LGG	N	N-back task	2-back congruent conditions	Attention shifting	FPN plastic capacity plays a role in cognitive deficits	N	
[48]	1	II	Y	Stroop task	Informative feedback blocks	Executive functions Attention Working memory Inhibition Mental flexibility Set shifting abilities Insular-related functioning (empathy scale and emotion recognition)	Role of the insula in self-monitoring	Y	Awake mapping multimodal protocol (modified version of the Stroop task)
[41]	16	11 HGG 5 LGG	Y	N-back task	Difference between 0-back and 2-back congruent conditions	National Institutes of Health Cognitive Battery	FPN connectivity is associated with cognitive performance	N	
[46]	46	25 HGG 21 LGG	N	N-back task	Difference between 0-back and 2-back congruent conditions	Cognitive performance	Cognitive deficits associated with reduced DMN	N	
[43]	23	NA	N	Working memory, executive function	N-back task	MoCa	Simultaneous cerebellar activation across different cognitive domains (except visual)	N	

Legend: HGG, High-grade glioma. IQ, Intelligence quotient. LGG, Low-grade glioma. MMSE, Mini mental state examination. N, No. NA, Not available. WHO, World Health Organization. Y, Yes. MoCa, Montreal Cognitive Assessment.

3.4. Additional Studies (PRISMA Qualitative Analysis)

In addition to this review, two studies reporting postoperative-only neuropsychological assessment together with fMRI mapping within the above-cited cognitive domains were found and, therefore, were qualitatively included.

In greater detail, Amiez et al. studied the sensorimotor function and specifically the role of the rostral part of the left dorsal premotor cortex in four patients with low-grade tumors close to that region [50]. The fMRI task's experimental design undertaken pre- and postoperatively was based on a visuomotor conditional task (i.e., the ability to select between competing responses based on previously defined conditional rules). The employment of the task enabled the localization of the functional activation area in the proximity of the tumor area in all four patients, enabling the exact delineation of premotor regions and aiding the planning of the surgery efficiently. This was further corroborated not only by an optimization of the EOR, but also by the absence of postoperative deficits in the visuomotor conditional task [50].

Quirarte et al. reported a case report of a left superior frontal glioma exhibiting linguistic impairments after surgery (i.e., language supplementary motor area syndrome). This study demonstrated the potential of fMRI mapping of the linguistic domain not only for localization purposes (i.e., surgical mapping), but also to gain better insights about plasticity processing occurring after surgery, and how these can be related to neuropsychological deficits exhibited by the patients [51].

4. Discussion

Function-preserving, maximally safe resection for brain tumors relies upon changes in the clinical approach. Indeed, not only does it consist of applying intraoperative mapping methods, but also improving the preoperative evaluation of patients and performing accurate presurgical planning to identify the relationship between the tumor and cortical and subcortical eloquent areas [15]. FMRI is a widely employed technology to identify the functional involvement of the cortical regions at rest or during different tasks [16,17,21]. The same tasks used during fMRI can be applied during awake surgery to obtain consistent results during direct cortical stimulation (DCS) [46]. Sensorimotor and language tasks are commonly used, but other functions (e.g., executive and attentive functions) are evaluated sporadically [40–42,46,48], and the task's experimental design is not standardized among different centers [18,19,22,26]. Furthermore, integrating fMRI results with neuropsychological assessment increases the variability of protocols and reduces the comparability of the outcome [29].

In this review, we found only 15 studies reporting both the fMRI tasks and neuropsychological tests included in the preoperative evaluation of patients. Previous evidence provides insights into the following cognitive domains: (i) sensorimotor functions, which can be efficiently mapped during standard motor paradigms and conserved by pre-surgically investigating visuospatial abilities; (ii) language, which historically is the most largely mapped cognitive function in glioma patients, with the recent efficient brain localization of different linguistic sub-components and related improvements in the recovery of the linguistic function after surgery; (iii) executive abilities, which might depend on resilient plastic processes of large-scale networks' connectivity, especially in the case of working memory. Nevertheless, the correlation between the results of the two examinations has yet to be extensively evaluated.

4.1. Anatomo-Functional Correlations of fMRI Mapping and Cognitive Performance

The first clinical implication of the investigated studies reporting fMRI data and neuropsychological assessment is the definition of correlations between areas with neoplastic involvement, functional brain sites, and cognitive status. These correlations may help in designing a specific multimodal presurgical planning strategy to improve the functional recovery of the patients. Currently, a broader part of the available literature focuses on language. But in this review, we demonstrated how, even for the same cognitive function (i.e., language), there is considerable variability across studies evaluating different eloquent areas as sub-parts of the cognitive network without a precise domain localization. Therefore, fMRI mapping, together with assessing neuropsychological correlates of a certain cognitive function, becomes influential if the tumor infiltrates one or more areas belonging to the supposed spatial localization of the corresponding neural network. In this context, language function is one of the most suitable candidates across cognitive domains, given the ease of its preoperative evaluation, intraoperative testing, and monitoring during follow-up [19,22]. On the contrary, evidence regarding other functions, such as sensorimotor, perceptual, and executive functions, is scarce and anecdotal. Therefore, more studies are needed to determine anatomo-functional correlations of cognitive domains at the single-subject level in glioma patients to precisely define a tailored surgical procedure.

4.2. Integration of fMRI Data and Neuropsychological Assessment

Secondly, only a few of the included studies emphasized the relationship between neuropsychological evaluation and fMRI tasks, reporting heterogeneous results. Some authors reported a complementary role of the two preoperative assessments; for example, regarding speech disturbances, fMRI-positive areas correlated with presurgical neuropsychological language tests were found in the frontal operculum [37,45], besides the well-known involvement of the arcuate fasciculus. Indeed, Schouwenaars et al. corroborated this hypothesis by showing how lower fMRI in-scanner performances in glioma patients compared to controls were associated with the same cognitive impairment during neuropsychological testing [46].

Further plastic evidence about specific compensatory and resilient fMRI activations, consequent to tumor presence, was given by Mitolo and colleagues, who assessed how they are positively correlated with a better cognitive performance in tests measuring general cognitive abilities (especially in semantic fluency, verbal short-term memory, and executive functions) [47]. Therefore, a correlation between fMRI and neuropsychological assessment is confirmed based on the currently available data. In cases where the two methods disagree, they should be seen as complementary tools, integrating their results for better surgical planning.

4.3. Role of fMRI-Positive Regions during Surgery

The most apparent transposition of functional data to surgical procedures is exhibited in the correlating of fMRI data and intraoperative monitoring with DCS to detect the spatial distribution of "function-positive" spots [45].

In this review, the only work reporting correspondence between DCS and fMRI was that by Leote, which was equal to 100% for the precentral gyrus for motor function and 84% regarding the opercular frontal inferior gyrus for language function. They found a correlation between worse presurgical neuropsychological performance and decreased DCS duration. Nevertheless, they stated that presurgical language disturbances limited the applicability of DCS mapping in awake surgery [45]. Indeed, according to the author's interpretation, this was due to the surgeon's decision to proceed with the tumor resection, having considered the systematic errors in language tasks instead of performing more efficient cortical mapping. Errors in language tasks without applying DCS were seen; therefore, the patients were engaged in spontaneous conversation, which demands a lower cognitive load but was not evaluated in fMRI paradigms [45].

Additionally, the surgeon focused more on negative cortical regions after DCS than positive fMRI regions, probably because of an unconscious preference for DCS for brain eloquent function mapping [45]. In addition to preoperative assessments with fMRI and comparisons with DCS results, some papers suggested the direct implementation of functional data intraoperatively [35,45,50]. This process can be done by fusing fMRI data into morphological MRI sequences and employing intraoperative navigation. Several pieces of evidence have highlighted how the intraoperative use of fMRI data in the context of neuronavigation can add highly informative and integrated knowledge about tumor features during resection [19]. Indeed, neuronavigation with fMRI increases both the neurosurgeon's accuracy and the identification of target regions for resection. Furthermore, information can be combined with neuronavigation in a multimodal manner, taking advantage also of the white matter fiber-tracking visualization [52,53]. The superimposition of white matter tracts on eloquent functional brain areas during neuronavigation may support both the evaluation of better functional limits for the resection and intraoperative neurophysiological mapping. As noted by Leote et al., the usage of navigated fMRI data during the surgery seems to not influence or improve DCS in practice [42], and therefore, further multimodal studies are still needed.

Finally, as a potential future development, Argiris et al. reported that patients performing worse on the mental rotation hand fMRI task had lower MEP amplitudes in the upper limb motor region during transcranial magnetic stimulation [44]; this may also have an impact on DCS during surgery if tested systematically.

4.4. Patient-Tailored Protocols According to the Lesion Site and Plasticity Evaluation

Different authors suggested that selecting well-adjusted and individualized neuropsychological tasks pre- and intraoperatively may enable the development of personalized and more efficient brain mapping protocols [39]. Yamamoto et al. pushed this concept forward, suggesting that fMRI findings, also regarding brain plasticity, may have important implications for the surgical management of patients with brain tumors [49]. Kamada et al. highlighted the same evidence, identifying how fMRI can successfully map the dissociation of cognitive functions, particularly within the language domain. Indeed, they probed that fMRI activation during abstract/concrete categorization tasks was located within their patients' right superior temporal region, in contrast to the activation of the left superior temporal and left supramarginal gyri in controls [35]. Moreover, knowledge of the areas showing functional reorganization may help avoid unintentional damage during neurosurgery [49]. Future prospective studies are needed to define fMRI and neuropsychological protocols specific to tumor locations and types, which will help clinicians select the most appropriate assessment method for the individual patient. In this context, preoperative task-based fMRI is a feasible and highly sensitive tool for localizing eloquent cortical areas in patients affected by brain tumors. Nevertheless, its prognostic role, regarding reduced morbidity and improved oncologic outcome, still needs to be definitively addressed and clarified.

4.5. Limitations of the Current Study

The major limitation of the current study is the limited number of studies that met the inclusion criteria. This paucity reflects the heterogeneity of the literature regarding fMRI studies, behavioral experimental design, neuropsychological tests, and specific aims investigated.

This systematic review included only brain gliomas, discarding secondary nervous system tumors. Here, we did not specifically focus on precise brain tumor sites and grading, given our aim of including the vast majority of relevant neuro-oncologic works in the field. Nevertheless, this contributed to a lack of straightforward conclusions; findings from the literature underlined this complexity. They were characterized by high variability in concordance rates, with sensitivity and specificity ranging from 59 to 100% and 0 to 97% [54,55]. Moreover, the difficulty in the generalizability of the results not only relates the hypothesis of each study, but is also caused by the nature of neural network connections related to cognitive functions, which are highly complex and require several heterogeneous tasks and tests in extensively mapping them. In addition, diverse acquisition protocols and image post-processing techniques may also significantly impact the areas identified as related to a specific cognitive function in fMRI studies. The heterogeneity of data further dilutes the strength of any current recommendations made in the review.

4.6. Future Perspectives

Based on the results of this systematic review, further studies focusing on the relationship between the preoperative mapping of eloquent brain areas and neuropsychological profiles in gliomas are needed. Attention should be primarily set on defining acquisition protocols that can be tailored to single patients, but assessing standardized and reproducible parameters, ideally including:

- Preoperative and postoperative evaluations, as longitudinal comparisons have not been made extensively in the previous literature;
- fMRI tasks focused on specific cognitive functions and put in perspective with relative neuropsychological assessments, also taking into account the tumor site and hemisphere;
- Correlations of specific neuropsychological tests with experimental fMRI tasks' results to identify clinical criteria for the indication of preoperative fMRI.

5. Conclusions

The results of this review re quite heterogeneous, and only a few papers satisfied the inclusion criteria, significantly impacting the generalizability of the findings. Most studies considered language as the most commonly evaluated cognitive function in clinical practice. In the literature, fMRI and neuropsychological assessments differed in almost every paper regarding the studied functions, but also the protocols fused to assess a specific function, impacting the chance of drawing broad conclusions and protocol suggestions. Furthermore, the fMRI and neuropsychological results demonstrated high variability even when evaluating the same cognitive domain [37]. Nevertheless, the correlation between neuropsychological and fMRI results reported by some authors suggests that altered functions during the neuropsychological assessment may help identify patients who could benefit from an fMRI evaluation in general and, possibly, which specific function should be tested [45]. However, this should follow a decision based also on tumor site and suspected grading, as low-grade tumors seem to be associated with higher levels of plasticity processes and better cognitive functions [47]. Based on the current literature, the neuropsychological evaluation and fMRI complement each other in preoperative patient assessment, surgical planning, and within the surgical procedure itself through neuronavigation and awake testing. They could potentially even improve post-surgical outcomes in the future. However, clinicians must consider multiple patient-related and tumor factors to determine the most appropriate protocol.

Supplementary Materials: The following supporting information can be downloaded at: https://www.mdpi.com/article/10.3390/brainsci13101380/s1.

Author Contributions: Conceptualization, D.C. and P.P.P.; methodology, M.P., F.B., D.C. and P.P.P.; validation, D.C., F.S., R.G. and M.M.F.; formal analysis, M.P., F.B. and D.C.; investigation, M.P., F.B. and D.C.; resources, P.P.P.; data curation, M.P. and F.B.; writing—original draft preparation, M.P.; writing—review and editing, F.B., F.S., D.C. and F.S.; visualization, F.B., D.C. and F.S.; supervision, R.G., F.S., M.M.F. and P.P.P.; project administration, M.P. All authors have read and agreed to the published version of the manuscript.

Funding: This research received no external funding.

Institutional Review Board Statement: Not applicable.

Informed Consent Statement: Not applicable.

Data Availability Statement: Data are contained within the article.

Conflicts of Interest: The authors declare no conflict of interest.

Abbreviations

BOLD, blood oxygen level-dependent; CEN, central executive network; DCS, direct cortical stimulation; DMN, default mode network; EOR, extent of resection; fMRI, functional magnetic resonance imaging; GTR, FPN, fronto-parietal network, gross total resection; HGG, high-grade glioma; IQ, intelligence quotient; LGG, low-grade glioma; MEP, motor evoked potential; MMSE, mini mental state examination; OSF, open science framework; PRISMA, Preferred Reporting Items for Systematic Reviews and Metanalyses; ROBIS, risk of bias in systematic reviews; WHO, World Health Organization.

References

1. Ius, T.; Pignotti, F.; Della Pepa, G.M.; La Rocca, G.; Somma, T.; Isola, M.; Battistella, C.; Gaudino, S.; Polano, M.; Dal Bo, M.; et al. A Novel Comprehensive Clinical Stratification Model to Refine Prognosis of Glioblastoma Patients Undergoing Surgical Resection. *Cancers* **2020**, *12*, 386. [CrossRef]
2. Lemée, J.-M.; Clavreul, A.; Menei, P. Intratumoral Heterogeneity in Glioblastoma: Don't Forget the Peritumoral Brain Zone. *Neuro-Oncology* **2015**, *17*, 1322–1332. [CrossRef]
3. Weller, M.; Van Den Bent, M.; Preusser, M.; Le Rhun, E.; Tonn, J.C.; Minniti, G.; Bendszus, M.; Balana, C.; Chinot, O.; Dirven, L.; et al. EANO Guidelines on the Diagnosis and Treatment of Diffuse Gliomas of Adulthood. *Nat. Rev. Clin. Oncol.* **2021**, *18*, 170–186. [CrossRef]
4. Brandes, A.A.; Tosoni, A.; Franceschi, E.; Reni, M.; Gatta, G.; Vecht, C. Glioblastoma in Adults. *Crit. Rev. Oncol./Hematol.* **2008**, *67*, 139–152. [CrossRef]
5. Chuang, D.F.; Lin, X. Targeted Therapies for the Treatment of Glioblastoma in Adults. *Curr. Oncol. Rep.* **2019**, *21*, 61. [CrossRef]
6. Kim, J.Y.; Park, J.E.; Jo, Y.; Shim, W.H.; Nam, S.J.; Kim, J.H.; Yoo, R.-E.; Choi, S.H.; Kim, H.S. Incorporating Diffusion- and Perfusion-Weighted MRI into a Radiomics Model Improves Diagnostic Performance for Pseudoprogression in Glioblastoma Patients. *Neuro-Oncology* **2019**, *21*, 404–414. [CrossRef]
7. Öz, G.; Alger, J.R.; Barker, P.B.; Bartha, R.; Bizzi, A.; Boesch, C.; Bolan, P.J.; Brindle, K.M.; Cudalbu, C.; Dinçer, A.; et al. Clinical Proton MR Spectroscopy in Central Nervous System Disorders. *Radiology* **2014**, *270*, 658–679. [CrossRef]
8. Mazzucchi, E.; La Rocca, G.; Ius, T.; Sabatino, G.; Della Pepa, G.M. Multimodality Imaging Techniques to Assist Surgery in Low-Grade Gliomas. *World Neurosurg.* **2020**, *133*, 423–425. [CrossRef]
9. Li, Y.M.; Suki, D.; Hess, K.; Sawaya, R. The Influence of Maximum Safe Resection of Glioblastoma on Survival in 1229 Patients: Can We Do Better than Gross-Total Resection? *JNS* **2016**, *124*, 977–988. [CrossRef]
10. Sanai, N.; Berger, M.S. Surgical Oncology for Gliomas: The State of the Art. *Nat. Rev. Clin. Oncol.* **2018**, *15*, 112–125. [CrossRef]
11. Wende, T.; Hoffmann, K.-T.; Meixensberger, J. Tractography in Neurosurgery: A Systematic Review of Current Applications. *J. Neurol. Surg. A Cent. Eur. Neurosurg.* **2020**, *81*, 442–455. [CrossRef] [PubMed]
12. Katsevman, G.A.; Turner, R.C.; Urhie, O.; Voelker, J.L.; Bhatia, S. Utility of Sodium Fluorescein for Achieving Resection Targets in Glioblastoma: Increased Gross- or near-Total Resections and Prolonged Survival. *J. Neurosurg.* **2020**, *132*, 914–920. [CrossRef] [PubMed]
13. Lacroix, M.; Abi-Said, D.; Fourney, D.R.; Gokaslan, Z.L.; Shi, W.; DeMonte, F.; Lang, F.F.; McCutcheon, I.E.; Hassenbusch, S.J.; Holland, E.; et al. A Multivariate Analysis of 416 Patients with Glioblastoma Multiforme: Prognosis, Extent of Resection, and Survival. *J. Neurosurg.* **2001**, *95*, 190–198. [CrossRef] [PubMed]
14. Gogos, A.J.; Young, J.S.; Morshed, R.A.; Hervey-Jumper, S.L.; Berger, M.S. Awake Glioma Surgery: Technical Evolution and Nuances. *J. Neurooncol.* **2020**, *147*, 515–524. [CrossRef] [PubMed]
15. Roland, J.L.; Hacker, C.D.; Leuthardt, E.C. A Review of Passive Brain Mapping Techniques in Neurological Surgery. *Neurosurgery* **2021**, *88*, 15–24. [CrossRef]
16. Raffa, G.; Conti, A.; Scibilia, A.; Sindorio, C.; Quattropani, M.C.; Visocchi, M.; Germanò, A.; Tomasello, F. Functional Reconstruction of Motor and Language Pathways Based on Navigated Transcranial Magnetic Stimulation and DTI Fiber Tracking for the Preoperative Planning of Low Grade Glioma Surgery: A New Tool for Preservation and Restoration of Eloquent Networks. In *Trends in Reconstructive Neurosurgery*; Visocchi, M., Mehdorn, H.M., Katayama, Y., Von Wild, K.R.H., Eds.; Acta Neurochirurgica Supplement; Springer International Publishing: Cham, Switzerland, 2017; Volume 124, pp. 251–261; ISBN 978-3-319-39545-6.
17. Spena, G.; Nava, A.; Cassini, F.; Pepoli, A.; Bruno, M.; D'Agata, F.; Cauda, F.; Sacco, K.; Duca, S.; Barletta, L.; et al. Preoperative and Intraoperative Brain Mapping for the Resection of Eloquent-Area Tumors. A Prospective Analysis of Methodology, Correlation, and Usefulness Based on Clinical Outcomes. *Acta Neurochir.* **2010**, *152*, 1835–1846. [CrossRef]
18. Manan, H.A.; Franz, E.A.; Yahya, N. Utilization of Functional MRI Language Paradigms for Preoperative Mapping: A Systematic Review. *Neuroradiology* **2020**, *62*, 353–367. [CrossRef]
19. Castellano, A.; Cirillo, S.; Bello, L.; Riva, M.; Falini, A. Functional MRI for Surgery of Gliomas. *Curr. Treat. Opt. Neurol.* **2017**, *19*, 34. [CrossRef]
20. Dierker, D.; Roland, J.L.; Kamran, M.; Rutlin, J.; Hacker, C.D.; Marcus, D.S.; Milchenko, M.; Miller-Thomas, M.M.; Benzinger, T.L.; Snyder, A.Z.; et al. Resting-State Functional Magnetic Resonance Imaging in Presurgical Functional Mapping. *Neuroimaging Clin. N. Am.* **2017**, *27*, 621–633. [CrossRef]
21. Silva, M.A.; See, A.P.; Essayed, W.I.; Golby, A.J.; Tie, Y. Challenges and Techniques for Presurgical Brain Mapping with Functional MRI. *NeuroImage Clin.* **2018**, *17*, 794–803. [CrossRef]

22. Black, D.F.; Vachha, B.; Mian, A.; Faro, S.H.; Maheshwari, M.; Sair, H.I.; Petrella, J.R.; Pillai, J.J.; Welker, K. American Society of Functional Neuroradiology–Recommended FMRI Paradigm Algorithms for Presurgical Language Assessment. *AJNR Am. J. Neuroradiol.* **2017**, *38*, E65–E73. [CrossRef] [PubMed]
23. Puglisi, G.; Sciortino, T.; Rossi, M.; Leonetti, A.; Fornia, L.; Conti Nibali, M.; Casarotti, A.; Pessina, F.; Riva, M.; Cerri, G.; et al. Preserving Executive Functions in Nondominant Frontal Lobe Glioma Surgery: An Intraoperative Tool. *J. Neurosurg.* **2019**, *131*, 474–480. [CrossRef] [PubMed]
24. DeYoe, E.A.; Raut, R.V. Visual Mapping Using Blood Oxygen Level Dependent Functional Magnetic Resonance Imaging. *Neuroimaging Clin. N. Am.* **2014**, *24*, 573–584. [CrossRef] [PubMed]
25. Pujol, J.; Deus, J.; Acebes, J.J.; Villanueva, A.; Aparicio, A.; Soriano-Mas, C.; Ortiz, H.; Conesa, G. Identification of the Sensorimotor Cortex with Functional MRI: Frequency and Actual Contribution in a Neurosurgical Context. *J. Neuroimaging* **2008**, *18*, 28–33. [CrossRef] [PubMed]
26. Bizzi, A. Presurgical Mapping of Verbal Language in Brain Tumors with Functional MR Imaging and MR Tractography. *Neuroimaging Clin. N. Am.* **2009**, *19*, 573–596. [CrossRef]
27. Acevedo-Vergara, K.; Perez-Florez, M.; Ramirez, A.; Torres-Bayona, S.; Dau, A.; Salva, S.; Maloof, D.; Garcia, C.; Luque, M.; Guillen-Burgos, H.F. Cognitive Deficits in Adult Patients with High-Grade Glioma: A Systematic Review. *Clin. Neurol. Neurosurg.* **2022**, *219*, 107296. [CrossRef]
28. Wolf, J.; Campos, B.; Bruckner, T.; Vogt, L.; Unterberg, A.; Ahmadi, R. Evaluation of Neuropsychological Outcome and "Quality of Life" after Glioma Surgery. *Langenbecks Arch. Surg.* **2016**, *401*, 541–549. [CrossRef]
29. Romero-Garcia, R.; Owen, M.; McDonald, A.; Woodberry, E.; Assem, M.; Coelho, P.; Morris, R.C.; Price, S.J.; Santarius, T.; Suckling, J.; et al. Assessment of Neuropsychological Function in Brain Tumor Treatment: A Comparison of Traditional Neuropsychological Assessment with App-Based Cognitive Screening. *Acta Neurochir.* **2022**, *164*, 2021–2034. [CrossRef]
30. MacDonald, M.E.; Pike, G.B. MRI of Healthy Brain Aging: A Review. *NMR Biomed.* **2021**, *34*, e4564. [CrossRef]
31. Odusami, M.; Maskeliūnas, R.; Damaševičius, R.; Misra, S. Machine Learning with Multimodal Neuroimaging Data to Classify Stages of Alzheimer's Disease: A Systematic Review and Meta-Analysis. *Cogn. Neurodyn.* **2023**, 1–20. [CrossRef]
32. Moher, D.; Liberati, A.; Tetzlaff, J.; Altman, D.G.; for the PRISMA Group. Preferred Reporting Items for Systematic Reviews and Meta-Analyses: The PRISMA Statement. *BMJ* **2009**, *339*, b2535. [CrossRef] [PubMed]
33. Page, M.J.; McKenzie, J.E.; Bossuyt, P.M.; Boutron, I.; Hoffmann, T.C.; Mulrow, C.D.; Shamseer, L.; Tetzlaff, J.M.; Akl, E.A.; Brennan, S.E.; et al. The PRISMA 2020 Statement: An Updated Guideline for Reporting Systematic Reviews. *BMJ* **2021**, *372*, n71. [CrossRef]
34. Whiting, P.; Savović, J.; Higgins, J.P.T.; Caldwell, D.M.; Reeves, B.C.; Shea, B.; Davies, P.; Kleijnen, J.; Churchill, R. ROBIS: A New Tool to Assess Risk of Bias in Systematic Reviews Was Developed. *J. Clin. Epidemiol.* **2016**, *69*, 225–234. [CrossRef] [PubMed]
35. Kamada, K.; Takeuchi, F.; Kuriki, S.; Todo, T.; Morita, A.; Sawamura, Y. Dissociated Expressive and Receptive Language Functions on Magnetoencephalography, Functional Magnetic Resonance Imaging, and Amobarbital Studies: Case Report and Review of the Literature. *JNS* **2006**, *104*, 598–607. [CrossRef] [PubMed]
36. Papagno, C.; Miracapillo, C.; Casarotti, A.; Romero Lauro, L.J.; Castellano, A.; Falini, A.; Casaceli, G.; Fava, E.; Bello, L. What Is the Role of the Uncinate Fasciculus? Surgical Removal and Proper Name Retrieval. *Brain* **2011**, *134*, 405–414. [CrossRef] [PubMed]
37. Bizzi, A.; Nava, S.; Ferrè, F.; Castelli, G.; Aquino, D.; Ciaraffa, F.; Broggi, G.; DiMeco, F.; Piacentini, S. Aphasia Induced by Gliomas Growing in the Ventrolateral Frontal Region: Assessment with Diffusion MR Tractography, Functional MR Imaging and Neuropsychology. *Cortex* **2012**, *48*, 255–272. [CrossRef]
38. Kristo, G.; Raemaekers, M.; Rutten, G.-J.; De Gelder, B.; Ramsey, N.F. Inter-Hemispheric Language Functional Reorganization in Low-Grade Glioma Patients after Tumour Surgery. *Cortex* **2015**, *64*, 235–248. [CrossRef]
39. Sierpowska, J.; Gabarrós, A.; Fernandez Coello, A.; Camins, À.; Castañer, S.; Juncadella, M.; De Diego-Balaguer, R.; Rodriguez-Fornells, A. Morphological Derivation Overflow as a Result of Disruption of the Left Frontal Aslant White Matter Tract. *Brain Lang.* **2015**, *142*, 54–64. [CrossRef]
40. Arbula, S.; Pacella, V.; De Pellegrin, S.; Rossetto, M.; Denaro, L.; D'Avella, D.; Della Puppa, A.; Vallesi, A. Addressing the Selective Role of Distinct Prefrontal Areas in Response Suppression: A Study with Brain Tumor Patients. *Neuropsychologia* **2017**, *100*, 120–130. [CrossRef]
41. Lang, S.; Gaxiola-Valdez, I.; Opoku-Darko, M.; Partlo, L.A.; Goodyear, B.G.; Kelly, J.J.P.; Federico, P. Functional Connectivity in Frontoparietal Network: Indicator of Preoperative Cognitive Function and Cognitive Outcome Following Surgery in Patients with Glioma. *World Neurosurg.* **2017**, *105*, 913–922.e2. [CrossRef]
42. De Baene, W.; Jansma, M.J.; Schouwenaars, I.T.; Rutten, G.-J.M.; Sitskoorn, M.M. Task-Evoked Reconfiguration of the Fronto-Parietal Network Is Associated with Cognitive Performance in Brain Tumor Patients. *Brain Imaging Behav.* **2020**, *14*, 2351–2366. [CrossRef] [PubMed]
43. Zacharia, T.T.; Eslinger, P.J. Functional MRI Activation Patterns of Cerebellum in Patients with Epilepsy and Brain Tumors. *Clin. Anat.* **2019**, *32*, 1053–1060. [CrossRef] [PubMed]
44. Argiris, G.; Budai, R.; Maieron, M.; Ius, T.; Skrap, M.; Tomasino, B. Neurosurgical Lesions to Sensorimotor Cortex Do Not Impair Action Verb Processing. *Sci. Rep.* **2020**, *10*, 523. [CrossRef]
45. Leote, J.; Loução, R.; Viegas, C.; Lauterbach, M.; Perez-Hick, A.; Monteiro, J.; Nunes, R.G.; Ferreira, H.A. Impact of Navigated Task-Specific FMRI on Direct Cortical Stimulation. *J. Neurol. Surg. A Cent. Eur. Neurosurg.* **2020**, *81*, 555–564. [CrossRef]

46. Schouwenaars, I.T.; De Dreu, M.J.; Rutten, G.-J.M.; Ramsey, N.F.; Jansma, J.M. A Functional MRI Study of Presurgical Cognitive Deficits in Glioma Patients. *Neuro-Oncol. Pract.* **2021**, *8*, 81–90. [CrossRef]
47. Mitolo, M.; Zoli, M.; Testa, C.; Morandi, L.; Rochat, M.J.; Zaccagna, F.; Martinoni, M.; Santoro, F.; Asioli, S.; Badaloni, F.; et al. Neuroplasticity Mechanisms in Frontal Brain Gliomas: A Preliminary Study. *Front. Neurol.* **2022**, *13*, 867048. [CrossRef] [PubMed]
48. Gomez-Andres, A.; Cunillera, T.; Rico, I.; Naval-Baudin, P.; Camins, A.; Fernandez-Coello, A.; Gabarrós, A.; Rodriguez-Fornells, A. The Role of the Anterior Insular Cortex in Self-Monitoring: A Novel Study Protocol with Electrical Stimulation Mapping and Functional Magnetic Resonance Imaging. *Cortex* **2022**, *157*, 231–244. [CrossRef]
49. Yamamoto, A.K.; Sanjuán, A.; Pope, R.; Parker Jones, O.; Hope, T.M.H.; Prejawa, S.; Oberhuber, M.; Mancini, L.; Ekert, J.O.; Garjardo-Vidal, A.; et al. The Effect of Right Temporal Lobe Gliomas on Left and Right Hemisphere Neural Processing During Speech Perception and Production Tasks. *Front. Hum. Neurosci.* **2022**, *16*, 803163. [CrossRef]
50. Amiez, C.; Kostopoulos, P.; Champod, A.-S.; Collins, D.L.; Doyon, J.; Maestro, R.D.; Petrides, M. Preoperative Functional Magnetic Resonance Imaging Assessment of Higher-Order Cognitive Function in Patients Undergoing Surgery for Brain Tumors. *JNS* **2008**, *108*, 258–268. [CrossRef]
51. Quirarte, J.A.; Kumar, V.A.; Liu, H.-L.; Noll, K.R.; Wefel, J.S.; Lang, F.F. Language Supplementary Motor Area Syndrome Correlated with Dynamic Changes in Perioperative Task-Based Functional MRI Activations: Case Report. *J. Neurosurg.* **2021**, *134*, 1738–1742. [CrossRef]
52. Alexopoulos, G.; Cikla, U.; El Tecle, N.; Kulkarni, N.; Pierson, M.; Mercier, P.; Kemp, J.; Coppens, J.; Mahmoud, S.; Sehi, M.; et al. The Value of White Matter Tractography by Diffusion Tensor Imaging in Altering a Neurosurgeon's Operative Plan. *World Neurosurg.* **2019**, *132*, e305–e313. [CrossRef] [PubMed]
53. Romano, A.; D'Andrea, G.; Minniti, G.; Mastronardi, L.; Ferrante, L.; Fantozzi, L.M.; Bozzao, A. Presurgical Planning and MR-Tractography Utility in Brain Tumour Resection. *Eur. Radiol.* **2009**, *19*, 2798–2808. [CrossRef]
54. Kuchcinski, G.; Mellerio, C.; Pallud, J.; Dezamis, E.; Turc, G.; Rigaux-Viode, O.; Malherbe, C.; Roca, P.; Leclerc, X.; Varlet, P.; et al. Three-Tesla Functional MR Language Mapping: Comparison with Direct Cortical Stimulation in Gliomas. *Neurology* **2015**, *84*, 560–568. [CrossRef] [PubMed]
55. Giussani, C.; Roux, F.-E.; Ojemann, J.; Sganzerla, E.P.; Pirillo, D.; Papagno, C. Is Preoperative Functional Magnetic Resonance Imaging Reliable for Language Areas Mapping in Brain Tumor Surgery? Review of Language Functional Magnetic Resonance Imaging and Direct Cortical Stimulation Correlation Studies. *Neurosurgery* **2010**, *66*, 113–120. [CrossRef] [PubMed]

brain sciences

MDPI

Article

Hand Motions Reveal Attentional Status and Subliminal Semantic Processing: A Mouse-Tracking Technique

Kunchen Xiao [1,*], Anqi Zhang [1], Jingke Qu [1], Feifei Deng [1], Chenyan Guo [1] and Takashi Yamauchi [2,*]

[1] Institute of Brain and Psychological Sciences, Sichuan Normal University, Chengdu 610066, China
[2] Department of Psychological and Brain Sciences, Texas A&M University, College Station, TX 77843-4235, USA
* Correspondence: xiao.kun.chong@163.com (K.X.); takashi-yamauchi@tamu.edu (T.Y.)

Abstract: Theories of embodied cognition suggest that hand motions and cognition are closely interconnected. An emerging technique of tracking how participants move a computer mouse (i.e., the mouse-tracking technique) has shown advantages over the traditional response time measurement to detect implicit cognitive conflicts. Previous research suggests that attention is essential for subliminal processing to take place at a semantic level. However, this assumption is challenged by evidence showing the presence of subliminal semantic processing in the near-absence of attention. The inconsistency of evidence could stem from the insufficient sensitivity in the response time measurement. Therefore, we examined the role of attention in subliminal semantic processing by analyzing participants' hand motions using the mouse-tracking technique. The results suggest that subliminal semantic processing is not only enhanced by attention but also occurs when attention is disrupted, challenging the necessity of facilitated top-down attention for subliminal semantic processing, as claimed by a number of studies. In addition, by manipulating the color of attentional cues, our experiment shows that the cue color per se could influence participants' response patterns. Overall, the current study suggests that attentional status and subliminal semantic processing can be reliably revealed by temporal–spatial features extracted from cursor motion trajectories.

Keywords: mouse-tracking; cursor motion; attention; congruency effects; subliminal semantic processing; area under the curve

Citation: Xiao, K.; Zhang, A.; Qu, J.; Deng, F.; Guo, C.; Yamauchi, T. Hand Motions Reveal Attentional Status and Subliminal Semantic Processing: A Mouse-Tracking Technique. *Brain Sci.* **2023**, *13*, 1267. https://doi.org/10.3390/brainsci13091267

Academic Editor: Daniele Corbo

Received: 29 July 2023
Revised: 22 August 2023
Accepted: 29 August 2023
Published: 31 August 2023

1. Introduction

1.1. Background

Regarding the relationship between consciousness and attention, a classic theory is that consciousness and unconsciousness have distinct features: conscious processing is elaborate, flexible, and controlled by attention, while unconscious processing is superficial, stereotypical, and independent of attention [1,2]. On the contrary, recent Global Workspace Theory argues that unconscious processing is more elaborate and flexible than previously regarded [3]. Research on masked priming suggests that semantic processing, which is conventionally considered relatively elaborate, can take place at a subliminal level. And subliminal semantic processing is modulated by top-down attention [4]. However, the role of attention in unconsciousness is controversial because of inconsistent results; the unsatisfactory reproducibility could be attributed to the lack of sensitivity in the traditional reaction time measure to reveal subtle unconscious cognition [5]. Thus, the current study proposes a novel methodology, which tracks participants' hand motions in choice-reaching tasks, to investigate the influence of attention on subliminal semantic processing.

According to theories of embodied cognition, cognition and body motion are closely interconnected [6], such as the interplay of visual perception and motion processing [7]. In particular, choice-reaching by hand (i.e., moving hands to select an intended target) is a fluid process of decision making, where muscles and neurons coordinate dynamically in a continuous feedback loop: higher cortical systems make a coarse motor plan, and local sensorimotor subsystems adjust the hand motion by adapting to sensory feedback in real-time [8,9]. If these sensorimotor coordination processes are intervened by cognitive conflicts, the motor plan for choice-reaching will be restructured (e.g., speed change or making a turn), modifying temporal–spatial features of the hand motion path [10–12]. Thus, hidden cognitive states, such as implicit semantic processing, can be revealed by how people move their hands during choice-reaching [13,14].

An emerging technique has been developed to track users' hand motion by recording the cursor movements of a computer mouse [15,16]. Temporal–spatial features of the cursor motion paths reflect participants' dynamic decision-making processes [17,18]. The mouse-tracking technique has been widely used in research on semantic processing [19], social judgment [20], and affective computing [21,22]. Recent research further integrates mouse-tracking into neuropsychological tests and finds that mouse-tracking benefits psychopathological assessment of stress [23,24], impulsivity [25], and attention deficit hyperactivity disorder [26–28]. Particularly, the cursor motion features are powerful in uncovering hidden cognitive conflicts [29–31], such as the implicit social stereotype [32], attitudinal ambivalence [33], uncertainty in economic choices [34], and syntactic incongruency [35]. Thus, the mouse-tracking technique provides a promising methodology to shed light on unconscious cognition.

1.2. Subliminal Semantic Processing and Attention

A typical unconscious cognitive conflict is semantic incongruency. For example [36], participants judged if a target digit was larger or smaller than five, preceded by a subliminal priming digit (i.e., the "prime"), which was masked (Figure 1). Although the prime was invisible, it influenced participants' semantic judgment: the response time was shorter for "congruent" trials (i.e., priming and target digits were both smaller or both larger than five) than "incongruent" trials (e.g., prime = 4, target = 9), known as the "congruency effect".

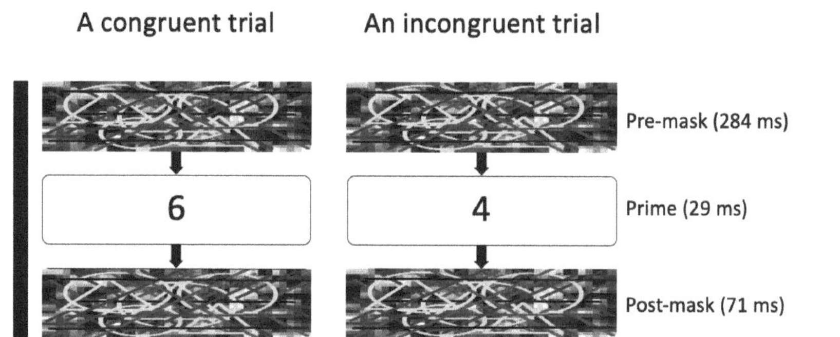

Figure 1. An illustration of the "congruency effect". The participants' task was to judge whether the target number was larger or smaller than 5. Before presenting the target, a priming digit was displayed briefly for 29 ms, sandwiched by masks to make the priming digit subliminal. The response time was shorter for congruent trials than incongruent trials, which is the "congruency effect".

The congruency effect demonstrates that semantic information of the subliminal stimuli can been processed without participants' awareness [37,38]. Furthermore, participants' attentional status was manipulated by occasionally presenting a cue: in the "cued" condition, an attentional cue (i.e., a green square) forecasting the impending target facilitated top-down attention, while in the "uncued" condition, the pre-target duration varied randomly with no cue presented to disrupt participants' temporal attention window [39]. The congruency effect was observed only in the cued condition, suggesting that top-down attention enhanced relevant types of subliminal semantic processing [40,41].

However, the exact role of attention in unconscious semantic processing is controversial [42,43]. Some research argued that assistance from attention was necessary for semantic processing to occur at a subliminal level [44–46]. Meanwhile, other studies only confirmed that attention modulated semantic priming but found no solid evidence for the necessity of attention [47,48]. On the other hand, some studies failed to replicate the modulation effect of attention on unconscious processing [49–51]. And subliminal congruency effects were also observed in the near-absences of attention [52,53]. For example, the gender of masked faces could be reliably identified with little attention, whereas non-face stimuli could not be unconsciously distinguished [54–56]. And masked pictures of natural scenes could be correctly categorized when the focal attention was suppressed [57,58].

The inconsistency of evidence regarding the role of attention in subliminal semantic processing may arise from the way that congruency effects have been evaluated. The traditional method of measuring response time captures only two data points for each trial: the onset and end times. This approach overlooks any cognitive process that takes place in between. Relying solely on the temporal measurement provides limited information about the intricate cognitive processes that unfold rapidly in real-time, which are yet fundamental for subliminal semantic processing to occur [59]. To probe these dynamic processes, a measurement recording fine-tuned data points corresponding to the continuous cognition-sensorimotor coordination is needed. In this case, the mouse-tracking data are beneficial because they integrate temporal–spatial information that characterizes participants' dynamic cognitive processes [60,61]. Though temporal data alone may underestimate semantic congruency effects, the temporal–spatial features of cursor motion help tease apart subliminal semantic processing [62,63].

1.3. Summary of Previous Research

A number of studies suggest that subliminal semantic congruency effects measured by response time are generally small and difficult to replicate [64–66]. One reason for the mediocre reproducibility could be the inadequate sensitivity in the measurement to detect the nuance of unconscious processing [67,68]. Indeed, substantial research corroborates the advantage of mouse-tracking measurement over response time, particularly for uncovering implicit cognitive conflicts [31,59,62]. For example, when participants judged whether a face belonged to a white or black person, their cursor motion trajectories were attracted to the unintended label "white" without awareness if an atypical black face was presented [20]. Similarly, semantic congruency effects can be reliably measured by the attraction of cursor trajectories toward the unintended alternative option, which is quantified by the "area under the curve" (AUC) of the cursor motion curves [59,69]. Research comparing mouse-tracking and response time shows that semantic congruency effects measured by AUC were significantly larger than those by response time [62,70,71]. Thus, the absence of semantic congruency effects in the uncued condition in the Naccache et al. [39] and follow-up studies could stem from the insufficient resolution in the response time measurement to reliably expose subliminal processing.

Another unresolved issue is the role of the attentional cue. It is possible that the cue not only facilitated top-down attention but also worked as another indirect prime. Previous research shows that the color of a visual cue is closely associated with semantic priming [72]; for example, a green-color priming picture facilitates participants' positive responses while a red-color picture polarizes negative responses [69,71,73]. Because the green/red colors are often associated with go/no-go signals, the cue color per se could influence the pattern of responses. Thus, the exact role of the attentional cue needs further clarification.

1.4. Purposes of the Current Study

To address these unresolved issues, the current study employs the mouse-tracking paradigm rather than the response time method to examine the role of attention in subliminal semantic processing. Following the experimental procedures in the Naccache et al. study [39], participants' attentional status was manipulated by occasionally presenting an attentional cue. To investigate the influence of cue color, the experiment contrasted two between-participant conditions: one group of participants received only a green cue (i.e., the green condition), while the other group received only a red cue (i.e., the red condition). The experiment seeks to clarify (1) whether subliminal semantic processing occurs when attention is disrupted, (2) the influence of cue color on subliminal semantic processing, and (3) whether subliminal semantic processing and attentional status can be revealed by characteristics of cursor motion trajectories.

2. Materials and Methods

2.1. Overview of the Experiment

We combined the procedure of the second experiment in the Naccache et al. (2002) research and the mouse-tracking technique [39]. Participants indicated if a target digit was larger or smaller than 5, and a masked priming digit was presented prior to the target. In some trials, an attentional cue (a green square) was shown to signal impending digits. Unlike the original study, however, we added a between-subject condition to manipulate the cue color—in addition to the green square cue, participants received a red square cue in another condition. In this manner, we investigated the role of the cue in a 2 × 2 × 2 factorial design, in which congruency (congruent vs. incongruent—within-subject factor), cue (cued vs. uncued—within-subject factor), and cue color (green vs. red—between-subject factor) were contrasted. If top-down attention enhances subliminal processing of the digits, both red and green cues should elicit more substantial priming effects relative to uncued conditions. If the cue color influences semantic priming, red and green cues should lead to different patterns of congruency effects.

2.2. Participants

Eighty undergraduate students participated in the experiment for course credit. They were randomly assigned to the green (n = 40) or red (n = 40) condition. All participants' response accuracy was above 80%. Three participants failed to finish the experiment, and data from 77 participants (n = 37 in the green condition; n = 40 in the red condition) were analyzed. There were 18 males and 19 females in the green condition while 18 males and 22 females were in the red condition. Their ages ranged from 17 to 23 (M_{green} = 19.68, SD_{green} = 2.12; M_{red} = 20.08, SD_{red} = 1.95). A meta-analysis shows that 23 studies on masked semantic priming (within-subject design) had a mean sample size of twenty and a mean effect size of 0.8 with a 95% confidence interval [0.60, 1.00] [5]. Using this pooled estimate of effect size, our prospective power analysis indicates that a sample size above 37 for each condition is sufficient to detect a congruency effect size of 0.6 at an alpha level of 0.05 with a power of 0.90.

2.3. Materials and Apparatus

We employed single-digit numbers (i.e., 1, 4, 6, or 9) as the prime and target. The display refresh rate of the monitor was 70 Hz, and the resolution was 1280 × 720. The computer program was developed using Microsoft Visual Studio v.15.0, and the computer mouse was a Logitech M100 Corded Mouse.

2.4. Experimental Procedure

In the mouse-tracking paradigm, participants move a computer mouse to select a choice, and the area under the curve (AUC) is measured in each trial (Figure 2). A smaller AUC denotes a more straightforward and certain response, while a larger AUC reflects uncertainty during the response and more distraction from the unselected option [15,30,59].

To depict the curve, the computer program obtains the location of the cursor on the monitor screen every 15 ms; all data points in each trial are standardized into 101 steps with a linear interpolation method.

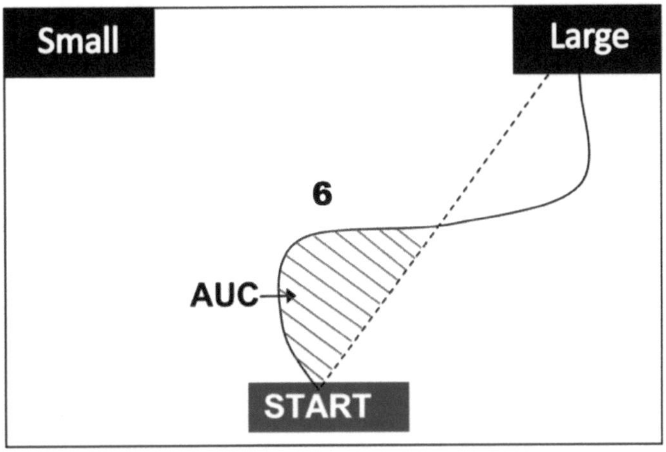

Figure 2. An illustration of the area under the curve (AUC). In each trial, the AUC (shaded area) is measured as the number of pixels enclosed by the dashed direct line connecting the starting and ending points and the actual cursor curve that goes over the direct path toward the unselected option. Any area that goes over the direct path toward the selected choice is subtracted as negative AUC. The cursor always starts from the center of the "START" button and ends where participants click one of the two options.

The experiment had two phases: a number judgment task and an awareness test. In the number judgment task, participants were assigned randomly to one of the two conditions: the green condition (i.e., the cue color is green) or the red condition. The procedure followed Experiment 2 in the Naccache et al. study [39], except that the red condition was added.

In each trial, a mask (i.e., a black-white rectangle doodled with circles and lines) was displayed for a random number of frames (i.e., 15 to 25 frames) in the center of the screen; each frame lasted for a fixed duration of 71 ms. Thus, this pre-mask lasted randomly from 1065 ms to 1775 ms with a 71 ms increment. Succeeding the pre-mask, a cue (i.e., a 200 ms green/red square) in the cued trials or one frame of the mask (71 ms) in the uncued trials was presented. The cue was green for participants assigned to the green condition and red for those to the red condition. Following the cue (or the 71 ms mask), 4 frames of 71 ms masks were displayed, succeeded by a 29 ms priming digit (i.e., 1, 4, 6, or 9), and then a 71 ms post-mask was presented. Following that, a target number (i.e., 1, 4, 6, or 9) was presented for 200 ms (Figure 3). Participants were required to judge if the target number was smaller or larger than five and move a computer mouse to select the "Small" or "Large" button on the top left/right locations on the screen (Figure 2). The locations of the "Small" and "Large" buttons were counterbalanced among participants.

In this number judgment task, there were 96 cued trials and 192 uncued trials, presented in random order. Half of the trials were congruent (i.e., the target and prime were both smaller or larger than five; e.g., prime = 4, target = 1), and the remaining half were incongruent (e.g., prime = 4, target = 6). Before the number judgment task, there was a practice session, where 96 trials (32 cued and 64 uncued trials) were performed. Trials with a response time longer than 5000 ms were excluded from the data analysis (0.6% of total trials).

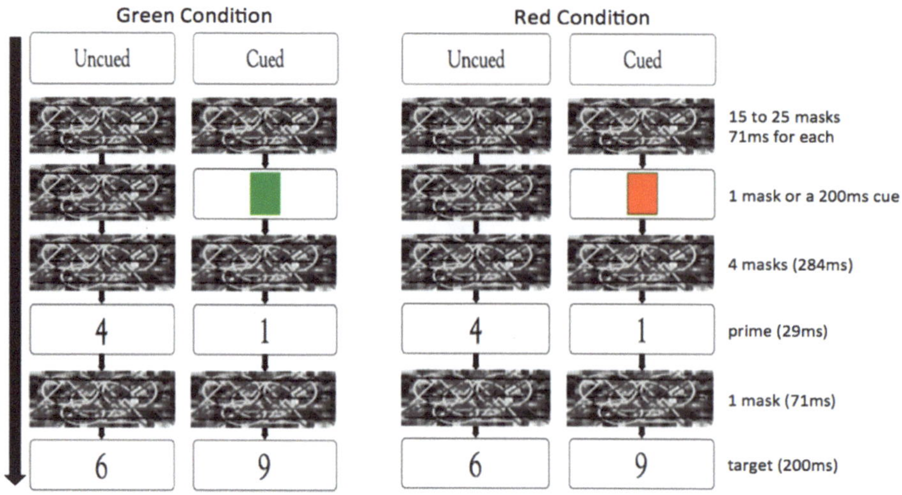

Figure 3. The procedure in the green and red conditions. The two conditions differed only in the cue color.

Following the number judgment task, an awareness test was given, where trials (192 trials in total) were the same as in the number judgment task. Participants were explicitly informed that a priming digit would flash briefly and were required to judge if the prime was visible and whether it was larger or smaller than 5, instead of judging the target digit [37]. The *d'* was calculated for each participant to check the visibility of priming digits: correctly indicating a prime as larger than five was defined as "hit" and choosing "Large" when the prime was smaller than five as "false alarm". We applied linear regression analyses for congruency effects on *d*'s to examine the extent to which the visibility of primes would contribute to priming effects [37].

3. Results

3.1. The Three-Way ANOVA

The experiment was a 2 (cue: cued, uncued; within-subject) × 2 (cue color: green, red; between-subject) × 2 (congruency: congruent, incongruent; within-subject) factorial design. The dependent variable was the area under the curve (AUC).

A three-way ANOVA (cue color × cue × congruency) revealed significant impacts of cue color ($F(1, 75) = 5.62$, $MSE = 48{,}246{,}665.40$, $p = 0.02$, partial $\eta^2 = 0.07$), congruency ($F(1, 75) = 53.43$, $MSE = 953{,}709.97$, $p < 0.001$, partial $\eta^2 = 0.42$), and interaction between congruency and cue ($F(1, 75) = 13.54$, $MSE = 618{,}393.72$, $p < 0.001$, partial $\eta^2 = 0.15$). The three-way interaction was not significant ($F < 1$).

The congruency effects were prominent, suggesting that the masked priming numbers can be processed at a semantic level without much awareness. Below, we report the impacts of presenting an attentional cue and those of the cue color on congruency effects, respectively.

3.2. The Impact of Presenting an Attentional Cue

The significant interaction between congruency and cue ($F(1, 75) = 13.54$, $MSE = 618{,}393.72$, $p < 0.001$, partial $\eta^2 = 0.15$) shows that the congruency effects were larger in the cued than the uncued conditions (Figure 4), which means presenting an attentional cue enhanced the congruency effects. It is noticeable that congruency effects were still robust in uncued conditions (Figure 4), suggesting that the masked numbers elicited semantic processing even when the temporal attentional window was disrupted (i.e., in the uncued condition).

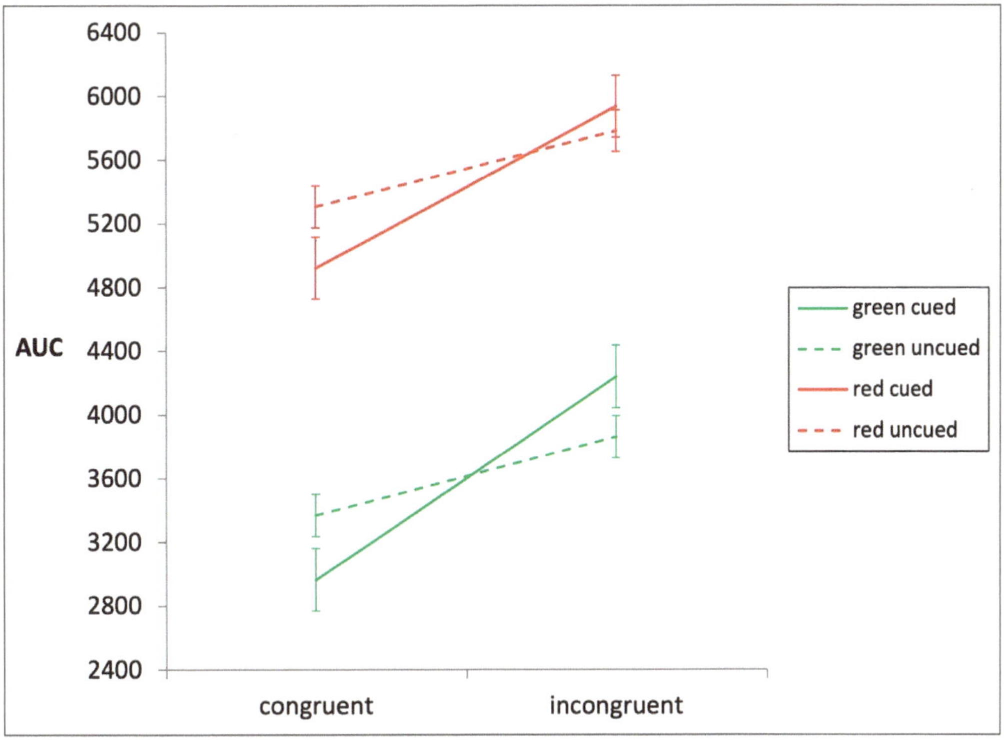

Figure 4. The average AUCs of congruent and incongruent trials are shown for cued and uncued conditions in the green and red conditions, respectively. The AUCs were measured by amounts of pixels on the screen.

3.3. The Impact of Cue Color

The cue color influenced overall AUCs. The average AUC in the red condition (M = 3611.62, SD = 3383.87) was larger than that in the green condition (M = 5506.54, SD = 3724.76); (F(1, 75) = 5.62, MSE = 4,8246,665.40, p = 0.02, partial η^2 = 0.07). Meanwhile, the interaction between cue and congruency was significant in the green condition (F(1, 36) = 11.00, MSE = 515,318.83, p = 0.002, partial η^2 = 0.23) but not in the red condition (F(1, 39) = 4.04, MSE = 713,539.77, p = 0.051, partial η^2 = 0.09).

3.4. Characteristic of Average Motion Trajectories: Congruent vs. Incongruent Trials in Green and Red Conditions

Figure 5 below shows the average cursor motion trajectories in the cued/uncued conditions as well as in the green (a) and red (b) conditions. Blue trajectories are for congruent trials while brown trajectories are for incongruent trials. The congruency effect is illustrated by the distance between the congruent and the incongruent trajectories toward the same ending location, which corresponds to the difference in average AUCs between congruent and incongruent trials (i.e., the size of congruency effects). The congruency effect is larger in the cued than the uncued conditions yet still present in the uncued condition (Figure 5), suggesting that subliminal semantic processing occurs in the near-absence of top-down attention.

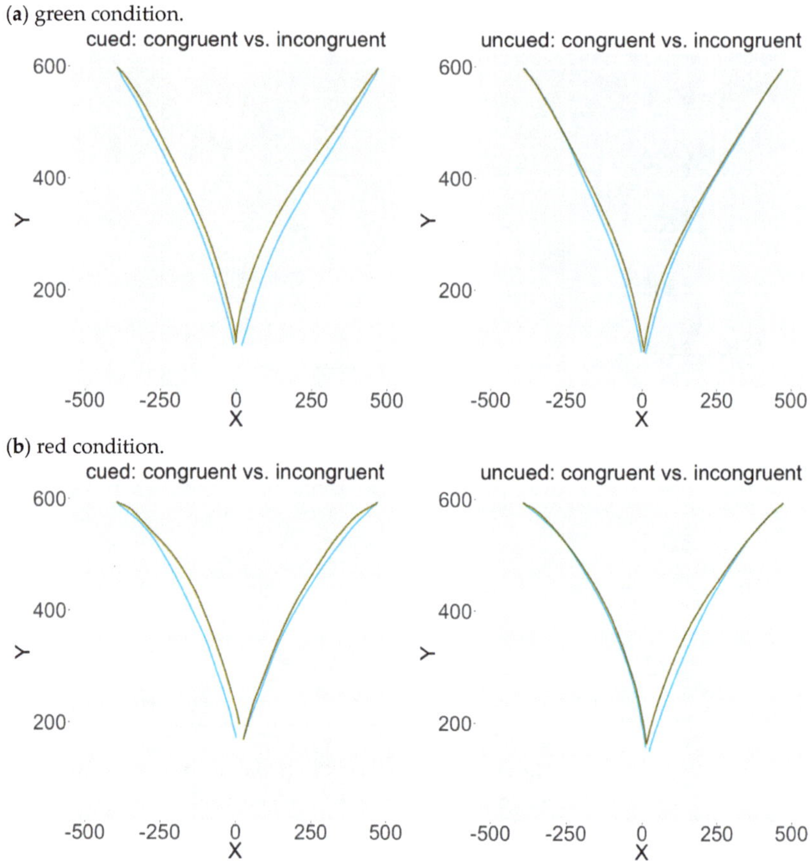

Figure 5. Average trajectories in the green condition (**a**) and the red condition (**b**), with the cued condition on the left panel and the uncued condition on the right panel. Trajectories in light blue represent congruent trials while those in light brown represent incongruent trials. All trajectories start from the lower-middle location and travel to one of the two options either on the upper-left corner (e.g., "Small") or on the upper-right corner (e.g., "Large") of the screen, depending on participants' choices. The congruency effect is illustrated by the distance between the congruent and the incongruent trajectories toward the same ending location, which corresponds to the difference in average AUCs between congruent and incongruent trials. The X and Y axes denote amounts of pixels on the screen.

3.5. Awareness Analyses

In the awareness test, no participant could correctly report any priming digit. To further examine the visibility of primes, we calculated the d's in the awareness test [37]. The d's were statistically indistinguishable from zero: in the green condition, $M = 0.001$, $t(36) = 0.10$, and $p = 0.923$; in the red condition, $M = 0.004$, $t(39) = 0.49$, and $p = 0.625$; these results suggest that participants had little awareness of primes. Linear regressions showed that d's could not predict congruency effects; thus, congruency effects were unlikely to be impacted by the visibility of primes (Tables 1 and 2). The intercepts were higher than zero at null d', indicating that congruency effects were significant at a subliminal level [37].

Table 1. Regressions with congruency effects [1] on *d*'s in the green condition.

	Predictor	*b*	*SE*	95% CI	*t*	*p*
Cued	(intercept)	1251.03	198.07	[862.81, 1639.25]	6.32 ***	<0.001
	d'	0.17	0.16	[−0.16, 0.50]	1.02	0.313
Uncued	(intercept)	479.97	133.92	[217.49, 742.45]	3.58 **	0.001
	d'	0.13	0.17	[−0.19, 0.46]	0.80	0.429

[1] The congruency effect = $AUC_{incongruent} - AUC_{congruent}$. ** $p < 0.01$, two-tailed. *** $p < 0.001$, two-tailed.

Table 2. Regressions with congruency effects on *d*'s in the red condition.

	Predictor	*b*	*SE*	95% CI	*t*	*p*
Cued	(intercept)	1014.28	297.93	[430.34, 1598.22]	3.40 **	0.002
	d'	−0.01	0.17	[−0.35, 0.34]	−0.03	0.973
Uncued	(intercept)	474.07	132.30	[214.76, 733.38]	3.57 **	0.001
	d'	0.04	0.16	[−0.28, 0.36]	0.25	0.807

** $p < 0.01$, two-tailed.

4. Discussion

The overall AUCs in the red condition were larger than those in the green condition, suggesting that cue color per se influenced subliminal semantic processing. Given that color is closely associated with semantic priming (e.g., the red color is often associated with negative connotations like "alarm" and "stop") [72,73], the red cue could make participants more cautious and hesitant to respond, resulting in larger AUCs. The same rationale could explain the smaller AUC in the green condition—positive meanings implied by the green color, such as "safe" and "go", might have prompted more straightforward responses [69,71]. The results suggest that researchers manipulating attentional status with visual cues shall proceed with caution and minimize the influence of cue color on priming effects.

The relationship between attention and consciousness is a long-debated core issue in human cognition. The classic theory claims that conscious processing is guided by attention while unconscious processing is autonomous, stereotypical, and independent of attention [74,75]. In contrast, recent research suggests that attention can orient to invisible stimuli and influence unconscious processing. Accumulated evidence suggests that top-down attention can modulate subliminal processing at a semantic level [76]. According to the Global Workspace Theory (GWT), unconscious processing needs assistance from top-down attention to reach a semantic level [77,78]. More specifically, subliminal semantic information is automatically coded by specialized neural modules in the peripheral workspace and sent into the global workspace; the top-down control mechanism in the global workspace allocates more cognitive resources to the attended semantic information, leading to enhanced subliminal priming, while unattended subliminal processing quickly fades away [79].

Challenging this view, the current study finds significant congruency effects in the uncued conditions, suggesting that unconscious information is processed at a semantic level even with little support from attention. Thus, unconscious processing need not be sustained by attention to produce semantic priming [62,70]. On the other hand, congruency effects are larger in the cued than in the uncued conditions, consistent with previous research on GWT [39,80]. Therefore, the current study rather proposes a revision than disproves the GWT: in the global workspace, the top-down control mechanism does enhance attended unconscious information; meanwhile, at least some types of semantic processing, though not sustained by attention, persist for a notable duration in a bottom-up manner instead of immediately vanishing.

The current study adds solid evidence to the robustness and flexibility of subliminal semantic processing, contrary to the traditional opinion that unconscious processing is short-lived and stereotypical [81,82]. Given the close interplay of top-down attention and subliminal information, the boundary between conscious control and unconscious processing becomes vague [83]. Since hand movements reflected subliminal processing, the underlying mechanism linking overt body motions and implicit cognition deserves further clarification to enrich the embodied cognition theory.

Overall, the mouse-tracking measure (e.g., AUC) shows advantages over response time to detect implicit cognitive conflicts [71]. Conventionally, decoding real-time cognitive activities relies on brain imaging techniques (e.g., EEG and fMRI), which require costly devices and cumbersome settings. In contrast, the mouse-tracking technique is an affordable, easy-to-use method adaptive to a broad range of user environments [59]. A computer mouse is one of the most ubiquitous devices and the cursor motion is traced by time-stamped x-y coordinates, minimizing costs for data collection and analysis. Given that subthreshold motor control processes cannot be revealed by button responses yet can be detected by BOLD signals [84], it is also possible that similar mechanisms can be measured with mouse-tracking. Thus, the mouse-tracking technique provides a promising tool to dissect dynamic cognitive processes and contribute to brain sciences in future research [16,85].

Admittedly, the current study comes with limitations. First, since disrupted attention in the uncued condition did not guarantee a complete absence of attention, we cannot rule out the possibility that minimal attention still played a role there. Rather, because semantic priming was present in both cued and uncued conditions, assistance from attention is unlikely to be essential for subliminal semantic processing. Second, the main effect of cue color was based on between-group comparisons, which could be confounded with potential between-group individual differences in movement styles, if any. Although the between-group design helps prevent red and green colors from working against each other, future research shall consider within-group design to re-examine the color effect.

5. Conclusions

In conclusion, the current study investigates the role of attention in subliminal semantic processing with the mouse-tracking technique and provides three critical messages. First, while attention enhances subliminal semantic processing, the cue color per se influences semantic priming effects. Second, top-down attention is unlikely to be a prerequisite for subliminal semantic processing, which occurs in the near-absence of attention. The findings add solid evidence to the robustness and flexibility of unconscious processing and propose a revision of the Global Workspace Theory. Third, subliminal semantic processing, attentional status, and the influence of cue color can be revealed by cursor motion features, such as the curvature of trajectories, demonstrating that the mouse-tracking technique is a promising tool to reveal dynamic implicit cognition on a broad range of topics in future research.

Author Contributions: Conceptualization, K.X. and T.Y.; methodology, K.X. and T.Y.; software, K.X., A.Z. and T.Y.; validation, K.X., J.Q., C.G. and T.Y.; formal analysis, K.X., A.Z., F.D. and T.Y.; investigation, K.X. and A.Z.; resources, K.X., F.D. and T.Y.; data curation, K.X., J.Q., C.G. and T.Y.; writing—original draft preparation, K.X.; writing—review and editing, K.X., J.Q., C.G. and T.Y.; visualization, K.X., F.D. and T.Y.; supervision, K.X. and T.Y.; project administration, K.X. and T.Y.; funding acquisition, K.X. All authors have read and agreed to the published version of the manuscript.

Funding: This research was funded by the Young Scholars Program of the National Social Science Fund of China for Education Sciences ("The Brain Mechanism underlying the Intervention of Hand Tracking Techniques on Children's Digital Reading—Based on an Embodied Cognition Perspective"), grant number CBA220316.

Institutional Review Board Statement: The study was conducted in accordance with the Declaration of Helsinki and approved by the Institutional Review Board of Texas A&M University (IRB No. 2012-0451M).

Informed Consent Statement: Informed consent was obtained from all subjects involved in the study.

Data Availability Statement: The data presented in this study are available on request from the corresponding author.

Conflicts of Interest: The authors declare no conflict of interest. The funders had no role in the design of the study; in the collection, analyses, or interpretation of data; in the writing of the manuscript, or in the decision to publish the results.

References

1. Collins, A.M.; Loftus, E.F. A spreading-activation theory of semantic processing. *Psychol. Rev.* **1975**, *82*, 407–428. [CrossRef]
2. Anderson, J.R. A spreading activation theory of memory. *J. Verbal Learn. Verbal Behav.* **1983**, *22*, 261–295. [CrossRef]
3. Dehaene, S.; Changeux, J.P.; Naccache, L.; Sackur, J.; Sergent, C. Conscious, preconscious, and subliminal processing: A testable taxonomy. *Trends Cogn. Sci.* **2006**, *10*, 204–211. [CrossRef]
4. Koch, C.; Tsuchiya, N. Attention and consciousness: Two distinct brain processes. *Trends Cogn. Sci.* **2007**, *11*, 16–22. [CrossRef]
5. Van den Bussche, E.; Van den Noortgate, W.; Reynvoet, B. Mechanisms of masked priming: A meta-analysis. *Psychol. Bull.* **2009**, *135*, 452–477. [CrossRef]
6. Barsalou, L.W.; Niedenthal, P.M.; Barbey, A.K.; Ruppert, J.A. Social embodiment. In *Psychology of Learning and Motivation*; Ross, B.H., Ed.; Academic Press: San Diego, CA, USA, 2003; Volume 43, pp. 43–92.
7. Xiao, K.; Gao, Y.; Imran, S.A.; Chowdhury, S.; Commuri, S.; Jiang, F. Cross-modal motion aftereffects transfer between vision and touch in early deaf adults. *Sci. Rep.* **2021**, *11*, 4395. [CrossRef]
8. Thelen, E.; Spencer, J.P. Postural control during reaching in young infants: A dynamic systems approach. *Neurosci. Biobehav. Rev.* **1998**, *22*, 507–514. [CrossRef]
9. Keysers, C.; Paracampo, R.; Gazzola, V. What neuromodulation and lesion studies tell us about the function of the mirror neuron system and embodied cognition. *Curr. Opin. Psychol.* **2018**, *24*, 35–40. [CrossRef]
10. Coombes, S.A.; Janelle, C.M.; Duley, A.R. Emotion and motor control: Movement attributes following affective picture processing. *J. Mot. Behav.* **2005**, *37*, 425–436. [CrossRef]
11. Harris, C.M.; Wolpert, D.M. Signal-dependent noise determines motor planning. *Nature* **1998**, *394*, 780–784. [CrossRef]
12. Gallivan, J.P.; Bowman, N.A.; Chapman, C.S.; Wolpert, D.M.; Flanagan, J.R. The sequential encoding of competing action goals involves dynamic restructuring of motor plans in working memory. *J. Neurophysiol.* **2016**, *115*, 3113–3122. [CrossRef] [PubMed]
13. Song, J.H.; Nakayama, K. Hidden cognitive states revealed in choice reaching tasks. *Trends Cogn. Sci.* **2009**, *13*, 360–366. [CrossRef]
14. Mollo, G.; Pulvermüller, F.; Hauk, O. Movement priming of EEG/MEG brain responses for action-words characterizes the link between language and action. *Cortex* **2016**, *74*, 262–276. [CrossRef] [PubMed]
15. Spivey, M.J.; Dale, R. Continuous dynamics in real-time cognition. *Curr. Dir. Psychol. Sci.* **2006**, *15*, 207–211. [CrossRef]
16. Dale, R.; Kehoe, C.; Spivey, M.J. Graded motor responses in the time course of categorizing atypical exemplars. *Mem. Cogn.* **2007**, *35*, 15–28. [CrossRef] [PubMed]
17. Wolpert, D.M.; Landy, M.S. Motor control is decision-making. *Curr. Opin. Neurobiol.* **2012**, *22*, 996–1003. [CrossRef] [PubMed]
18. Dotan, D.; Pinheiro-Chagas, P.; Al Roumi, F.; Dehaene, S. Track it to crack it: Dissecting processing stages with finger tracking. *Trends Cogn. Sci.* **2019**, *23*, 1058–1070. [CrossRef]
19. Yamauchi, T.; Kohn, N.; Yu, N.Y. Tracking mouse movement in feature inference: Category labels are different from feature labels. *Mem. Cogn.* **2007**, *35*, 852–863. [CrossRef]
20. Freeman, J.B.; Pauker, K.; Apfelbaum, E.P.; Ambady, N. Continuous dynamics in the real-time perception of race. *J. Exp. Soc. Psychol.* **2010**, *46*, 179–185. [CrossRef]
21. Yamauchi, T.; Xiao, K. Reading emotion from mouse cursor motions: Affective computing approach. *Cogn. Sci.* **2018**, *42*, 771–819. [CrossRef]
22. Yamauchi, T.; Bowman, C. Mining cursor motions to find the gender, experience and feelings of computer users. In Proceedings of the IEEE International Conference on Data Mining (ICDM 2014): Workshop on Domain Driven Data Mining, Shenzhen, China, 14 December 2014; IEEE Computer Society: Piscataway, NJ, USA; pp. 221–230.
23. Freihaut, P.; Göritz, A.S.; Rockstroh, C.; Blum, J. Tracking stress via the computer mouse? Promises and challenges of a potential behavioral stress marker. *Behav. Res. Methods* **2021**, *53*, 2281–2301. [PubMed]
24. Freihaut, P.; Göritz, A.S. Using the computer mouse for stress measurement—An empirical investigation and critical review. *Int. J. Hum. Comput. Stud.* **2021**, *145*, 102520. [CrossRef]
25. Leontyev, A.; Yamauchi, T. Discerning Mouse Trajectory Features with the Drift Diffusion Model. *Cogn. Sci.* **2021**, *45*, e13046. [CrossRef]
26. Leontyev, A.; Sun, S.; Wolfe, M.; Yamauchi, T. Augmented Go/No-Go Task: Mouse Cursor Motion Measures Improve ADHD Symptom Assessment in Healthy College Students. *Front. Psychol.* **2018**, *9*, 496. [CrossRef] [PubMed]
27. Leontyev, A.; Yamauchi, T. Mouse movement measures enhance the stop-signal task in adult ADHD assessment. *PLoS ONE* **2019**, *14*, e0225437. [CrossRef]
28. Leontyev, A.; Yamauchi, T.; Razavi, M. Machine Learning Stop Signal Test (ML-SST): ML-based Mouse Tracking Enhances Adult ADHD Diagnosis. In Proceedings of the 2019 8th International Conference on Affective Computing and Intelligent Interaction Workshops and Demos (ACIIW), Cambridge, UK, 3–6 September 2019; IEEE Computer Society: Piscataway, NJ, USA.

29. Freeman, J.B.; Johnson, K.L. More than meets the eye: Split-second social perception. *Trends Cogn. Sci.* **2016**, *20*, 362–374. [CrossRef]
30. Freeman, J.B. Doing psychological science by hand. *Curr. Dir. Psychol. Sci.* **2018**, *27*, 315–323. [CrossRef] [PubMed]
31. Tafuro, A.; Vallesi, A.; Ambrosini, E. Cognitive brakes in interference resolution: A mouse-tracking and EEG co-registration study. *Cortex* **2020**, *133*, 188–200. [CrossRef]
32. Freeman, J.B.; Ambady, N. Motions of the hand expose the partial and parallel activation of stereotypes. *Psychol. Sci.* **2009**, *20*, 1183–1188. [CrossRef] [PubMed]
33. Wojnowicz, M.T.; Ferguson, M.J.; Dale, R.; Spivey, M.J. The self-organization of explicit attitudes. *Psychol. Sci.* **2009**, *20*, 1428–1435. [CrossRef] [PubMed]
34. Calluso, C.; Committeri, G.; Pezzulo, G.; Lepora, N.; Tosoni, A. Analysis of hand kinematics reveals inter-individual differences in intertemporal decision dynamics. *Exp. Brain Res.* **2015**, *233*, 3597–3611. [CrossRef]
35. Farmer, T.A.; Anderson, S.E.; Spivey, M.J. Gradiency and visual context in syntactic garden-paths. *J. Mem. Lang.* **2007**, *57*, 570–595. [CrossRef]
36. Dehaene, S.; Naccache, L.; Le Clec'H, G.; Koechlin, E.; Mueller, M.; Dehaene-Lambertz, G.; van de Moortele, P.F.; Le Bihan, D. Imaging unconscious semantic priming. *Nature* **1998**, *395*, 597–600. [CrossRef] [PubMed]
37. Greenwald, A.G.; Sean, C.D.; Richard, L.A. Three cognitive markers of unconscious semantic activation. *Science* **1996**, *273*, 1699–1702. [CrossRef] [PubMed]
38. Berkovitch, L.; Dehaene, S. Subliminal syntactic priming. *Cogn. Psychol.* **2019**, *109*, 26–46. [CrossRef]
39. Naccache, L.; Blandin, E.; Dehaene, S. Unconscious masked priming depends on temporal attention. *Psychol. Sci.* **2002**, *13*, 416–424. [CrossRef]
40. Prasad, S.; Mishra, R.K. The nature of unconscious attention to subliminal cues. *Vision* **2019**, *3*, 38. [CrossRef]
41. Mao, X.; Xie, C.; Shi, J.; Huang, Q.; Jiang, R.; Meng, F.; Shen, H.; Miao, L.; Cui, S.; Li, A. The essential role of executive attention in unconscious visuomotor priming. *Front. Psychol.* **2022**, *13*, 800781. [CrossRef]
42. Van Boxtel, J.J.; Tsuchiya, N.; Koch, C. Consciousness and attention: On sufficiency and necessity. *Front. Psychol.* **2010**, *1*, 217. [CrossRef]
43. Cohen, M.A.; Dennett, D.C.; Kanwisher, N. What is the bandwidth of perceptual experience? *Trends Cogn. Sci.* **2016**, *20*, 324–335. [CrossRef] [PubMed]
44. Mashour, G.A.; Roelfsema, P.; Changeux, J.P.; Dehaene, S. Conscious processing and the global neuronal workspace hypothesis. *Neuron* **2020**, *105*, 776–798. [CrossRef]
45. Sim, E.J.; Harpaintner, M.; Kiefer, M. Is subliminal face processing modulated by attentional task sets? Evidence from masked priming effects in a gender decision task. *Open Psychol.* **2020**, *2*, 76–89. [CrossRef]
46. Khalid, S.; Finkbeiner, M.; König, P.; Ansorge, U. Subcortical human face processing? Evidence from masked priming. *J. Exp. Psychol. Hum. Percept. Perform.* **2013**, *39*, 989–1002. [CrossRef]
47. Wang, J.; JIAO, R.; Zhang, M. The mechanism of the effect of task setting on negative compatibility effect: The effect of top-down cognition control on subliminal prime processing. *Acta Psychol. Sin.* **2016**, *48*, 1370–1378. [CrossRef]
48. Kiefer, M.; Trumpp, N.M.; Schaitz, C.; Reuss, H.; Kunde, W. Attentional modulation of masked semantic priming by visible and masked task cues. *Cognition* **2019**, *187*, 62–77. [CrossRef]
49. Block, N. Perceptual consciousness overflows cognitive access. *Trends Cogn. Sci.* **2011**, *15*, 567–575. [CrossRef]
50. Block, N. Rich conscious perception outside focal attention. *Trends Cogn. Sci.* **2014**, *18*, 445–447. [CrossRef]
51. Handschack, J.; Rothkirch, M.; Sterzer, P.; Hesselmann, G. No effect of attentional modulation by spatial cueing in a masked numerical priming paradigm using continuous flash suppression (CFS). *PeerJ* **2023**, *11*, e14607. [CrossRef] [PubMed]
52. Eo, K.; Cha, O.; Chong, S.C.; Kang, M.S. Less is more: Semantic information survives interocular suppression when attention is diverted. *J. Neurosci.* **2016**, *36*, 5489–5497. [CrossRef]
53. Rahnev, D.A.; Huang, E.; Lau, H. Subliminal stimuli in the near absence of attention influence top-down cognitive control. *Atten. Percept. Psychophys.* **2012**, *74*, 521–532. [CrossRef] [PubMed]
54. Finkbeiner, M.; Palermo, R. The role of spatial attention in nonconscious processing: A comparison of face and nonface stimuli. *Psychol. Sci.* **2009**, *20*, 42–51. [CrossRef] [PubMed]
55. Reddy, L.; Reddy, L.; Koch, C. Face identification in the near-absence of focal attention. *Vis. Res.* **2006**, *46*, 2336–2343. [CrossRef]
56. Reddy, L.; Patrick, W.; Christof, K. Face-gender discrimination is possible in the near-absence of attention. *J. Vis.* **2004**, *4*, 4. [CrossRef]
57. Li, F.F.; VanRullen, R.; Koch, C.; Perona, P. Rapid natural scene categorization in the near absence of attention. *Proc. Natl. Acad. Sci. USA* **2002**, *99*, 9596–9601. [CrossRef] [PubMed]
58. Underwood, G.; Templeman, E.; Lamming, L.; Foulsham, T. Is attention necessary for object identification? Evidence from eye movements during the inspection of real-world scenes. *Conscious. Cogn.* **2008**, *17*, 159–170. [CrossRef]
59. Stillman, P.E.; Shen, X.; Ferguson, M.J. Ferguson. How mouse-tracking can advance social cognitive theory. *Trends Cogn. Sci.* **2018**, *22*, 531–543. [CrossRef]
60. Freeman, J.B.; Ambady, N. MouseTracker: Software for studying real-time mental processing using a computer mouse-tracking method. *Behav. Res. Methods* **2010**, *42*, 226–241. [CrossRef] [PubMed]
61. Freeman, J.B.; Ambady, N. A dynamic interactive theory of person construal. *Psychol. Rev.* **2011**, *118*, 247–279. [CrossRef]

62. Xiao, K.; Yamauchi, T. Subliminal semantic priming in near absence of attention: A cursor motion study. *Conscious. Cogn.* **2015**, *38*, 88–98. [CrossRef]
63. Lowry, M.; Trivedi, N.; Boyd, P.; Julian, A.; Treviño, M.; Lama, Y.; Heley, K.; Perna, F. Making decisions about health information on social media: A mouse-tracking study. *Cogn. Res. Princ. Implic.* **2022**, *7*, 68. [CrossRef]
64. Vadillo, M.A.; Konstantinidis, E.; Shanks, D.R. Underpowered samples, false negatives, and unconscious learning. *Psychon. Bull. Rev.* **2016**, *23*, 87–102. [CrossRef]
65. Yang, Y.H.; Tien, Y.H.; Yang, P.L.; Yeh, S.L. Role of consciousness in temporal integration of semantic information. *Cogn. Affect. Behav. Neurosci.* **2017**, *17*, 954–972. [CrossRef]
66. Avneon, M.; Lamy, D. Reexamining unconscious response priming: A liminal-prime paradigm. *Conscious. Cogn.* **2018**, *59*, 87–103. [CrossRef]
67. Schneider, I.K.; van Harreveld, F.; Rotteveel, M.; Topolinski, S.; van der Pligt, J.; Schwarz, N.; Koole, S.L. The path of ambivalence: Tracing the pull of opposing evaluations using mouse trajectories. *Front. Psychol.* **2015**, *6*, 996. [CrossRef] [PubMed]
68. Yamamoto, N.; Incera, S.; McLennan, C.T. A reverse Stroop task with mouse tracking. *Front. Psychol.* **2016**, *7*, 670. [CrossRef] [PubMed]
69. Xiao, K.; Yamauchi, T. Semantic priming revealed by mouse movement trajectories. *Conscious. Cogn.* **2014**, *27*, 42–52. [CrossRef]
70. Xiao, K.; Yamauchi, T. The role of attention in subliminal semantic processing: A mouse tracking study. *PLoS ONE* **2017**, *12*, e0178740. [CrossRef]
71. Xiao, K.; Yamauchi, T.; Bowman, C. Assessing Masked Semantic Priming: Cursor Trajectory versus Response Time Measures. In Proceedings of the 37th Annual Meeting of the Cognitive Science Society (CogSci 2015), Pasadena, CA, USA, 22–25 July 2015.
72. Kareklas, I.; Muehling, D.D.; King, S. The effect of color and self-view priming in persuasive communications. *J. Bus. Res.* **2019**, *98*, 33–49. [CrossRef]
73. Argiris, G.; Rumiati, R.I.; Crepaldi, D. No fruits without color: Cross-modal priming and EEG reveal different roles for different features across semantic categories. *PLoS ONE* **2021**, *16*, e0234219. [CrossRef]
74. Schneider, W.; Shiffrin, R.M. Controlled and automatic human information processing: I. Detection, search, and attention. *Psychol. Rev.* **1977**, *84*, 1–66. [CrossRef]
75. Shiffrin, R.M.; Schneider, W. Controlled and automatic human information processing: II. Perceptual learning, automatic attending and a general theory. *Psychol. Rev.* **1977**, *84*, 127–190. [CrossRef]
76. Cohen, M.A.; Cavanagh, P.; Chun, M.M.; Nakayama, K. The attentional requirements of consciousness. *Trends Cogn. Sci.* **2012**, *16*, 411–417. [CrossRef] [PubMed]
77. Dehaene, S.; Kerszberg, M.; Changeux, J.P. A neuronal model of a global workspace in effortful cognitive tasks. *Proc. Natl. Acad. Sci. USA* **1998**, *95*, 14529–14534. [CrossRef] [PubMed]
78. Dehaene, S.; Naccache, L. Towards a cognitive neuroscience of consciousness: Basic evidence and a workspace framework. *Cognition* **2001**, *79*, 1–37. [CrossRef]
79. Baars, B.J.; Franklin, S.; Ramsoy, T.Z. Global workspace dynamics: Cortical "binding and propagation" enables conscious contents. *Front. Psychol.* **2013**, *4*, 200. [CrossRef]
80. Kiefer, M.; Brendel, D. Attentional modulation of unconscious "automatic" processes: Evidence from event-related potentials in a masked priming paradigm. *J. Cogn. Neurosci.* **2006**, *18*, 184–198. [CrossRef] [PubMed]
81. Mudrik, L.; Faivre, N.; Koch, C. Information integration without awareness. *Trends Cogn. Sci.* **2014**, *18*, 488–496. [CrossRef] [PubMed]
82. Mudrik, L.; Breska, A.; Lamy, D.; Deouell, L.Y. Integration without awareness: Expanding the limits of unconscious processing. *Psychol. Sci.* **2011**, *22*, 764–770. [CrossRef]
83. Koch, C.; Massimini, M.; Boly, M.; Tononi, G. Neural correlates of consciousness: Progress and problems. *Nat. Rev. Neurosci.* **2016**, *17*, 307–321. [CrossRef]
84. Gavazzi, G.; Rossi, A.; Orsolini, S.; Diciotti, S.; Giovannelli, F.; Salvadori, E.; Pantoni, L.; Mascalchi, M.; Viggiano, M.P. Impulsivity trait and proactive cognitive control: An fMRI study. *Eur. J. Neurosci.* **2019**, *49*, 1171–1179. [CrossRef]
85. Quétard, B.; Spatola, N.; Parris, B.A.; Ferrand, L.; Augustinova, M. A mouse-tracking study of the composite nature of the Stroop effect at the level of response execution. *PLoS ONE* **2023**, *18*, e0279036. [CrossRef] [PubMed]

brain sciences

MDPI

Review

The Effect of Physical Exercise on People with Psychosis: A Qualitative Critical Review of Neuroimaging Findings

Francesca Saviola [1], Giacomo Deste [1,2], Stefano Barlati [1,2], Antonio Vita [1,2], Roberto Gasparotti [1,3] and Daniele Corbo [1,*]

1 Department of Medical and Surgical Specialties, Radiological Sciences and Public Health, University of Brescia, 25123 Brescia, Italy; francesca.saviola@unibs.it (F.S.); stefano.barlati@unibs.it (S.B.); roberto.gasparotti@unibs.it (R.G.)
2 Department of Mental Health and Addiction Services, ASST Spedali Civili of Brescia, 25123 Brescia, Italy
3 Neuroradiology Unit, ASST Spedali Civili of Brescia, 25123 Brescia, Italy
* Correspondence: daniele.corbo@unibs.it

Abstract: Recently, genuine motor abnormalities have been recognized as prodromal and predictive signs of psychosis onset and progression. Therefore, physical exercise could represent a potentially relevant clinical tool in promoting the reshaping of neural connections in motor circuitry. The aim of this review is to provide an overview of the literature on neuroimaging findings as a result of physical treatment in psychosis cohorts. Twenty-one studies, all research articles, were included and discussed in this narrative review. Here, we first outlined how the psychotic brain is susceptible to structural plastic changes after aerobic physical training in pathognomic brain areas (i.e., temporal, hippocampal and parahippocampal regions). Secondly, we focused on functional changes, both region-specific and in terms of connections, to gain insights into the involvement of distant but inter-related neural regions in the plastic process occurring after treatment. Third, we attempted to bridge neural plastic changes occurring after physical interventions with clinical and cognitive outcomes of psychotic patients in order to assess the relevance of such neural reshaping in the psychiatric rehabilitation field. In conclusion, we suggest that the current state of the art is presenting physical intervention as effective in promoting neural changes for patients with psychosis; it is not only useful at the onset of the pathology but also in improving the course of the illness and its functional outcome. However, more evidence is needed to improve our knowledge of the efficacy of physical exercise in plastically reorganizing the psychotic brain in the long term, especially within regions lacking specific investigations, such as motor circuitry.

Keywords: physical exercise; neural plasticity; brain connectomics; psychosis; genuine movement abnormalities

check for updates

Citation: Saviola, F.; Deste, G.; Barlati, S.; Vita, A.; Gasparotti, R.; Corbo, D. The Effect of Physical Exercise on People with Psychosis: A Qualitative Critical Review of Neuroimaging Findings. *Brain Sci.* **2023**, *13*, 923. https://doi.org/10.3390/brainsci13060923

Academic Editor: Notger G. Müller

Received: 13 May 2023
Revised: 31 May 2023
Accepted: 5 June 2023
Published: 7 June 2023

1. Introduction

Psychosis is a pathological condition characterized by a set of severe symptoms affecting different levels of everyday life (i.e., social, cognitive and perceptual domains) and is related to poor functional outcomes [1]. Signs of altered motor development are increasingly recognized as an important marker of risk for psychosis [2,3]. Indeed, a growing body of literature suggests that genuine movement abnormalities (GMA; e.g., akathisia, catatonia, dyskinesia, parkinsonism and psychomotor slowing) [4] are present long before the first signs of thought disorder, and prospective studies of youth at risk show that GMA may predict the transition to psychosis [5–7]. Therefore, it is alleged that motor circuitry has a relevant role in the pathophysiology of psychosis [4].

At a neural level, several brain regions were shown to be structurally implicated in the neuropathology of psychosis [8–12], with a main focus on the medial temporal lobe (together with hippocampal and parahippocampal areas), parietal lobe and subcortical areas involvement. However, there is still limited information about how functional motor areas disruption co-occur with GMA in psychosis and how different motor circuits may contribute to the understanding of the pathology [13].

Specifically, the functional somatomotor network (SMN) is a large-scale functional network usually involving both (i) motor circuitry within the precentral gyrus together with supplementary motor areas and (ii) somatosensory regions relying on the postcentral gyrus [14].

The SMN connections were found to be functional transdiagnostic hubs across several psychiatric disorders [15,16]. Indeed, dysconnectivity within the SMN is believed to be a common feature across multiple psychiatric conditions; whereas its connections to frontal areas are rather disrupted, with an illness-specific pattern, when considering psychotic disorders [16]. These findings highlight that psychiatric disorders share common impairments linked to perception and motor output (i.e., related to the bottom-up networking within regions of the motor circuitry); whereas unique brain connectivity profiles can be found for each pathology mediating more distinctive features of the disorder (i.e., related to the top-down networking between motor circuitry and frontal areas) [4,17].

Hence, in the case of psychosis, efficiently targeting motor circuitry connections could be of foremost importance in treating and acting on overall symptomatology. Clinical treatment of psychosis is generally carried out through the application of a pharmacological therapy combined with several non-pharmacological psychosocial interventions [18,19] consistently varying in types across different guidelines. Evidence-based psychosocial treatments are aimed at improving the quality of life and at a more efficient management of the illness; among them, the ones recognized as most common are cognitive-behavioral therapy, social skills training and cognitive remediation. Recently, physical exercise has also appeared as a valid tool for psychotic rehabilitation [20], not only given its potential to target GMA from the beginning of the pathology—or even in high-risk subjects with a preventive role—but also as a non-pharmacological treatment acknowledged to improve symptoms and cognitive performance in schizophrenia [21]. Furthermore, physical training is characterized by high practical feasibility, which makes it extremely appealing for psychiatric rehabilitation. Thus, a better understanding of its expected neural effects on the multifaceted symptomatology of psychosis and its potential outcomes may represent an important clinical topic.

Aims

Therefore, the general aim of this review is to provide an overview of brain connectomics literature (both structural and functional) on physical exercise plasticity effects as a treatment for mental illness, with a particular focus on psychosis.

The first more specific aim is to evaluate how volumetric properties of diverse brain regions are affected by physical training, considering both areas thought to be pathogenic for the psychotic phenomena and ones believed to be targeted during the treatment (i.e., within the neural motor circuitry). The second specific aim is to obtain a better look at effects on the implicated regions from a functional point of view in order to gain insights about the reshaping of functional connections as a result of aerobic training, even between remote regions or areas not structurally affected. The third aim tries to combine neural plastic changes resulting from physical training with behavioral outcomes of the patient, potentially ameliorating prognosis. In addition, our last specific aim is to understand if physical treatment via promoting neural changes can be considered a potentially interesting clinical tool among the everyday rehabilitation techniques proposed for psychotic patients.

2. Materials and Methods

This narrative review was conducted by following a re-adaptation of the PRISMA flow [22].

2.1. Inclusion Criteria

As previously stated, the aim of this narrative review was to include all research studies fulfilling the following criteria: MRI-based neuroimaging studies investigating the effects of physical training in psychotic populations. To be included, research studies had to provide enough methodological details with regard to both the MRI acquisition protocol and strategy of analysis and the type of physical training applied. No age limit was applied to the samples included as psychotic patients. Studies had to index psychotic populations via a validated measure (e.g., Positive and Negative Syndrome Scale, Structured Interview for Psychosis-risk Syndromes, Structured Clinical Interviews for DSM-5, etc.). Samples consisting of patients potentially experiencing psychosis (i.e., bipolar disorders) but lacking diagnosis and/or clinical information on the nature and specificity of the phenomena in the study manuscripts were excluded.

2.2. Search Strategy and Selection of Studies

In order to evaluate the relevant literature in the field, works published up to April 2023 were included in this qualitative critical review, chosen from a search conducted in open databases (PubMed, Scopus and Google Scholar). The initial search and all databases used a combination of the following terms: [physical AND (exercise OR activity), neuroimaging AND (fMRI OR DWI OR anatomy OR Voxel-based morphometry OR cortical thickness)] AND (psychosis OR schizophrenia OR psychotic disorder) without specifying a certain period of publication dates. References from the detected articles, further sources and pertinent literature reviews were also assessed for potential inclusion in the reviews. Database inclusion, exclusion, secondary searches and final inclusion, summarized in the flow diagram (Figure 1), were conducted by one of the authors (FS).

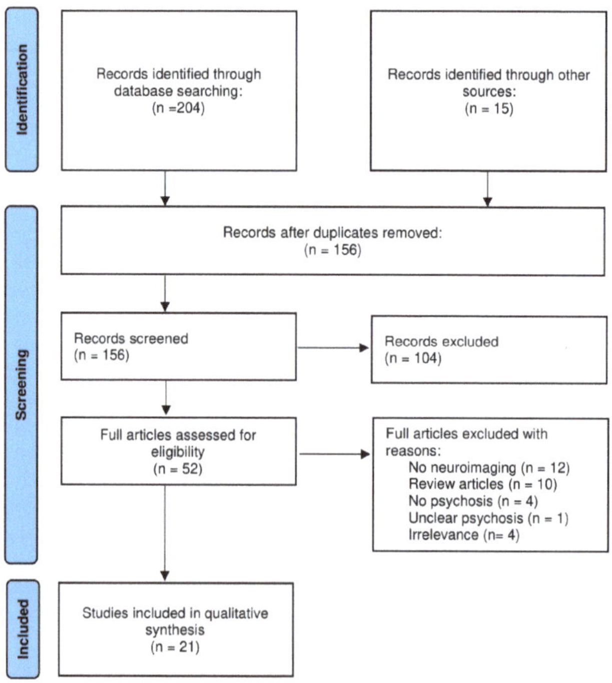

Figure 1. Flowchart of the narrative review process.

We found 204 hits from our initial search, among which, after adjusting for duplicates, 141 articles were retained and screened for title and abstract. Additionally, 104 of these were discarded as not meeting the inclusion criteria. The full texts of the remaining 37 articles were assessed for eligibility and 18 of these papers were included in the narrative review. Moreover, the reference lists of these articles were searched for relevant records, and this search yielded 15 additional articles. Of these secondary searchers, three articles were included in the review.

Thus, out of 52 full-text screened papers, the total number of articles that met the criteria for inclusion was 21. As a last step, the entire manuscript texts of the final full list of 21 final included articles was further reassessed to ensure that they meet inclusion/exclusion criteria.

The following information was extracted from each included study: (1) authors and year of publication; (2) sample size; (3) participant characteristics (demographic information of the sample, clinical diagnosis, clinical assessment scores, cognitive scores); (4) study characteristics (design, control condition); (5) physical exercise protocol characteristics (type, frequency, duration); (6) neuroimaging study characteristics (MRI scanner, MRI sequences, outcome measure); and (7) main conclusions.

3. Results

In recent decades, a growing body of literature has assessed the association between neural connectivity changes and exercise intervention in patients with psychotic disorders. The search identified 21 neuroimaging studies investigating physical training effects in psychosis, 20 studies looking at structural brain plasticity and 8 at functional ones.

These studies proposed several techniques for physical activities as treatments and investigated, by means of MRI, the neural plastic variations in brain connections both on a structural anatomical level and on a functional one. The type of physical training investigated in the above cited studies included mainly aerobic exercise (e.g., cycling, endurance training or yoga), ranging from two to three times a week, with sessions of variable duration (i.e., from 30 min up to 120 min) for an observational period of a minimum of 12 weeks. A large portion of the studies (N = 14) also included a control condition for the physical treatment frequently consisting of the absence of exercise (e.g., as waitlist or recreational/occupational activities such as table soccer). Furthermore, three of the selected studies considered only passive assessments of physical exercise without promoting physical treatment.

3.1. *Exercise Effects on Structural Plasticity in Psychosis*

3.1.1. Anatomical MRI

The largest portion of selected studies (N = 19) investigated the structural modification of cortical grey matter as an effect of physical exercise (Table 1).

Table 1. Structural plasticity studies: research articles investigating neural structural plasticity changes due to the effect of physical exercise in psychosis.

Author	Participant Characteristics						Study Characteristics		Exercise Protocol Characteristics			Neuroimaging Characteristics			Outcomes
	N	Age (Mean)	Gender (F)	Diagnosis	Clinical Scores	Cognitive Scores	Design	Control Condition	Exercise	Frequency	Duration	Scanner	Sequence	Outcome Measures	
Damme et al., 2022 [23]	25	21	11	CHR	SIPS	WRAT, RISE	RCT	Waitlist	AE	2/week	3 months	3T	T1-MPRAGE	HP and subfields GMV	(1) Stable HP volumes; (2) Decreased HP subfield volume with no exercise.
Dean et al., 2017 [24]	12	19.4	6	UHR	SIPS, GFS	MCCB	RCT	None	AE	3–2/week	12 weeks	3T	T1-MPRAGE	HP GMV	(1) No changes in HP volume.
Maurus et al., 2022a [25]	48	37.4	19	SCZ	N.A.	VLMT	CS	N.A.	AE assessment	N.A.	N.A.	3T	T1-MPRAGE	GMV of HP and subfields	(1) Positive associations of aerobic fitness levels and HP subfields volumes.
McEwen et al., 2017 [26]	37	N.A.	N.A.	FEP	N.A.	MCCB	LS	CT	CT and AE	N.A.	N.A.	N.A.	N.D.	CTh	(1) Increased CTh were in prefrontal regions.
Roell et al., 2023 [27]	92	36.8	31	SCZ	PANSS, GAF, FROGS	VLMT, Digit Span Test, TMT	RCT	NA-T	AE	N.A.	3 months	3T	T1-MPRAGE	GMV of HP and subfields	(1) In the AE group volumes within the HP formation increased.
van der Stouwe et al., 2021 [28]	31	34.3	12	PSY	PANSS, BNSS	N.A.	LRCT	Befriending	BEATVIC	1/week (75 min)	20 sessions	3T	T1-w image	VBM	(1) No differences at VBM.
Falkai et al., 2013 [29]	16	35.2	16	SCZ	PANSS	VLMT, Corsi block Tapping test	RCT	Table football	AE (cycling)	3/week	3 months	1.5 T	T1-MPRAGE; T2-w gradient echo	GMD, CSE	(1) Cortical changes in healthy controls; (2) No effect in cortical regions for SCZ.
Lin et al., 2015 [30]	124	24.5	124	SCZ, SCZA, PSY	PANSS, CDSS, QoL, FRS, CRS	VA, VR, Digit Span Test, LC Q score, Stroop task	RCT	Waitlist	Yoga or AE	3/week	12 weeks	3T	T1-MPRAGE	HP GMV	(1) AE increases HP volume.
Malchow et al., 2016 [31]	20	35.8	12	SCZ	PANSS	N.A.	LS	Table football	Endurance training	3/week	3 months	3T	T1-MPRAGE	VBM, manual and automatic segmentation of HP and subfields.	(1) No volume increase in HP and its subfields; (2) Endurance training increased volume of the left superior, middle and inferior anterior temporal gyri; (3) Table soccer increased volumes in the motor and anterior cingulate cortices; (4) Differences were no longer present after inactivity.

Table 1. *Cont.*

Author	Participant Characteristics						Study Characteristics		Exercise Protocol Characteristics			Neuroimaging Characteristics			Outcomes
	N	Age (Mean)	Gender (F)	Diagnosis	Clinical Scores	Cognitive Scores	Design	Control Condition	Exercise	Frequency	Duration	Scanner	Sequence	Outcome Measures	
Maurus et al., 2022b [32]	69	36.94	23	SCZ	PANSS, CDSS	TMT, Digit Span Test, VLMT, B-CATS, DSST, ERT	CS	N.A.	AE assessment	N.A.	N.A.	3T	T1-MPRAGE	Anatomical parcellation for GMV and WMV	(1) AE is associated to increased right HP GMV and para-HP WMV.
McEwen et al., 2023 [33]	37	22.7	10	FEP	N.A.	MCCB	LRCT	CT (Brain HQ)	CT + AE	3/week	24 weeks	3T	T1-MPRAGE; T2-weighted turbo spin-echo	CTh	(1) CT and AE increase CTh within the left anterior cingulate cortex over treatment period; (2) Directional tendencies were similar in the left dorsolateral prefrontal cortex.
Mittal et al., 2013 [34]	29	18.5	11	UHR	SIPS, SCID	N.A.	CS	N.A.	wristwatch recording (ActiLife scoring)	N.A.	5 days	3T	T1-MPRAGE; T2-weighted turbo spin-echo	GMV of HP and para-HP gyri	(1) UHR greater percentage of time in sedentary behavior; (2) Trend UHR group showed less total physical activity; (3) UHR smaller medial temporal volumes; (4) Inactivity is associated with medial temporal lobe health.
Pajonk et al., 2010 [35]	16	35.2	0	SCZ	PANSS, CGI	VLMT, Corsi block Tapping Test	RCT	Table football	AE (cycling)	3/week	3 months	1.5T	T1-MPRAGE; T2-weighted gradient echo; spin-echo MRS	GMV HP; neuro-metabolites of left HP	(1) AE increase HP volume in patients with no change in the non-exercise group of patients; (2) HP volume in the exercise group were correlated with improvements in aerobic fitness; (3) SCZ exercise group changes in HP volume were associated with a 35% increase in NAA/Cr in HP;
Rosenbau et al., 2015 [36]	5	N.A.	0	SCZ, SCZA	SAPS, SANS, WHOQOL-BREF	VLMT, WMS	Pilot	None	Stationary bike	2/week	12 weeks	1.5T	T1-MPRAGE	GMV of HP	(1) No significant HP volume increase.

Table 1. *Cont.*

Author	Participant Characteristics						Study Characteristics		Exercise Protocol Characteristics			Neuroimaging Characteristics			Outcomes
	N	Age (Mean)	Gender (F)	Diagnosis	Clinical Scores	Cognitive Scores	Design	Control Condition	Exercise	Frequency	Duration	Scanner	Sequence	Outcome Measures	
Scheewe et al., 2013 [37]	32	29.8	6	SCZ	PANSS	WAIS IQ	RCT	OC	AE	1/week	6 months	3T	T1-weighted FFE	GMV, CTh	(1) AE does not affect global brain and HP volume or CTh in patients and controls; (2) CRF improvement was related to increased cerebral matter volume and lateral and third ventricle volume decrease in patients and to thickening in the left hemisphere in large areas of the frontal, temporal and cingulate cortex irrespective of diagnosis; (3) One to two hours' exercise therapy did not elicit significant brain volume changes in patients or controls.
Takahashi et al., 2020 [38]	21	36.2	12	SCZ	GAF, SAS, PANSS, CGI, CDSS	VLMT, TMT, WCST	LS	Table soccer	AE (cycling)	1/week	12 weeks	3T	T1-MPRAGE	CTh of entorhinal, para-HP and lateral/medial PFC	(1) AE showed increase in CTh in right entorhinal cortex; (2) No significant longitudinal change in CTh in control groups.
Woodward et al., 2018 [39]	17	30.1	11	SCZ, SCZA	PANSS	WAR	LRCT	Weight-bearing exercise program	AE	3/week	12 weeks	3T	T1-MPRAGE; SWI	GMV HP and its subfields; SWI mapping of HP vasculature volume	(1) HP increased in the left CA-1 field; (2) HP vascular volume unchanged.
Woodward et al., 2020 [40]	51	24.5	51	FEP	PANSS	N.A.	RCT	Waitlist	AE/ Hatha yoga	3/week	12 weeks	3T	T1-MPRAGE	GMV and CTh of whole-brain and HP	(1) Increases in GMV and CTh in the medial temporal cortical regions for AE.

Table 1. *Cont.*

Author	Participant Characteristics						Study Characteristics		Exercise Protocol Characteristics			Neuroimaging Characteristics			Outcomes
	N	Age (Mean)	Gender (F)	Diagnosis	Clinical Scores	Cognitive Scores	Design	Control Condition	Exercise	Frequency	Duration	Scanner	Sequence	Outcome Measures	
Svatkova et al., 2015 [41]	33	30	6	SCZ	PANSS	IQ	LS	Life-as-usual	AE + NAE	1/week	6 months	3T	DWI	DTI and TBSS of major tracts	(1) Physical exercise increases the integrity of white matter fiber tracts; (2) Life-as-usual decreases fiber integrity.

Abbreviations: N.A. (not applicable); N.D. (not defined); N (number of participants); F (female); CHR (presence of diagnosed attenuated psychosis syndrome); UHR (ultra high-risk for psychosis); SCZ (schizophrenia); FEP (first-episode psychosis); PSY (psychotic spectrum); SCZA (schizoaffective disorder); SIPS (Structured Interview for Psychosis-risk Syndromes); GFS (global level of functioning); PANSS (Positive and Negative Syndrome Scale); GAF (Global Assessment of Functioning,); FROGS (Functional Remission of General Schizophrenia); BNSS (Brief Negative Symptom Scale); CDSS (Calgary Depression Scale for Schizophrenia); QoL (Quality of Life Scale); FRS (first-rank symptoms); CRS (Compliance Rating Scale); SCID (Structured Clinical Interview for DSM Disorders), CGI (Clinical Global Impression); SAPS (Scale for the Assessment of Positive Symptoms); SANS(Scale for the Assessment of Negative Symptoms); SAS (Self-Rating Anxiety Scale); WHOQOL-BREF (World Health Organization Quality of-Life Scale); WRAT (Wide Range Achievement Test); RISE (Relational and Item-Specific Encoding Task); MCCB (MATRICS Consensus Cognitive Battery); VLMT (Verbal learning and Memory test); TMT (Trial Making Test); VA (verbal acquisition); VR (verbal retention); LC Q score (Letter Cancellation Test); B-CATS (The Brief Cognitive Assessment Tool for Schizophrenia); DSST (Digit Symbol Substitution Test); ERT (Emotion Recognition Test); WMS (Weschler Memory Scale); WAIS IQ (Wechsler Adult Intelligence Scale); WCST (Wisconsin Card Sorting Test); WAR (Weschler Test of Adult Reading); IQ (intelligence quotient); RCT (randomized controlled trial); LRCT (longitudinal randomized controlled trial); LS (longitudinal study); CS (cross-sectional study); CT (cognitive training); NA-T (flexibility, strengthening and balance training non aerobic); AE (aerobic exercise); NAE (non-aerobic exercise); BEATVIC (session led by a therapist trained in body and movement-oriented interventions); T1-MEMPRAGE (T1-Multi-echo-MPRAGE); MRS (magnetic resonance spectroscopy); SWI (susceptibly weighted imaging); DWI (diffusion-weighted imaging); HP (hippocampus); GMV (grey matter volume); CTh (cortical thickness); VBM (voxel-based morphometry); GMD (grey matter density); CSE (cortical surface expansion); WMV (white matter volume); PFC (prefrontal cortex); DTI (diffusion tensor imaging); TBSS (tract-based spatial statistics); N-acetylaspartate-to-creatine ratio (NAA/Cr); CRF (cardiorespiratory fitness).

Most of the studies focused on precise a priori regions where the expected structural changes (Figure 2A) should take place based rather on the pathophysiology of psychosis than supposed motor neural networking enhancement.

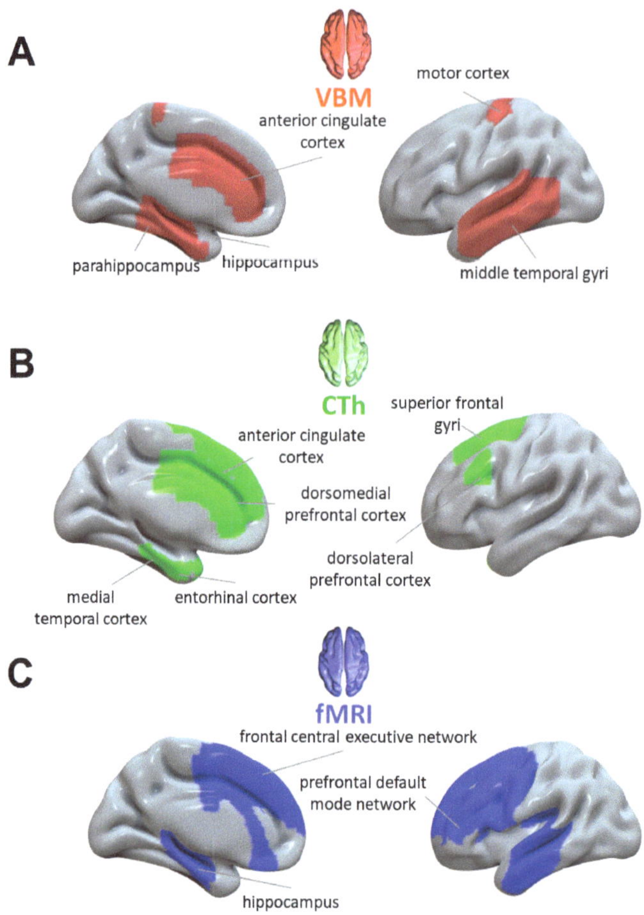

Figure 2. Neural plasticity processes. Evidence from neuroimaging studies on (**A**) structural morphometric (Red: VBM), (**B**) structural cortical thinning (Green: CTh) and (**C**) functional (Blue: fMRI) plasticity changes occurring in brain regions of psychotic patients as a function of physical activity. The figure was generated with Surf Ice toolbox (https://www.nitrc.org/projects/surfice/) (accessed on 29 April 2023) by rendering on the surface mesh previously selected and atlas-derived regions of interest (Harvard–Oxford cortical and subcortical structural atlases [42]; Anatomical labelling Atlas [43] (AAL3); Atlas of Intrinsic Connectivity of Homotopic Areas [44] (AICHA)) found to be implicated in previous studies. Abbreviations: VBM (voxel-based morphometry); CTh (cortical thickness); fMRI (functional MRI).

Therefore, the utmost investigated brain region is the hippocampus, together with its subfields and the parahippocampal areas. Volumetric studies reported mixed results consequent to aerobic fitness, with evidences for either an increased cortical volume of the hippocampus in schizophrenic patients [25,27,30,35,39] or stable hippocampal grey matter both in clinical high-risk subjects [23,24] and later stages of psychosis [31,36,37]. The same pattern is found also while looking at the hippocampal subfields, with Damme et al. [23] reporting a negative correlation of its volume in the absence of aerobic activity and Maurus et al. [25] reporting an increased volume associated with physical activity; whereas Malchow et al. [31] describe no structural changes. Only two studies instead focused on parahippocampal regions, highlighting that in ultra-high-risk subjects for psychosis. a structural smaller volume of the area is associated with lower physical activity [34] can develop an increased white matter volume promoted by aerobic training in cases of schizophrenia [25].

Similarly, focusing on whole brain volumetric changes, other works highlighted the absence of cortical region differences in psychosis due to treatment [28,29]. Only one study exhibited subtle localized differences in temporal gyri [31] in schizophrenic patients undergoing aerobic training and also found increased volumetric grey matter in the motor and anterior cingulate cortex for the control condition (i.e., table soccer), both of which fade after physical inactivity.

Differently, as concerns effects on cortical thickness (Figure 2B), physical exercise was found to enhance prefrontal and temporal regions both in early psychosis [26,33,40] and chronic schizophrenia [38].

3.1.2. Diffusion MRI

Only one study in the context of structural plasticity investigated the white matter structures' modifications in schizophrenia due to physical exercise [41], showing an improvement in white matter integrity (in terms of fractional anisotropy (FA)) in motor fiber tracts thanks to training and regardless of the presence of diagnosis (i.e., both in schizophrenia and in healthy controls). Concerning patients, this structural connectivity study also corroborates previous findings about the beneficial effect of physical training on overall mental health with a peculiar alleviation of positive psychotic symptoms.

3.2. *Exercise Effects on Functional Plasticity in Psychosis*

Functional MRI

Looking at the total number of studies included in this narrative review, a smaller portion of the studies (N = 8) investigated the effect of exercise on functional connections in psychotic patients compared to those dedicated to structural plasticity (Table 2, Figure 2C).

Table 2. Functional plasticity studies: research articles investigating neural functional plasticity changes due to the effect of physical exercise in psychosis.

Author	Participant Characteristics						Study Characteristics		Exercise Protocol Characteristics			Neuroimaging Characteristics			Outcomes
	N	Age (Mean)	Gender (F)	Diagnosis	Clinical Scores	Cognitive Scores	Design	Control Condition	Exercise	Frequency	Duration	Scanner	Sequence	Outcome Measures	
Damme et al., 2022 [23]	25	21	11	CHR	SIPS	WRAT, RISE	RCT	Waitlist	AE	2/week	3 months	3T	T1-MPRAGE; rs-fMRI	ROI-ROI FC (HP-occipital lobe)	(1) Increased HP connectivity.
Dean et al., 2017 [24]	12	19.4	6	UHR	SIPS, GFS	MCCB	RCT	None	AE	3–2/week	12 weeks	3T	T1-MPRAGE; rs-fMRI	ROI-ROI FC (left HP—right HP—bilateral occipital cortices)	(1) Increased FC between left HP and occipital cortex.
Lin et al., 2018 [45]	124	N.A.	124	FEP	PANSS	Working memory	RCT	Waitlist	Yoga/AE	N.A.	12 weeks	N.A.	N.D.	ALFF	(1) ALFF decreases in precuneus in yoga group and correlates with better negative symptoms.
Maurus et al., 2022a [25]	48	37.4	19	SCZ	N.A.	VLMT	CS	N.A.	AE assessment	N.A.	N.A.	3T	T1-MPRAGE; rs-fMRI	FC matrices of HP and subfields	(1) No associations of HP subfields FC or mediation effects on verbal memory.
McEwen et al., 2017 [26]	37	N.A.	N.A.	FEP	N.A.	MCCB	LS	CT	CT and AE	N.A.	N.A.	N.A.	N.D.	FC	(1) CT and exercise improved FC between right CEN and ventral attention network and also between left CEN and right CEN.
Roell et al., 2023 [27]	92	36.8	31	SCZ	PANSS, GAF. FROGS	VLMT, Digit Span Test, TMT	RCT	NA-T	AE	N.A.	3 months	3T	T1-MPRAGE; rs-fMRI	FC from the HP subfields, the striatum, the amygdala and thalamus, the DLPFC and CC	(1) No effects of exercise on HP formation connectivity were observed.
Takahashi et al., 2012 [46]	23	41.7	11	SCZ	PANSS	N.A.	LS	HC	AE	2/day (60–120 min)	3 months	1.5 T	T1-w image; Task-fMRI	Task-based fMRI GLM	(1) Activation of the body-selective EBA in the posterior temporal-occipital cortex during observation of sports-related actions was increased in the program group

Table 2. *Cont.*

Author	Participant Characteristics						Study Characteristics		Exercise Protocol Characteristics			Neuroimaging Characteristics			Outcomes
	N	Age (Mean)	Gender (F)	Diagnosis	Clinical Scores	Cognitive Scores	Design	Control Condition	Exercise	Frequency	Duration	Scanner	Sequence	Outcome Measures	
van der Stouwe et al., 2021 [28]	31	34.3	12	PSY	PANSS, BNSS	N.A.	LRCT	Befriending	BEATVIC	1/week (75 min)	20 sessions	3T	T1-weighted image; Task-fMRI	Task-based fMRI GLM and ICA FC	(1) GLM no differences between groups over time; (2) ICA increased activation of the salience network to angry and fearful faces in BEATVIC; (3) Increased activation of the salience network may suggest an increased alertness for potentiallydangerous faces.

Abbreviations: N.A. (not applicable); N.D. (not defined); N (number of participants); F (female); CHR (presence of diagnosed attenuated psychosis syndrome); UHR (ultra high-risk for psychosis); SCZ (schizophrenia); FEP (first-episode psychosis); PSY (psychotic spectrum); SCZA (schizoaffective disorder); SIPS (Structured Interview for Psychosis-risk Syndromes); GFS (global level of functioning); PANSS (Positive and Negative Syndrome Scale); GAF (Global Assessment of Functioning;); FROGS (Functional Remission of General Schizophrenia); BNSS (Brief Negative Symptom Scale); WRAT (Wide Range Achievement Test); RISE (Relational and Item-Specific Encoding Task); MCCB (MATRICS Consensus Cognitive Battery); VLMT (Verbal learning and Memory test; TMT (Trial Making Test); RCT (randomized controlled trial); LRCT (longitudinal randomized controlled trial); LS (longitudinal study); CS (cross-sectional study); CT (cognitive training); NA-T (flexibility, strengthening and balance training non aerobic); HC (healthy controls); Befriending (social interaction group intervention); AE (aerobic exercise); NAE (non-aerobic exercise); BEATVIC (session led by a therapist trained in body and movement-oriented interventions); rs-fMRI (resting-state fMRI); ROI (region of interest); HP (hippocampus); ALFF (amplitude of low-frequency fluctuations); FC (functional connectivity); DLPFC (dorsolateral prefrontal cortex); CC (corpus callosum); GLM (general linear model); ICA (independent component analysis); CEN (central executive network); EBA (extra-striate body Area).

Indeed, the vast majority of functional studies based their hypotheses on previous structural findings and aimed at shining new light on hippocampal changes consequent to exercise training by looking at its functional connectivity [23,24,27,32]. Functional connectivity of the hippocampus seemed to increase with the occipital lobe after physical exercise both in ultra-high risks subjects for psychosis [24] and in attenuated forms of psychosis [23] compared to control conditions. However, while looking at patients with schizophrenia, no changes were found in hippocampal functional connections to more neighboring structures (e.g., subfields, parahippocampal regions, amygdala, thalamus, middle/frontal and cingulate gyrus) after physical activity [27,32].

As regards whole-brain findings, only two studies investigated physical activity effects in psychosis on functional brain activation during tasks [28,46]. The study from Takahashi et al. is the first study to investigate functional activity changes after aerobic exercise. It demonstrated that a precise part of the occipital cortex, the extra-striate body area (EBA), increased its activity in schizophrenic patients in relation to exercise and this seemed to facilitate an improvement in psychotic symptoms. However, no further studies looked at how this increase in EBA BOLD activation relates to functional connections with other brain areas. The second task-based fMRI study, on the contrary, tried to depict a more comprehensive, connectome-based picture of brain activity changes after body-oriented interventions for psychotic patients. Their results showed a connectivity normalization of the aberrant salience network activation usually found in psychotic patients, potentially driven also by the aerobic kick-boxing training present in the body-oriented therapy. Furthermore, other functional connectivity studies highlighted an involvement of large-scale brain networking in plasticity after aerobic training, such as within the default mode network (DMN) (precuneal amplitude of low-frequency fluctuations (ALFF) decrease with yoga [45]) and the central executive network (CEN) (improved connectivity in combination with cognitive training [26]).

However, no studies have investigated the functional connectivity changes between hippocampus, DMN and SMN networking in relation to physical treatment in psychosis.

3.3. Behavioral Correlation with Neuroimaging Findings

A portion of the selected studies (N = 17; Table 3) concerning both structural and functional plasticity, also investigated the association of neural changes with behavioral variables that can be grouped into three main domains, namely, cognitive performance, psychotic symptomatology and social functioning.

Table 3. Behavioral correlations in neuroimaging studies: research articles investigating how neural plastic changes due to the effect of physical exercise in psychosis are correlated with behavioral/cognitive and clinical outcomes.

Author	Participant Characteristics						Study Characteristics		Exercise Protocol Characteristics			
	N	Age (Mean)	Gender (F)	Diagnosis	Clinical Scores	Cognitive Scores	Design	Control Condition	Exercise	Frequency	Duration	
Damme et al., 2022 [23]	25	21	11	CHR	SIPS	WRAT, RISE	RCT	Waitlist	AE	2/week	3 months	(1) Improved fitness; (2) Increased cognitive performance; (3) Decrease in positive symptoms.
Dean et al., 2017 [24]	12	19.4	6	UHR	SIPS, GFS	MCCB	RCT	None	AE	3–2/week	12 weeks	(1) Improved positive and negative symptoms; (2) Improved social functioning; (3) Improved cognition.
Maurus et al., 2022a [25]	48	37.4	19	SCZ	N.A.	VLMT	CS	N.A.	AE assessment	N.A.	N.A.	(1) No associations of HP subfields FC or mediation effects on verbal memory.
McEwen et al., 2017 [26]	37	N.A.	N.A.	FEP	N.A.	MCCB	LS	CT	CT and AE	N.A.	N.A.	(1) Improved FC between left and right CEN was associated with cognitive gains in reasoning and problem solving at 6-month follow-up.
Falkai et al., 2013 [29]	16	35.2	16	SCZ	PANSS	VLMT, Corsi block Tapping test	RCT	Table football	AE (cycling)	3/week	3 months	(1) Improved short-term memory; (2) Improved PANSS.
Lin et al., 2015 [30]	124	24.5	124	SCZ, SCZA, PSY	PANSS, CDSS, QoL, FRS, CRS	VA, VR, Digit Span Test, LC Q score, Stroop task	RCT	Waitlist	Yoga or AE	3/week	12 weeks	(1) Yoga and AE improved working memory; (2) Yoga improved verbal acquisition and attention; (3) Yoga and AE improved overall and depressive symptoms;
Malchow et al., 2016 [31]	20	35.8	12	SCZ	PANSS	N.A.	LS	Table football	Endurance training	3/week	3 months	(1) Psychopathological symptoms did not change.
Maurus et al., 2022b [32]	69	36.94	23	SCZ	PANSS, CDSS	TMT, Digit Span Test, VLMT, B-CATS, DSST, ERT	CS	N.A.	AE assessment	N.A.	N.A.	(1) No association between cognition or symptoms and AE.
McEwen et al., 2023 [33]	37	22.7	10	FEP	N.A.	MCCB	LRCT	CT (Brain HQ)	CT + AE	3/week	24 weeks	(1) Increased CTh in the left ACC was improved work/school functioning.
Mittal et al., 2013 [34]	29	18.5	11	UHR	SIPS, SCID	N.A.	CS	N.A.	wristwatch recording (ActiLife scoring)	N.A.	5 days	(1) Total level of physical activity in UHR correlated with smaller para-HP gyri bilaterally and with occupational functioning.
Pajonk et al., 2010 [35]	16	35.2	0	SCZ	PANSS, CGI	VLMT, Corsi block Tapping Test	RCT	Table football	AE (cycling)	3/week	3 months	(1) Short-term memory improvement in SCZ was correlated with change in HP volume.

Table 3. *Cont.*

Author	Participant Characteristics						Study Characteristics		Exercise Protocol Characteristics			
	N	Age (Mean)	Gender (F)	Diagnosis	Clinical Scores	Cognitive Scores	Design	Control Condition	Exercise	Frequency	Duration	
Takahashi et al., 2020 [38]	21	36.2	12	SCZ	GAF, SAS, PANSS, CGI, CDSS	VLMT, TMT, WCST	LS	Table soccer	AE (cycling)	1/week	12 weeks	(1) Significant correlation between CTh of right lateral PFC at baseline and improvement of social adaptation
Woodward et al., 2018 [39]	17	30.1	11	SCZ, SCZA	PANSS	WAR	LRCT	weight-bearing exercise program	AE	3/week	12 weeks	(1) Changes in HP volume and vascular volume were not significantly correlated with changes in symptom severity, nor did they affect scores.
Woodward et al., 2020 [40]	51	24.5	51	FEP	PANSS	N.A.	RCT	Waitlist	AE/ Hatha yoga	3/week	12 weeks	(1) AE increases in the entorhinal and fusiform/temporal gyri associated with reduced symptom severity; (2) Increased fusiform CTh associated with increased HP volume for all psychosis participants.
Svatkova et al., 2015 [41]	33	30	6	SCZ	PANSS	IQ	LS	Life-as-usual	AE + NAE	1/week	6 months	(1) Exercise improves brain structural connectivity and positive symptoms.
Lin et al., 2018 [45]	124	N.A.	124	FEP	PANSS	Working memory	RCT	Waitlist	Yoga/AE	N.A.	12 weeks	(1) Yoga and aerobic exercise improved working memory and psychotic symptoms; (2) ALFF decreases in precuneus in yoga group and correlates with better negative symptoms.
Takahashi et al., 2012 [46]	23	41.7	11	SCZ	PANSS	N.A.	LS	HC	AE	2/day (60–120 min)	3 months	(1) Increase in EBA activation was associated with PANSS improvement.

Abbreviations: N.A. (not applicable); N.D. (not defined); N (number of participants); F (female); CHR (presence of diagnosed attenuated psychosis syndrome); UHR (ultra high-risk for psychosis); SCZ (schizophrenia); FEP (first-episode Psychosis); PSY (psychotic spectrum); SCZA (schizoaffective disorder); SIPS (Structured Interview for Psychosis-risk Syndromes); GFS (global level of functioning); PANSS (Positive and Negative Syndrome Scale); GAF (Global Assessment of Functioning;); FROGS (Functional Remission of General Schizophrenia); BNSS (Brief Negative Symptom Scale); CDSS (Calgary Depression Scale for Schizophrenia); QoL (Quality of Life Scale); FRS (First Rank Symptoms); CRS (Compliance Rating Scale); SCID (Structured Clinical Interview for DSM Disorders); CGI (Clinical Global Impression); SAPS (Scale for the Assessment of Positive Symptoms); SANS (Scale for the Assessment of Negative Symptoms); SAS (Self-Rating Anxiety Scale); WHOQOL-BREF (World Health Organization Quality of Life Scale); WRAT (Wide Range Achievement Test); RISE (Relational and Item-Specific Encoding Task); MCCB (MATRICS Consensus Cognitive Battery); VLMT (Verbal learning and Memory test); TMT (Trial Making Test); VA (verbal acquisition); VR (verbal retention); LC Q score (Letter Cancellation Test); B-CATS (The Brief Cognitive Assessment Tool for Schizophrenia); DSST (Digit Symbol Substitution Test); ERT (Emotion Recognition Rest); WMS (Weschler Memory Scale); WAIS IQ (Wechsler Adult Intelligence Scale); WCST (Wisconsin Card Sorting Test); WAR (Weschler Test of Adult Reading); IQ (intelligence quotient); RCT (randomized controlled trial); LRCT (longitudinal randomized controlled trial); LS (longitudinal study; CS (cross-sectional study); CT (cognitive training); NA-T (flexibility, strengthening and balance training non aerobic); AE (aerobic exercise); NAE (non-aerobic exercise); BEATVIC (session led by a therapist trained in body and movement-oriented interventions); HP (hippocampus); FC (functional connectivity); CEN (central executive network); CTh (cortical thickness); ACC (anterior cingulate cortex); PFC (prefrontal cortex); EBA (extra-striate body area); ALFF (amplitude of low frequency fluctuations).

Regarding the cognitive profile, neural changes promoted by physical exercise were found to be associated with: (i) a better general cognitive outcome in high-risk subjects [23,24]; (ii) improved short term memory in schizophrenia [29,35]; (iii) enhanced working memory and reasoning/problem solving skills both in early and overt psychosis [45].

Similarly, as regards behavioral clinical variables, all studies demonstrated an improved outcome as a function of exercise. Concerning clinical scores, the evidence was found for better overall psychotic symptomatology in schizophrenia [29,30,40,41,46] and for negative symptoms in early psychosis [45] and positive symptoms in high-risk cohorts [23,24]. Even though the proven effects on psychotic symptoms, there is no such clear evidence showing either an investigation of GMA or an effect of physical training in delaying the worsening/improving the management of these symptoms in association to neural changes.

On the other hand, concerning scores of social functioning, a positive association of physical treatment with improved skills of occupation [34] and work/school [33] functioning, and social adaptation [24,38] was found in diverse conditions of psychosis.

4. Discussion

At the current state of the art, beside the given clinical importance of physical activity [47–50] in the context of psychosis, there is no clear consensus on the expected neural plastic changes occurring after such treatment in psychotic patients.

In this narrative review we showed the following (Figure 3): (i) hippocampal and parahippocampal structural and functional plasticity in schizophrenia and high-risk patients as a consequence of physical exercise; (ii) preliminary whole-brain functional plastic processes occurring after physical treatment and their associations with patients' outcome.

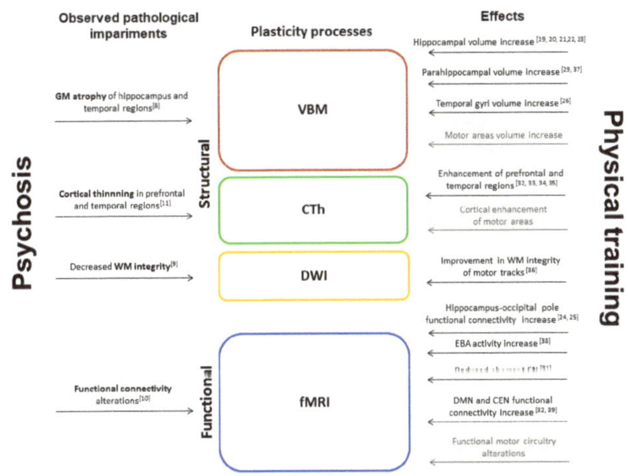

Figure 3. Schematic representation of physical training effects on multiple levels and relevant observed pathological impairments in psychosis. Observed effects are reported with the relevant reference, whereas the expected effects still lacking investigations are reported in grey. Abbreviations: VBM (voxel-based morphometry); CTh (cortical thickness); DWI (diffusion-weighted imaging); fMRI (functional MRI); GM (grey matter); WM (white matter); EBA (extra-striate body area); SN (salience network); DMN (default mode network); CEN (central executive network).

As regards structural plasticity, most studies reported results based on a priori hypothesized regions involved in the psychotic disorder. The focus was concentrated on hippocampal and medial temporal regions, but there was no clear consensus on the volumetric effect of physical exercise across the selected studies. The only region showing consistent directionality in structural plasticity is the parahippocampus, which seemed to have a protective role both in grey and white matter portions. Prior evidence obtained from both animal and human models demonstrated that the structure and function of the parahippocampal region susceptible to physical exercise [51] are critically implicated in psychosis [52]. Reductions in parahippocampal volume have been reported in high-risk individuals with transient or isolated psychotic symptoms, and high-risk patients who developed psychosis showed longitudinal reduction. Exercise has various effects on parahippocampal regions such as increasing neural excitability, increasing grey/white matter, volumetric reduction, enhancing regional glucose metabolism, increasing cerebral blood flow, augmenting various markers of synaptic plasticity and increasing the functional connectivity with other proximal brain structures. Therefore, it is reasonable to suppose a potential role of parahippocampal regions in the motor domain. Future work could investigate the effects of exercise on connections of these regions to other areas, such as (i) the motor circuitry [13] system and (ii) or temporal and frontal regions, which were previously found to be structurally affected by aerobic training in psychosis [31,38].

Concerning structural connections changes, we also highlighted that only one study investigated the effect of exercise on white matter motor bundles. Brain white matter alterations are well-documented in adults with psychotic disorders; the results indicate widespread lower FA, with the largest effect sizes in the anterior corona radiata and corpus callosum (CC). The involvement of the CC seems to be connected to the role of abnormal hemispheric specialization and abnormal interhemispheric communication in the etiology of the disease. Accumulating evidence suggests that aerobic exercise training and higher cardiorespiratory fitness are associated with improved microstructural organization of the CC; in particular, practicing aerobic exercise regularly or intense cardiorespiratory fitness leads to an increase in FA values in the CC. Despite the idea that temporary physical training might not be persistent or effective enough to provoke such structural changes, all these elements highlight the need of a more extensive investigation of structural connectivity.

In addition, a small number of studies looked at functional plasticity reporting based on regions structurally affected by exercise, changes in connectivity of the hippocampus and in large scale networking. However, the generalizability of these results is very poor due to power issues limiting the possibility to draw comprehensive conclusions; hence, further studies are needed.

Moreover, the selected studies investigating behavioral associations with neural changes due to physical treatment consistently reported a positive effect on multiple domains affected by the presence of psychosis (i.e., cognitive deficits, psychotic symptomatology and functional outcome). This poses aerobic exercise as a potentially effective treatment for schizophrenia and psychotic spectrum disorders in general, as previously hypothesized [48]. However, the level of persistence of plastic neural changes over time after the conclusion of the treatment, which is thought to re-organize the psychotic brain back to physiology, is still a matter of debate.

Nevertheless, there is another relevant point to raise regarding the investigation of motor circuitry changes. Previous investigations [13] demonstrated how three main networks involved in motor functions, namely, the basal ganglia, the cerebellar–thalamocortical and the corticomotor circuits, are relevant for the pathophysiology of psychosis. With this review, we have highlighted the lack of studies investigating how these connections can be affected or influenced by physical and motor exercise in psychosis. Indeed, the vast majority of studies included in this narrative review were focused mainly on brain changes occurring after physical treatment in regions allegedly expected to be affected [8] by psychosis symptomatology, rather than on motor connections. Furthermore, in this narrative review no evidence was found investigating the correlation between the rise in neural plasticity thanks to exercise and the effect on GMA symptoms, leaving the door open for upcoming investigations. Therefore, future experimental studies should examine changes in the inter-relationships of motor networks in psychosis as a result of aerobic exercise to better probe its potential as a targeted treatment on GMA.

Most of the studies focus on the effects of aerobic activity, although there is no literature that guarantees a better effect of this physical exercise over others. This choice is probably due to a greater simplicity in conducting the intervention compared to team sports or other sports such as contact sports, which could have a positive impact on the dopaminergic circuits but increase aggressive attitudes. It would be of interest to compare various modalities of exercise with other forms of intervention, including medication, cognitive behavioral therapy and imagined locomotion. In future studies, it will also be important to acquire information on the amount of previously played sports to correlate with the MRI acquired before treatment, as the outcome of the therapy could be different in relation to patient's habit of physical exercise.

However, the limitations of this study must be considered. Being a narrative review, it was beyond the scope of this work to systematically investigate the quantitative effects of physical treatment in psychosis; therefore, future meta-analyses could shed light on precise estimated effects.

Despite this limitation, this review is of scientific interest as it underlines how physical exercise can have a strong therapeutic impact on those suffering from psychosis.

5. Conclusions

In this review, we provide evidence of both structural and functional brain plasticity in psychosis as a result of physical treatment. Although there is no obvious directionality and the effects do not last once the exercise phase is over, physical training, especially aerobic exercise, promotes structural changes in the areas of the hippocampus. Physical exercise and brain changes altered the hippocampus' functional connectivity as well as that of other large-scale networks, and these changes were associated with improved functional outcomes for patients. To evaluate the long-term effectiveness of physical therapy within the psychotic spectrum and its reliability in the therapeutic setting, more research is still required. Future studies should concentrate on enhancing our understanding of how neural connections within motor circuitry alter in psychosis, emphasizing both cortical and subcortical connection changes.

Author Contributions: Conceptualization, F.S. and D.C.; methodology, F.S.; software, F.S.; formal analysis, F.S.; investigation, F.S. and D.C.; writing—original draft preparation, F.S. and D.C.; writing—review and editing, G.D., R.G., A.V. and S.B. All authors have read and agreed to the published version of the manuscript.

Funding: This research received no external funding.

Institutional Review Board Statement: Not applicable.

Informed Consent Statement: Not applicable.

Data Availability Statement: Not applicable.

Conflicts of Interest: The authors declare no conflict of interest.

References

1. Harvey, P.D.; Strassnig, M. Predicting the severity of everyday functional disability in people with schizophrenia: Cognitive deficits, functional capacity, symptoms, and health status. *World Psychiatry* **2012**, *11*, 73–79. [CrossRef] [PubMed]
2. Bernard, J.A.; Mittal, V.A. Updating the research domain criteria: The utility of a motor dimension. *Psychol. Med.* **2015**, *45*, 2685–2689. [CrossRef] [PubMed]
3. Mittal, V.A. Cross-Cutting Advancements Usher in a New Era for Motor Research in Psychosis. *Schizophr. Bull.* **2016**, *42*, 1322–1325. [CrossRef] [PubMed]
4. Walther, S.; Mittal, V.A. Motor Behavior is Relevant for Understanding Mechanism, Bolstering Prediction, And Improving Treatment: A Transdiagnostic Perspective. *Schizophr. Bull.* **2022**, *48*, 741–748. [CrossRef] [PubMed]
5. Callaway, D.A.; Perkins, D.O.; Woods, S.W.; Liu, L.; Addington, J. Movement abnormalities predict transitioning to psychosis in individuals at clinical high risk for psychosis. *Schizophr. Res.* **2014**, *159*, 263–266. [CrossRef]
6. Mittal, V.A.; Neumann, C.; Saczawa, M.; Walker, E.F. Longitudinal progression of movement abnormalities in relation to psychotic symptoms in adolescents at high risk of schizophrenia. *Arch. Gen. Psychiatry* **2008**, *65*, 165–171. [CrossRef]

7. Mittal, V.A.; Walker, E.F.; Bearden, C.E.; Walder, D.; Trottman, H.; Daley, M.; Simone, A.; Cannon, T.D. Markers of Basal Ganglia Dysfunction and Conversion to Psychosis: Neurocognitive Deficits and Dyskinesias in the Prodromal Period. *Biol. Psychiatry* **2010**, *68*, 93–99. [CrossRef]
8. Shenton, M.E.; Dickey, C.C.; Frumin, M.; McCarley, R.W. A review of MRI findings in schizophrenia. *Schizophr. Res.* **2001**, *49*, 1–52. [CrossRef]
9. Kubicki, M.; Park, H.; Westin, C.F.; Nestor, P.G.; Mulkern, R.V.; Maier, S.E.; Niznikiewicz, M.; Connor, E.E.; Levitt, J.J.; Frumin, M.; et al. DTI and MTR abnormalities in schizophrenia: Analysis of white matter integrity. *Neuroimage* **2005**, *26*, 1109–1118. [CrossRef]
10. Lynall, M.E.; Bassett, D.S.; Kerwin, R.; McKenna, P.J.; Kitzbichler, M.; Muller, U.; Bullmore, E. Functional connectivity and brain networks in schizophrenia. *J. Neurosci.* **2010**, *30*, 9477–9487. [CrossRef]
11. Kuperberg, G.R.; Broome, M.R.; McGuire, P.K.; David, A.S.; Eddy, M.; Ozawa, F.; Goff, D.; West, W.C.; Williams, S.C.R.; Van der Kouwe, A.J.W.; et al. Regionally localized thinning of the cerebral cortex in schizophrenia. *Arch. Gen. Psychiatry* **2003**, *60*, 878–888. [CrossRef]
12. De Peri, L.; Crescini, A.; Deste, G.; Fusar-Poli, P.; Sacchetti, E.; Vita, A. Brain Structural Abnormalities at the Onset of Schizophrenia and Bipolar Disorder: A Meta-analysis of Controlled Magnetic Resonance Imaging Studies. *Curr. Pharm. Des.* **2012**, *18*, 486–494. [CrossRef]
13. Mittal, V.A.; Bernard, J.A.; Northoff, G. What can different motor circuits tell us about psychosis? An RDoC perspective. *Schizophr. Bull.* **2017**, *43*, 949–955. [CrossRef]
14. Uddin, L.Q.; Yeo, B.T.T.; Spreng, R.N. Towards a Universal Taxonomy of Macro-scale Functional Human Brain Networks. *Brain Topogr.* **2019**, *32*, 926–942. [CrossRef]
15. Kebets, V.; Holmes, A.J.; Orban, C.; Tang, S.; Li, J.; Sun, N.; Kong, R.; Poldrack, R.A.; Yeo, B.T.T. Somatosensory-Motor Dysconnectivity Spans Multiple Transdiagnostic Dimensions of Psychopathology. *Biol. Psychiatry* **2019**, *86*, 779–791. [CrossRef]
16. Huang, C.C.; Luo, Q.; Palaniyappan, L.; Yang, A.C.; Hung, C.C.; Chou, K.H.; Zac Lo, C.Y.; Liu, M.N.; Tsai, S.J.; Barch, D.M.; et al. Transdiagnostic and Illness-Specific Functional Dysconnectivity Across Schizophrenia, Bipolar Disorder, and Major Depressive Disorder. *Biol. Psychiatry Cogn. Neurosci. Neuroimaging* **2020**, *5*, 542–553. [CrossRef]
17. Hirjak, D.; Meyer-Lindenberg, A.; Fritze, S.; Sambataro, F.; Kubera, K.M.; Wolf, R.C. Motor dysfunction as research domain across bipolar, obsessive-compulsive and neurodevelopmental disorders. *Neurosci. Biobehav. Rev.* **2018**, *95*, 315–335. [CrossRef]
18. Solmi, M.; Croatto, G.; Piva, G.; Rosson, S.; Fusar-Poli, P.; Rubio, J.M.; Carvalho, A.F.; Vieta, E.; Arango, C.; DeTore, N.R.; et al. Efficacy and acceptability of psychosocial interventions in schizophrenia: Systematic overview and quality appraisal of the meta-analytic evidence. *Mol. Psychiatry* **2023**, *28*, 354–368. [CrossRef]
19. Maj, M.; van Os, J.; De Hert, M.; Gaebel, W.; Galderisi, S.; Green, M.F.; Guloksuz, S.; Harvey, P.D.; Jones, P.B.; Malaspina, D.; et al. The clinical characterization of the patient with primary psychosis aimed at personalization of management. *World Psychiatry* **2021**, *20*, 4–33. [CrossRef]
20. Deste, G.; Corbo, D.; Nibbio, G.; Italia, M.; Dell'Ovo, D.; Calzavara-Pinton, I.; Lisoni, J.; Barlati, S.; Gasparotti, R.; Vita, A. Impact of Physical Exercise Alone or in Combination with Cognitive Remediation on Cognitive Functions in People with Schizophrenia: A Qualitative Critical Review. *Brain Sci.* **2023**, *13*, 320. [CrossRef]
21. Vita, A.; Gaebel, W.; Mucci, A.; Sachs, G.; Erfurth, A.; Barlati, S.; Zanca, F.; Giordano, G.M.; Birkedal Glenthøj, L.; Nordentoft, M.; et al. European Psychiatric Association guidance on assessment of cognitive impairment in schizophrenia. *Eur. Psychiatry* **2022**, *65*, E58. [CrossRef] [PubMed]
22. Page, M.J.; McKenzie, J.E.; Bossuyt, P.M.; Boutron, I.; Hoffmann, T.C.; Mulrow, C.D.; Shamseer, L.; Tetzlaff, J.M.; Akl, E.A.; Brennan, S.E.; et al. The PRISMA 2020 statement: An updated guideline for reporting systematic reviews. *BMJ* **2021**, *372*, n71. [CrossRef] [PubMed]
23. Damme, K.S.F.; Gupta, T.; Ristanovic, I.; Kimhy, D.; Bryan, A.D.; Mittal, V.A. Exercise Intervention in Individuals at Clinical High Risk for Psychosis: Benefits to Fitness, Symptoms, Hippocampal Volumes, and Functional Connectivity. *Schizophr. Bull.* **2022**, *48*, 1394–1405. [CrossRef] [PubMed]
24. Dean, D.J.; Bryan, A.D.; Newberry, R.; Gupta, T.; Carol, E.; Mittal, V.A. A supervised exercise intervention for youth at risk for psychosis: An open-label pilot study. *J. Clin. Psychiatry* **2017**, *78*, e1167–e1173. [CrossRef] [PubMed]
25. Maurus, I.; Roell, L.; Keeser, D.; Papazov, B.; Papazova, I.; Lembeck, M.; Roeh, A.; Wagner, E.; Hirjak, D.; Malchow, B.; et al. Fitness is positively associated with hippocampal formation subfield volumes in schizophrenia: A multiparametric magnetic resonance imaging study. *Transl. Psychiatry* **2022**, *12*, 388. [CrossRef]
26. McEwen, S.; Jarrahi, B.; Subotnik, K.; Ventura, J.; Nuechterlein, K. 396. Neuroplasticity Benefits of Adding Aerobic Exercise to Cognitive Training in First-Episode Schizophrenia Patients. *Biol. Psychiatry* **2017**, *81*, S161–S162. [CrossRef]
27. Roell, L.; Keeser, D.; Papazov, B.; Lembeck, M.; Papazova, I.; Muenz, S.; Schneider-axmann, T.; Sykorova, E.; Christina, E.; Vogel, B.O.; et al. Effects of Aerobic Exercise on Hippocampal Formation Volume and Connectivity in Patients with Schizophrenia—A Multicenter Randomized-Controlled Clinical Trial. 2023.
28. van der Stouwe, E.C.D.; Pijnenborg, G.H.M.; Opmeer, E.M.; de Vries, B.; Marsman, J.B.C.; Aleman, A.; van Busschbach, J.T. Neural changes following a body-oriented resilience therapy with elements of kickboxing for individuals with a psychotic disorder: A randomized controlled trial. *Eur. Arch. Psychiatry Clin. Neurosci.* **2021**, *271*, 355–366. [CrossRef]

29. Falkai, P.; Malchow, B.; Wobrock, T.; Gruber, O.; Schmitt, A.; Honer, W.G.; Pajonk, F.G.; Sun, F.; Cannon, T.D. The effect of aerobic exercise on cortical architecture in patients with chronic schizophrenia: A randomized controlled MRI study. *Eur. Arch. Psychiatry Clin. Neurosci.* **2013**, *263*, 469–473. [CrossRef]
30. Lin, J.; Chan, S.K.; Lee, E.H.; Chang, W.C.; Tse, M.; Su, W.W.; Sham, P.; Hui, C.L.; Joe, G.; Chan, C.L.; et al. Aerobic exercise and yoga improve neurocognitive function in women with early psychosis. *NPJ Schizophr.* **2015**, *1*, 15047. [CrossRef]
31. Malchow, B.; Keeser, D.; Keller, K.; Hasan, A.; Rauchmann, B.S.; Kimura, H.; Schneider-Axmann, T.; Dechent, P.; Gruber, O.; Ertl-Wagner, B.; et al. Effects of endurance training on brain structures in chronic schizophrenia patients and healthy controls. *Schizophr. Res.* **2016**, *173*, 182–191. [CrossRef]
32. Maurus, I.; Röll, L.; Keeser, D.; Karali, T.; Papazov, B.; Hasan, A.; Schmitt, A.; Papazova, I.; Lembeck, M.; Hirjak, D.; et al. Associations between aerobic fitness, negative symptoms, cognitive deficits and brain structure in schizophrenia—A cross-sectional study. *Schizophrenia* **2022**, *8*, 63. [CrossRef]
33. McEwen, S.C.; Jarrahi, B.; Ventura, J.; Subotnik, K.L.; Nguyen, J.; Woo, S.M.; Nuechterlein, K.H. A combined exercise and cognitive training intervention induces fronto-cingulate cortical plasticity in first-episode psychosis patients. *Schizophr. Res.* **2023**, *251*, 12–21. [CrossRef]
34. Mittal, V.A.; Gupta, T.; Orr, J.M.; Pelletier-Baldelli, A.; Dean, D.J.; Lunsford-Avery, J.R.; Smith, A.K.; Robustelli, B.L.; Leopold, D.R.; Millman, Z.B. Physical activity level and medial temporal health in youth at ultra high-risk for psychosis. *J. Abnorm. Psychol.* **2013**, *122*, 1101–1110. [CrossRef]
35. Pajonk, F.G.; Wobrock, T.; Gruber, O.; Scherk, H.; Berner, D.; Kaizl, I.; Kierer, A.; Müller, S.; Oest, M.; Meyer, T.; et al. Hippocampal plasticity in response to exercise in schizophrenia. *Arch. Gen. Psychiatry* **2010**, *67*, 133–143. [CrossRef]
36. Rosenbaum, S.; Lagopoulos, J.; Curtis, J.; Taylor, L.; Watkins, A.; Barry, B.K.; Ward, P.B. Aerobic exercise intervention in young people with schizophrenia spectrum disorders; improved fitness with no change in hippocampal volume. *Psychiatry Res.-Neuroimaging* **2015**, *232*, 200–201. [CrossRef]
37. Scheewe, T.W.; van Haren, N.E.M.; Sarkisyan, G.; Schnack, H.G.; Brouwer, R.M.; de Glint, M.; Hulshoff Pol, H.E.; Backx, F.J.G.; Kahn, R.S.; Cahn, W. Exercise therapy, cardiorespiratory fitness and their effect on brain volumes: A randomised controlled trial in patients with schizophrenia and healthy controls. *Eur. Neuropsychopharmacol.* **2013**, *23*, 675–685. [CrossRef]
38. Takahashi, S.; Keeser, D.; Rauchmann, B.S.; Schneider-Axmann, T.; Keller-Varady, K.; Maurus, I.; Dechent, P.; Wobrock, T.; Hasan, A.; Schmitt, A.; et al. Effect of aerobic exercise combined with cognitive remediation on cortical thickness and prediction of social adaptation in patients with schizophrenia. *Schizophr. Res.* **2020**, *216*, 397–407. [CrossRef]
39. Woodward, M.L.; Gicas, K.M.; Warburton, D.E.; White, R.F.; Rauscher, A.; Leonova, O.; Su, W.; Smith, G.N.; Thornton, A.E.; Vertinsky, A.T.; et al. Hippocampal volume and vasculature before and after exercise in treatment-resistant schizophrenia. *Schizophr. Res.* **2018**, *202*, 158–165. [CrossRef]
40. Woodward, M.L.; Lin, J.; Gicas, K.M.; Su, W.; Hui, C.L.M.; Honer, W.G.; Chen, E.Y.H.; Lang, D.J. Medial temporal lobe cortical changes in response to exercise interventions in people with early psychosis: A randomized controlled trial. *Schizophr. Res.* **2020**, *223*, 87–95. [CrossRef]
41. Svatkova, A.; Mandl, R.C.W.; Scheewe, T.W.; Cahn, W.; Kahn, R.S.; Hulshoff Pol, H.E. Physical Exercise Keeps the Brain Connected: Biking Increases White Matter Integrity in Patients with Schizophrenia and Healthy Controls. *Schizophr. Bull.* **2015**, *41*, 869–878. [CrossRef]
42. Makris, N.; Goldstein, J.M.; Kennedy, D.; Hodge, S.M.; Caviness, V.S.; Faraone, S.V.; Tsuang, M.T.; Seidman, L.J. Decreased volume of left and total anterior insular lobule in schizophrenia. *Schizophr. Res.* **2006**, *83*, 155–171. [CrossRef] [PubMed]
43. Rolls, E.T.; Huang, C.C.; Lin, C.P.; Feng, J.; Joliot, M. Automated anatomical labelling atlas 3. *Neuroimage* **2020**, *206*, 116189. [CrossRef] [PubMed]
44. Joliot, M.; Jobard, G.; Naveau, M.; Delcroix, N.; Petit, L.; Zago, L.; Crivello, F.; Mellet, E.; Mazoyer, B.; Tzourio-Mazoyer, N. AICHA: An atlas of intrinsic connectivity of homotopic areas. *J. Neurosci. Methods* **2015**, *254*, 46–59. [CrossRef] [PubMed]
45. Lin, J.J.; Geng, X.; Chen, E.Y.H. Yoga reduces the brain's amplitude of low-frequency fluctuations in patients with early psychosis. *Eur. Psychiatry* **2018**, *48*, S211.
46. Takahashi, H.; Sassa, T.; Shibuya, T.; Kato, M.; Koeda, M.; Murai, T.; Matsuura, M.; Asai, K.; Suhara, T.; Okubo, Y. Effects of sports participation on psychiatric symptoms and brain activations during sports observation in schizophrenia. *Transl. Psychiatry* **2012**, *2*, e96. [CrossRef]
47. Vancampfort, D.; Rosenbaum, S.; Ward, P.B.; Stubbs, B. Exercise improves cardiorespiratory fitness in people with schizophrenia: A systematic review and meta-analysis. *Schizophr. Res.* **2015**, *169*, 453–457. [CrossRef]
48. Martin, H.; Beard, S.; Clissold, N.; Andraos, K.; Currey, L. Combined aerobic and resistance exercise interventions for individuals with schizophrenia: A systematic review. *Ment. Health Phys. Act.* **2017**, *12*, 147–155. [CrossRef]
49. Firth, J.; Stubbs, B.; Rosenbaum, S.; Vancampfort, D.; Malchow, B.; Schuch, F.; Elliott, R.; Nuechterlein, K.H.; Yung, A.R. Aerobic exercise improves cognitive functioning in people with schizophrenia: A systematic review and meta-analysis. *Schizophr. Bull.* **2017**, *43*, 546–556. [CrossRef]
50. Firth, J.; Cotter, J.; Elliott, R.; French, P.; Yung, A.R. A systematic review and meta-Analysis of exercise interventions in schizophrenia patients. *Psychol. Med.* **2015**, *45*, 1343–1361. [CrossRef]

51. Loprinzi, P.D. The effects of physical exercise on parahippocampal function. *Physiol. Int.* **2019**, *106*, 114–127. [CrossRef]
52. Mechelli, A.; Riecher-Rössler, A.; Meisenzahl, E.M.; Tognin, S.; Wood, S.J.; Borgwardt, S.J.; Koutsouleris, N.; Yung, A.R.; Stone, J.M.; Phillips, L.J.; et al. Neuroanatomical abnormalities that predate the onset of psychosis: A multicenter study. *Arch. Gen. Psychiatry* **2011**, *68*, 489–495. [CrossRef]

brain sciences

MDPI

Systematic Review

Benefits of Treadmill Training for Patients with Down Syndrome: A Systematic Review

Karolina Kamińska [1], Michał Ciołek [1], Krzysztof Krysta [2,*] and Marek Krzystanek [2]

1 Students' Scientific Association, Department and Clinic of Rehabilitation Psychiatry, Faculty of Medical Sciences in Katowice, Medical University of Silesia in Katowice, 40-055 Katowice, Poland; karolina.kaminska311@gmail.com (K.K.); m.cioleksl@gmail.com (M.C.)

2 Department and Clinic of Rehabilitation Psychiatry, Faculty of Medical Sciences in Katowice, Medical University of Silesia in Katowice, 40-055 Katowice, Poland; krzystanekmarek@gmail.com

* Correspondence: krysta@mp.pl

Abstract: Background: The objective of this study was to evaluate the effectiveness of various results of treadmill training in children and adults with Down syndrome (DS). Methods: To provide an overview of this effectiveness, we conducted a systematic literature review of studies in which participants with DS from all age groups received treadmill training, alone or combined with physiotherapy. We also looked for comparisons with control groups of patients with DS who did not undergo treadmill training. The search was performed in medical databases: PubMed, PEDro, Science Direct, Scopus, and Web of Science, and included trials published until February 2023. Following PRISMA criteria, the risk of bias assessment was conducted using a tool developed by the Cochrane Collaboration for RCT. The selected studies presented multiple outcomes with differences in methodology; therefore, we were not able to conduct any sort of data synthesis, so we present measures of treatment effect as mean differences and corresponding 95% confidence intervals. Results: We selected 25 studies for the analysis with a total number of 687 participants, and identified 25 different outcomes which are presented in a narrative manner. In all outcomes we observed positive results favoring the treadmill training. Discussion: Introducing treadmill exercise into typical physiotherapy generates improvement in mental and physical health of people with DS.

Keywords: Down syndrome; treadmill training; physiotherapy; intellectual disability

check for updates

Citation: Kamińska, K.; Ciołek, M.; Krysta, K.; Krzystanek, M. Benefits of Treadmill Training for Patients with Down Syndrome: A Systematic Review. *Brain Sci.* **2023**, *13*, 808. https://doi.org/10.3390/brainsci13050808

Academic Editor: Daniele Corbo

Received: 22 April 2023
Revised: 11 May 2023
Accepted: 13 May 2023
Published: 16 May 2023

1. Introduction

Down syndrome (DS) is a genetic disorder that affects people of all races and societies, occurring in approximately 1 in every 1000–1100 children. It is caused by either partial or complete triplication of chromosome 21, and is considered to be one of the most prevalent causes of intellectual disability globally [1]. There are three genetic types of Down syndrome: trisomy 21, mosaicism, and Robertsonian translocation [2].

1.1. Health Complications Associated with Down Syndrome

It has been reported that people with Down syndrome struggle with many health problems and their life expectancy (approximately 55 years) is shorter than in neurotypical individuals. Furthermore, its incidence has been increasing over the years [3]. The reported health problems include congenital heart defects, hypothyroidism, leukemia, coeliac disease, muscle hypotonia, ligamentous laxity, atlantoaxial instability, epilepsy, obstructive sleep apnea, autoimmune diseases, recurrent respiratory infections, hearing and vision problems, early onset Alzheimer disease, and anxiety disorder [3,4].

1.2. Motor Development and Cognitive Function in Down Syndrome

Low muscle tone and ligamentous laxity results in delay of walking onset in infants with Down syndrome. They stand alone and walk independently when they are 9, 18, or

19 months old, whereas neurotypical infants usually walk when they are 6, 11, or 12 months old [5,6]. The atypical motor behavior of infants with Down syndrome is seen even in the first weeks of their life and leads to various implications for their future motor skills, for example, weaker grip strength and difficulty maintaining body posture [7,8]. Independent walking is also important because of its influence on cognitive and social development [9].

1.3. Obesity and Alzheimer's Disease in Down Syndrome

People with Down syndrome also have a higher risk of obesity, which is associated with a pro-inflammatory status [10]. Furthermore, they are susceptible to early progressive loss in executive functions (cognitive strategies to process adaptive, goal-directed actions) which is associated with early onset Alzheimer's disease in individuals with Down syndrome. Postmortem analyses showed the presence of beta-amyloid plaques, characteristic features of the development of Alzheimer's disease, in almost all of the neurological tissue of people with Down syndrome aged 35–40 years [11]. These plaques were also found in 8-year-old children [12].

1.4. Treadmill Training as a Physical Therapy Method

Studies have shown that adolescents who have Down syndrome exhibit weaker working memory, slower processing speed, and poor attention span and inhibitory control when compared to their peers who are typically developing. It surely degrades their quality of life [13,14]. Individuals with Down syndrome also have difficulty in productivity of words and switching in the semantic and the phonological fluency test compared with neurotypical persons [15]. It is widely accepted that treadmill training is a universal physical therapy method for supporting the treatment of many various conditions, for example, Parkinson's disease [16], chronic stroke [17], and cerebral palsy [18]. Physical therapy incorporates treadmill training as a means to enhance the locomotor functions of patients, which is made feasible by the plasticity of the central nervous system. The activation of trophic factors, neurogenesis, synaptogenesis, and angiogenesis during treadmill training leads to this improvement, especially in the early stages of development [19]. According to reports, a single 30 min exercise session could enhance the information processing speed and executive function of teenagers and young adults who have Down syndrome [20].

1.5. Evaluating the Effectiveness of Treadmill Training for Down Syndrome

It might be worthy of attention to evaluate the effectiveness of treadmill training, not only to improve locomotor performance, but also more universally, in various problems that people with Down syndrome struggle with, especially those related to the brain functions (executive and cognitive) or their inflammatory system. There are several existing systematic reviews that focus only on one of these functions or not particularly on treadmill training, or not only in patients with Down syndrome, and often describe only studies related to children [21–27]. The aim of this review is to address the existing gap in accessible analyses and evaluate the effectiveness of various results of treadmill training in children and adults with Down syndrome, because up until the present time there are no available systematic reviews on this topic.

2. Materials and Methods

This systematic review was conducted according to reporting standards of the PRISMA protocol [28]. The review was registered at the International Prospective Register of Systematic Reviews PROSPERO with the following number: CRD42023412948.

2.1. Research Question

The research question was formulated using the PICO format. We aimed to analyze the benefits of treadmill training alone or combined with physiotherapy in study groups compared to control groups, in which no interventions were applied, in patients with Down syndrome.

2.2. Objectives of Systematic Review

The objectives of this systematic review were to:

- Determine various effects of treadmill training, alone or combined with physiotherapy among children and adults with Down syndrome;
- Assess the quality of the included RCTs using the Cochrane risk-of-bias tool for randomized trials;
- Analyze and compare the results of the selected studies.

2.3. Search Strategy and Selection Process

The search was conducted on 21 February 2023 in five online medical databases: PubMed, PEDro, Science Direct, Scopus, and Web of Science. All databases were searched using the same search query: ("Down Syndrome") AND ("treadmill") in the title, abstract, and keywords fields. These search terms were developed based on preliminary searches carried out in each selected database. All identified records were documented and screened in Mendeley reference manager, which was also used for duplicate removal. Both authors independently screened titles and abstracts against the eligibility criteria of the review. Cases of disagreement were discussed until a consensus was reached. The search process is illustrated in Figure 1.

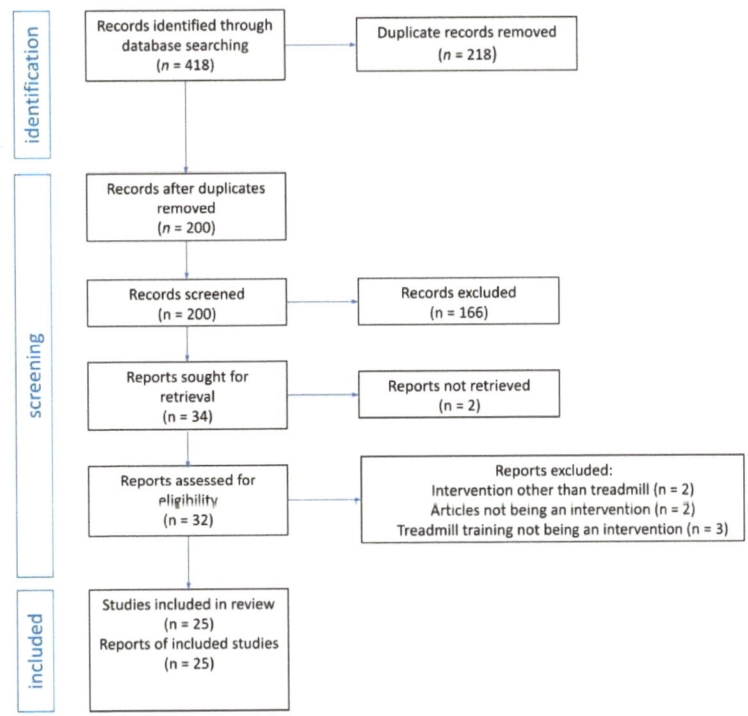

Figure 1. The search process.

2.4. Eligibility Criteria

- Types of studies: Randomized controlled, quasi-experimental, or clinical trial or pilot study published in English or Polish in peer-reviewed journals published from the inception of the database until 21st February 2023.
- Participants: Study participants with Down syndrome from all age groups.

- Intervention: Studies in which participants undergo treadmill training, alone or combined with physiotherapy.
- Comparison: Control groups formed only with patients with DS, who were offered only standard physiotherapy care or no therapeutic intervention.
- Outcome: In order to be included in the analysis, the study had to use a defined clinical outcome relating to mental or physical health in Down syndrome.

2.5. Risk of Bias Assessment

Following the Preferred Reporting Items for Systematic Reviews and Meta-Analyses (PRISMA) criteria [28], the risk of bias (RoB) assessment was conducted using a tool developed by the Cochrane Collaboration for randomized controlled trials [29].

2.6. Data Extraction

Data were extracted from each of the included reports using the Data Collection Form for Intervention Reviews for RCTs and non-RCTs template, consistent with the Cochrane Handbook for Systematic Reviews of Interventions [30].

2.7. Data Analysis

In the current review, we aim to summarize and analyze various types of outcome because in the selected articles we identified differences in study designs, effect measures, and clinical questions. Moreover, the age spectrum of the participants was wide, from infants to elderly people. Taking into consideration all the reasons presented above, we report all the data in a narrative review instead of in a meta-analysis. All outcome measures in this review are continuous. In order to measure the treatment effect, we used the mean difference and corresponding 95% confidence intervals.

3. Results

A total of 418 articles we originally identified. Within those results, duplicate removal was conducted using Mendeley reference manager and manual searching, which resulted in 200 records. Next, 166 articles were excluded based on title and abstract evaluation, mostly because the article was not an intervention study (specifically, reviews, book reviews, media reviews, editorials, obituaries, and case studies), or it did not have an eligible design, it recruited participants without DS, or a treadmill was not used as an intervention. Two articles were excluded because they could not be obtained. Therefore, 32 reports underwent full-text examination, after which 7 were excluded: 2 due to an intervention other than treadmill, 2 due to not being an intervention, and 3 due to treadmill training not being an intervention. Finally, 25 articles met the inclusion criteria.

The results of the search are presented in Table 1.

Table 1. Study characteristics and participants. Abbreviations: SG—study group, CG—control group, EG—experimental group, LG—lower-intensity-generalized training group, HI—higher-intensity-individualized training group, LG—generalized low intensive training, MI—moderate-intensity exercise.

No.	Study	Duration	Design	Intervention	Age	Co-Morbidities
1	Ulrich et al. (1992) [31]	11 months	Single group design	Supported treadmill stepping for infants with DS	7 months	Congenital heart defects (n = 5)
2	Ulrich et al. (1995) [32]	4–21 months	Single group design	Longitudinal supported treadmill stepping for infants with DS	8–11 months	Congenital heart defects (n = 5)

Table 1. *Cont.*

No.	Study	Duration	Design	Intervention	Age	Co-Morbidities
3	Ulrich et al. (2001) [33]	Until independent walking	Randomized controlled trial	Treadmill stepping practice for infants with DS	307.4 days	Congenital heart disease requiring surgery (SG: n = 7, CG: n = 2)
4	Carmeli et al. (2002) [34]	6 months	Parallel group design	Treadmill walking program for adults with DS	63 years	SG: Cardiac disease (n = 2)
5	Carmeli et al. (2004) [35]	15 weeks	Parallel group design	Treadmill walking program for ID adults with arterial occlusive disease	SG: 65.5 years, CG: 62 years	Arterial occlusive disease
6	Wu et al. (2007) [36]	Until 3 independent steps + 1 and 3-month follow-up	Randomized controlled trial	Different treadmill interventions for infants with DS	SG: LG—21.4 months, HI—19.2 months, CG: 23.9 months	Not reported
7	Wu et al. (2008 [37])	LG group—11 months, HI group—9.6 months + 1-year follow-up	Randomized trial (no control)	Treadmill interventions for newly walking toddlers with DS	HI group: 9.65 months, LG group: 10.40 months	Not reported
8	Ulrich et al. (2008) [38]	Until 3 independent steps	Randomized trial (no control)	Individualized, progressively intense treadmill training for infants with DS	HI group: 9.65 months, LG group: 10.40 months	Congenital heart defects (HI: n = 8, LG: n = 6)
9	Angulo-Barroso et al. (2008) [39]	15 months + 1-year follow-up	Randomized trial (no control)	Higher intensity, individualized TMT protocol for infants with DS	HI group: 9.65 months, LG group: 10.40 months	Not reported
10	Angulo-Barroso, Wu et al. (2008) [40]	15 months + 1-year follow-up	Randomized trial (no control)	Long-term effect of different treadmill interventions on gait patterns in infants with DS	HI group: 9.7 months, LG group: 10.40 months	Not reported
11	Mendonca et al. (2009) [41]	12 + 28 weeks	Single group design	28-week training program for DS males to improve aerobic capacity and locomotor economy	34.5 years	Not reported
12	Looper et al. (2010) [42]	Until 3 independent steps + 1-month follow-up	Randomized controlled trial	Early orthosis use combined with treadmill training in infants with DS vs. treadmill training alone	SG: 578 days, CG: 642 days	Not reported
13	Wu et al. (2010) [43]	Until 3 independent steps + 1-year gait follow-up	Randomized trial (no control)	Different treadmill interventions on joint kinematic patterns in infants with DS	LG: 35.7 months, HI: 75 months	Not reported

Table 1. *Cont.*

No.	Study	Duration	Design	Intervention	Age	Co-Morbidities
14	Ordonez et al. (2010) [44]	12-week training	Parallel group design with matched control	Aerobic training for reducing protein oxidation	16.3 years	Not reported
15	El- Meniawy et al. (2012) [45]	3 months intervention + 3-month follow-up	Randomized trial (no control)	Treadmill training vs. suspension therapy on balance in children with DS	9.34 years	Not reported
16	Ordonez et al. (2013) [46]	10-week aerobic training program	Randomized controlled trial	Aerobic training on plasma adipokines in obese women with DS	EG: 24.7 years, CG: 25.1 years	Obesity, mild ID
17	Lin et al. (2012) [47]	6-week program + 6-week follow-up	Randomized controlled trial	Strength and agility training for adolescents with DS	SG: 10.6 years, CG: 11.2 years	Not reported
18	Ordonez et al. (2014) [48]	10-week aerobic training program	Randomized controlled trial	Aerobic training on pro-inflammatory cytokines and acute phase proteins in women with DS	EG: 24.7 years, CG: 25.1 years	Obesity
19	Rodenbusch et al. (2013) [49]	Only intervention	Single group design	Effects of upward treadmill inclination on gait of children with DS	8.43 years	Not reported
20	Chen et al. (2014) [50]	~20 min (only intervention)	Randomized controlled trial	Relation between grip strength, anthropometric factors, and aerobic exercise impact on grip strength in young men with DS	EG: 21.76 years, CG: 17.77 years	Not reporte
21	Rosety-Rodriguez et al. (2014) [51]	10-week aerobic training program + 6-month follow-up	Parallel group design with matched control	Reduced inflammation effects maintenance after aerobic program completion	EG: 24.7 years, CG: 25.1 years	Obesity, mild ID
22	Chen et al. (2015) [52]	~20 min (only intervention)	Randomized controlled trial	Impact of single exercise intervention on executive function in young adults with DS	EG: 23.45 years, CG: 20.58 years	Not reported
23	Chen et al. (2016) [53]	~20 min (only intervention)	Parallel group design with matched control	Dose-response relationship between acute exercise intensity and cognitive performance	MI: 23.7 years, HI: 22.10 years, CG: 19.11 years	Not reported
24	Chen et al. (2019) [54]	~20 min intervention + 5–10 min rest + ~1.5 h verbal tests	Parallel group design with matched control	Relationship between acute exercise and verbal fluency	MI: 21.42 years, HI: 22.70 years, CG: 20.58 years	Not reported
25	Alsakhawi et al. (2019) [8]	4 months	Randomized controlled trial	Core stability training vs. treadmill exercises on balance in children with DS	4.59 years	Not reported

3.1. Study Characteristics of Included Studies

We included 25 studies with a total number of 687 participants [8,31–54]. The real number of subjects included in this analysis might be different because seven of the trials were performed with the same experimental group [36–40]. Papers by Ulrich et al. (1992) [31] and Ulrich et al. (1995) [32] had the same method of recruitment and intervention and the same number of participants. Similar situations happened in the case of trials by Ordonez et al. (2013) [46], Ordonez et al. (2014) [48], and Rosety-Rodriguez et al. (2014) [51]. Six studies [50,52–54] had the same method of recruitment and the same type of intervention. In the manuscripts by Wu et al. (2007) [33] and Ulrich et al. (2001) [36], there was the same control group.

3.1.1. Study Design

Nine studies were randomized controlled trials [8,33,36,42–44,50,52], six trials used a randomized parallel group design with no control group [37–40,44,47], six were quasi-experimental designs [34,35,44,51,53,54], and four were single-group designs [31,32,41,49].

3.1.2. Setting

Fourteen studies were conducted in the USA [31–33,36–40,42,43,50,52–54], four in Spain [44,47,48,51], two in Israel [34,35], two in Egypt [8,45], one in Portugal [41], one in Taiwan [46], and one in Brazil [49].

In eleven studies, interventions were provided at participants' homes [31–33,36–40,42,43]. In the study by Looper et al. (2010) [42], interventions were also located at the motor development laboratory. In the El-Meniawy et al. (2011) trial [45], the intervention took place in the outpatient clinic of The Faculty of Physical Therapy in Cairo University. Lin et al. (2012) [47] performed their study in a 48 m^2 room located at the Department of Occupational Therapy of Kaohsiung Medical University. Other studies did not include information about their settings.

3.1.3. Participants

Characteristics

All studies [8,31–54] focused on people diagnosed with Down syndrome, both in the experimental and control groups. The most common comorbidities were those associated with the cardiovascular system, for example, congenital heart defects [31–33,38], cardiac disease [34], arterial occlusive disease [35], and obesity [46,48,51]. The percentage of male participants was approximately 53.86%. In the studies of Ulrich et al. (2001) [36], Looper et al. (2010) [42], and Chen et al. (2016) [53], there was no information about this. In Alskhawi et al. (2019) [8] and El-Meniawy et al. (2011) [45], there was information indicating that both sexes participated in the study. The trials of Mendonca et al. (2009) [41], Ordonez et al. (2010) [44], and Chen et al. (2014) [50] included only males, while those of Ordonez et al. (2013) [46], Ordonez et al. (2014) [48], and Rosety-Rodriguez et al. (2014) [51] included only females. The age of patients varied substantially. Some of the studies referred to infants [31–33,36–40,42,43] from 8 months [32] to 75 months [43], and children and adolescents [44,45,47,49,54] from 4.5 years [8] to 16.3 years old [44]. Other studies [34,35,41,46,48,50–54] involved adults with Down syndrome, with a mean age of 24. The information about race or ethnicity was only known in 5 studies: Ulrich et al. (2001) [33] (2 mixed race and 28 white); Ulrich et al. (2007) [38] (2 African American, 26 white, and 2 biracial); Angulo-Barroso et al. (2008) [39] (2 African American, 26 Caucasian, and 2 other); Mendonca et al. (2009) [41].

Number of Participants, Method of Recruitment, and Other Relevant Information

Ulrich et al. (1992) [31] enrolled seven infants, all of them to the experimental group. They were recruited from the Down Syndrome Support Association of Central Indiana. These infants received physical therapy from 30 min to 1 h per week before the beginning of the study. Ulrich et al. (1995) [32] enrolled seven infants, all in the experimental group.

They were recruited from the same place as in the Ulrich 1992 trial. Two infants were born prematurely. Ulrich et al. (2001) [33] conducted another study, which involved recruiting a total of 30 participants from both parent support groups and DS clinics, with 15 participants in each of the experimental and control groups. Before the experiment, they received traditional physical therapy every week. Carmeli et al. (2002) [34] enlisted 26 inhabitants of an Israeli foster home, with 16 assigned to the experimental group and 10 assigned to the control group. It was ensured that the subjects did not consume any drugs that could have impeded their balance or strength performance. Additionally, Carmeli et al. (2004) [35] conducted another study with 26 residents of an Israeli foster home, in which 14 individuals were placed in the experimental group and 12 were placed in the control group. Six participants in the experimental group had symptoms of intermittent claudication. The paper by Wu et al. (2007) [36] involved the enrollment of 45 participants, with 30 assigned to the experimental group and 15 to the control group. The study included two cohorts of participants from two separate studies. The initial cohort was comprised of infants with DS who were recruited when they were capable of completing six steps per minute while receiving assistance from a treadmill. Participants were sourced from parent support groups in Lower Michigan. The control group for the analysis was taken from another study, by Ulrich et al. (2001) [33], who enrolled patients from parent support groups and DS clinics when they could sit alone for 30 s. Studies by Wu et al. (2008) [37], Ulrich et al. (2007) [38], Angulo-Barroso et al. (2008) [39], Angulo-Barroso, Wu et al. (2008) [40], and Wu et al. (2010) [43] each involved 30 participants, who were members only of the experimental groups, which were the same groups as in the paper by Wu et al. (2007) [36]. Mendonca et al. (2009) [41] enrolled 12 participants, all of them to the study group. They were recruited from a vocational center dedicated to the professional employability of individuals with intellectual disabilities. At the entry to the study, the subjects' average weight was 66.9 $\pm$ 8.9 kg and their height was 155.4 $\pm$ 9.5. In the study conducted by Looper et al. (2010) [42], a total of 17 participants with DS were included, with 10 assigned to the experimental group and 7 to the control group. The study by Ordonez et al. (2010) [44] included 38 participants (31 in the experimental group and 7 in the control group). Thirty participants took part in the study by El-Meniawy et al. (2012) [45]; all of them were members of the experimental group, but they were divided into two equal groups. In the work by Lin et al. (2012) [47], 92 individuals with disabilities from the Kaohsiung and Pingtung metropolitan areas met the inclusion criteria for the study (46 of them were in each of the experimental and control groups). School nurses, teachers, and directors selected individuals eligible for the experiment. In the study conducted by Ordonez et al. (2013) [46], they selected 20 women with Down syndrome who were obese, of which 11 were in the experimental group and 9 were in the control group. The participants did not have any harmful habits such as smoking or alcohol consumption, and were not taking any medication that could affect their appetite regulation or physical performance. To ensure that diet did not affect the results, the parents of the participants were carefully instructed to avoid any differences in the quantity or quality of food given to the control and experimental groups. In a similar study by Ordonez et al. (2014) [48], 20 obese women with DS were also enrolled (11 to the experimental group and 9 to the control group). They were recruited from different community groups for people with ID and their families. In the experiment by Rodenbush et al. (2013) [49], the participants were 16 non-probabilistically chosen subjects from 2 rehabilitation centers (the Association of Parents and Friends of Exceptional Children—APAE—and the Association to Assist Disabled People—ADOTE). There was no control group. Rosety-Rodriguez et al. (2014) [51] had the same experimental and study group as that of Ordonez et al. (2014) [48]. Chen et al. (2014) [50] recruited 20 young men—12 to the experimental group and 8 to the control group. In the studies by Chen et al. (2015 [52], 2016 [53], and 2019 [54]) the investigators invited participants recruited from local Special Olympic programs or DS organizations (e.g., Sharing Down syndrome Arizona, DS Network Arizona). Chen et al. (2015) [52] enrolled 20 young adults (10 in each of the experimental and study groups). In the study by Chen et al. (2016) [53] there were 18 young adults (12 in the experimental group and 6 in control group). Chen et al. (2019) [54] recruited 28 young adults (18 to the experimental group and 10 to the control group). Alsakhawi et al. (2019) [8] invited 45 participants recruited from an outpatient clinic at the Faculty of Physical Therapy in Cairo University. There were three equal groups: A, B and C. The inclusion criteria in studies focusing on infants referred to the ability to take a minimum of six spontaneous steps on a treadmill [37–40,43]. The exclusion criteria often refer to pre-existing conditions that may affect the ability to undergo the intervention, for example blindness, seizure disorder, musculoskeletal and auditory problems [34,36,38,40–42,45,46], mental age of participants lower than 3 years [50,52–54], or hypothyroidism [48]. Following the application of these criteria and occurrence of other conditions, there were some cases of withdrawing the participants from the studies described in the above table [32,36,38,39,42,43,46,49]. The information about ethical approval of the studies was included in all of them except for two [31,45]. The information about informed consent obtained from the participants was included in all the studies except for one [31].

3.1.4. Interventions

Treadmill Training Interventions in Various Studies

Treadmill training was the form of intervention applied in all of the twenty-six studies [8,31–54]. Specifications such as speed, duration, and inclination of treadmill exercise were adapted to the aims of the studies and the characteristics of the participants. In the study by Ulrich et al. (1992) [31], the treadmill training was performed 8 times for 30 s, and the speed of this training was 0.1 m/s, 0.15 m/s, and 0.2 m/s, in each set in 6 of 8 trials. Infants were supported by the investigators under the arms. The same method of intervention was used by Ulrich et al. (1995) [32], but additional information was added that the infants perform training trials monthly for several months, until every participant demonstrated consistent alternating step patterns across three successive testing sessions. Ulrich et al. (2001) [33] conducted an experiment in which the treadmill training was conducted for eight minutes a day, five days a week, until the participants displayed the capability to walk independently. The speed of the training was set at 2 m/s, and the infants were supported by their parents. Infants began participating in the study when they were able to sit independently for 30 s. The study was monitored by a team of researchers who visited all participants biweekly. Carmeli et al. (2002) [34] and Carmeli et al. (2004) [35] instructed the participants to walk at a comfortable pace, and if required, they could hold onto the handrails for balance adjustments while walking. Prior to the training, a warm-up session was conducted for 3 min, followed by knee extension and flexion exercises in the sitting position, and then 5 min of active stretching. The participants only walked between 9:30 a.m. and 11:30 a.m. indoors, and the environmental conditions were controlled to maintain a temperature of 23 °C and 40% humidity. In the study by Carmeli et al. (2002) [34], treadmill training was performed 3 times per week for 25 consecutive weeks, initially for 10–25 min, then increasing to 45 min. In the experiment of Carmeli et al. (2004) [35], treadmill training was performed 3 times per week, for 15 consecutive weeks, initially 5–15 min, then increasing to 40 min. The control groups did not undergo treadmill exercises. The study by Wu et al. (2007) [36] had an experimental group which consisted of two subgroups. The first was a lower-intensity-generalized (LG) training group, which included 14 participants, while the second subgroup was a higher-intensity-individualized (HI) training group, which was composed of 16 participants. The same groups of participants can be found in the publications by Wu et al. (2008) [37], Ulrich et al. (2007) [38], Angulo-Barroso et al. (2008) [39], Angulo-Barroso and Wu et al. (2008) [40], and Wu et al. (2010) [43]. In studies by Wu et al. (2007) [36], Wu et al. (2008) [37], Angulo-Barroso and Wu et al. (2008) [40], and Wu et al. (2010) [43], the LG group performed treadmill training for 6 min per day for 5 days a week; the HI group trained for 5 days per week, but the participants had individualized programs. The speed was 0.18 m/s. In works by Ulrich et al. (2007) [38] and Angulo-Barroso et al. (2008) [39], the LG group performed treadmill training for 8 min per day, 5 days per week; the HI group trained for 8–12 min a day. The speed for the LG group was 0.15 m/s and that for the HI group 0.15–0.3 m/s. Infants were supported by their parents in all studies [36–40,43]. Throughout the pre-walking phase of the study by Wu et al. (2007) [36], the research team made biweekly visits to all participants in order to measure their physical dimensions, record a 5 min period of treadmill stepping, and ensure that the training program was being followed as necessary. In a subsequent gait follow-up, participants were asked to walk at their own pace on the GAITRite mat, with measurements taken from an average of four walking trials. Participants in the control group did not engage in the treadmill training program but were instead asked to walk at their own pace across an 8-foot walkway that was covered with a long strip of 3-foot-wide butcher paper during the follow-up. The research performed by Wu et al. (2008) [37] also involved regular visits to the infants' homes by the staff during the treadmill intervention. The high-intensity (HI) group utilized ankle weights ranging from 0 to 115% of their calf mass. Thirteen individuals from each group completed a one-year follow-up of their gait after the treadmill intervention. The initial visit was scheduled as soon as parents reported that their child could walk 8–10 steps continuously at home, which took approximately 3 months for both groups. Ulrich et al. (2007) [38] conducted an experiment on the HI group where ankle weights were added, equivalent to 125% of the calf mass circumference, along with increased belt speed and daily duration to optimize the stepping response. The researchers visited all families every two weeks to ensure that the infants were following the treadmill training protocols. They also recorded five 1 min trials of the infants stepping while being supported on the treadmill, measured body weight, height, and shank length, and answered questions from the caregivers. In the work by Angulo-Barroso et al. (2008) [39], the HI group participated in the treadmill intervention, with progressively increasing treadmill belt speed, time, and ankle weights (to 125% of calf mass). In the study by Angulo-Barroso and Wu et al. (2008) [40], the research staff visited all the participants biweekly throughout the treadmill intervention. The HI group had an ankle weight that was 14–115% of calf mass. There was a one-year gait follow-up after the treadmill intervention—thirteen new walkers in the LG group (four male, nine female) and twelve in the HI group (nine male, three female) came into the laboratory immediately after walking onset. In the paper by Wu et al. (2010) [43], as soon as the parents informed the researchers that their child could take 8 to 10 independent steps at home, the initial visit was promptly arranged. The papers authored by Wu et al. (2007) [36], Angulo-Barroso et al. (2008) [39], Angulo-Barroso and Wu et al. (2008) [40], and Wu et al. (2010) [43] all had a similar scheduling pattern for the second, third, and fourth gait visits, which were set to occur three, six, and twelve months, respectively, after the first visit. In the experiment by Mendonca et al. (2009) [41], treadmill training was performed 8 min (submaximal protocol) and 12 min (graded maximal exercise protocol) 2 times per week for 28 weeks. The submaximal protocol was at 2.5 km/h and there was a 0% grade. The graded maximal exercise protocol was at 4 km/h and increased by 1.6 km/h every minute until exhaustion, and the grade increased by 2.5% every 2 min until 12.5% was reached. Before the study, participants received pre-training for 2 times a week for 12 weeks, as follows: 30 min (treadmill), 5 min warm-up dynamic activities, 40 min of continuous ergometer conditioning (treadmill, stepper, upright stationary cycle, and rowing ergometer), followed by a 5 min dynamic cool-down.

Treadmill Training with Orthoses and Virtual Reality

In Looper and colleagues' study (2010) [42], children wore orthoses throughout the entire study, which involved training sessions lasting 8 min per day, 5 days per week. The training continued until the children could take three independent steps, and the infants were assisted by their parents during the process. Between walking onset and follow-up testing, the children underwent treadmill training at a speed of 0.2 m/s while wearing the orthoses. After one month from the walking onset, the developmental tests were administered again. The control group did not wear orthoses. In the study by Ordonez et al.

(2010) [44], there was a 12-week treadmill training program (3 days per week for 20–35 min, increasing by 5 min each 3 weeks). The participants also received a warm-up (15 min) and cool-down (10 min). The work intensity on the treadmill was 60–75% of peak heart rate. The control group did not undergo treadmill training. In the study by El-Meniawy et al. (2012) [45], the duration of treadmill training was 20 min, 3 times a week for 3 months with 75% of over-ground speed. There was a warm-up for 5 min and cool-down for 5 min. Each group received an exercise program for balance and posture control for 30 min. Data were also collected three months after the treatment. In the experiment by Lin et al. (2012) [47], treadmill training took place for 5 min, 3 times a week for 6 weeks. The speed varied between 2.0 (0% incline) and 3.0 kph (5% elevation). Additionally, there was a warm-up (10 min), and a single virtual reality-based activity lasted for 20 min, administered 3 times per week for a duration of 6 weeks, with a 10 min break. There was a follow-up after 6 weeks. The control group did not undergo treadmill training.

Treadmill Training Duration and Intensity

Ordonez et al. (2013) [46], Ordonez et al. (2014) [48], and Rosety-Rodriguez et al. (2014) [51] conducted studies in which treadmill training lasted for 30–40 min, 3 times per week, over a period of 10 weeks. The training duration was gradually increased by 2.5 min every 2 weeks. The participants began with a 10–15 min warm-up period and ended with a 5–10 min cool-down period. The initial stage involved walking at a speed of 4.0 km/h for 2 min. The incline of the treadmill was then raised by 2.5% every 2 min until it reached a grade of 12.5%. The grade was then kept constant while the speed was increased by 1.6 km/hr every minute until the point of exhaustion. The work intensity on the treadmill was 55–65% of peak heart rate. Rosety-Rodriguez et al. (2014) [51] conducted re-evaluations of the parameters at 1, 3, and 6 months following the completion of the training program, during which time the participants did not engage in any further training program. In the experiment by Rodenbusch et al. (2013) [49], the treadmill training lasted 2 min for each inclination (0–10%). Speed was comfortable for participants. There was a 1 min rest period between changes in inclination. Data were captured for 30 s in each inclination.

Chen et al. (2014) [50], (2015) [52], (2016) [53], and (2019) [54] conducted studies where the training session was 20 min long, with a speed of 2.0–3.0 mph. The treadmill incline was increased by 2.5% every 4 min until the walking protocol was completed, which involved 5 stages and a 0–10% incline. Prior to the training, a warm-up was performed, where the speed was increased from 0.5 mph to 2.0 mph, with increments of 0.5 mph per minute. The walking protocol was stopped if the participant's heart rate exceeded 85% of their predicted maximum, the entire protocol was completed, or the participant reported feeling too tired or being unwilling to continue. In the work by Chen et al. (2015) [52], during the last 15 s of each stage, the participants were asked to select a picture on the rating of perceived exertion (RPE) scale in response to their perception of exertion. In the experiment by Chen et al. (2016) [54], the participants were assigned to high-intensity exercise (i.e., 75–85% of predicted maximum heart rate) (N = 6) or moderate-intensity exercise (i.e., 50–75% of predicted maximum heart rate) (N = 6). Chen et al. (2019) [55] conducted a study with two groups of participants who performed either high- or moderate-intensity exercise, with the former involving heart rates of 70–85% of predicted maximum and the latter involving heart rates of 50–69% of predicted maximum. After the intervention, participants were given 5–10 min of rest before taking the verbal fluency test again as a post-test measure. The total duration of the testing period was about an hour and a half. Control groups in all studies spent 20 min watching a video.

Combining Treadmill Training with Other Physical Therapy Interventions

Alsakhawi et al. (2019) [8] conducted a study with three groups. Group A received traditional physical therapy for 60 min to improve balance in the participating children. Group B underwent the same therapy as Group A for 30 min and additional core stability exercise training. Group C followed the same intervention strategies as Group A for 30 min, combined with a treadmill exercise program (20 min, three times a week, for eight weeks at 75% of over-ground speed and individually prescribed low-endurance walking). Before each walking session, the children in all groups engaged in 5 min of active stretching exercises that involved prolonged and progressive stretching of the hamstrings, quadriceps muscles, and Achilles tendon, followed by 30 min of physical therapy.

3.1.5. Outcome Assessment Tools

Walking Onset and Gait Patterns in Infants

The Bayley Scales of Infant Development was used in the studies by Ulrich et al. (1995) [32], Ulrich et al. (2001) [33], Ulrich et al. (2007) [38], Angulo-Barroso et al. (2008) [39], and Angulo-Barroso and Wu (2008) [40]. To measure the motor skills (four step types—alternating, single, parallel, and double) in Ulrich et al. (1992) [31] and Ulrich et al. (1995) [32], researchers used the camera. The GAITRite system and the camera were used in studies by Wu et al. (2007) [36], Wu et al. (2008) [37], and Angulo-Barroso et al. (2008) [39] to measure average velocity, stride length, step width, stride time, stance time, and dynamic base of infants. An ankle band with an activity monitor was used in the study by Angulo-Barroso and Wu et al. (2008) [40]. In the experiment by Wu et al. (2010) [43], the researchers used markers to measure the kinematics of the hip, knee, and ankle joints. The study conducted by Looper et al. (2010) [42] utilized Gross Motor Function as a tool for assessing gross motor skill development in various positions and movements, including lying and rolling, sitting, crawling and kneeling, and standing, as well as walking, running, and jumping.

Motor and Cardiovascular Function

In the study by Carmeli et al. (2002) [34], a dynamometer and the "timed get-up and go" test were applied to measure dynamic balance. In the study by Carmeli et al. (2004) [35], researchers used Vasculab PPG, Pain Physiopathology Index (PPI), and ABI to assess the volume of blood present in capillaries and pain levels. In the experiment by Mendonca et al. (2009) [41], a standardized body composition assessment was used (anthropometric measurements and bioelectrical impedance spectroscopy), and the following tools were applied: a resting protocol, submaximal steady-state exercise protocol, and a maximal graded exercise protocol to measure VO2 and pulmonary minute ventilation elicited by the selected submaximal treadmill exercise task and peak VO2. A Biodex instrument system was applied to assess stability level, feet angles, and heel coordinates in the study by El-Meniawy et al. (2011) [45]. In the experiment by Lin et al. (2012) [47], a handheld dynamometer, the Bruininks–Oseretsky Test of Motor Proficiency—Second Edition, a study questionnaire, and the Wechsler Intelligence Scale for Children—Third Edition were used to measure muscle strength (hip extensor, hip flexor, knee extensor, knee flexors, hip abductors, and ankle plantarflexor) and agility performance. In the work by Rodenbusch et al. (2013) [49], GMFCS, Berg Balance Scale (BBS), and the Qualisys Motion Capture System helped to evaluate spatial-temporal variables and angular variation of the hip, knee, and ankle in the sagittal plane. A hydraulic dynamometer was used in the study by Chen et al. (2014) [50] to measure grip strength. In the study by Alsakhawi et al. (2019) [8], the Berg Balance Scale and the Biodex Balance System were applied to assess functional balance and the steadiness of a circular force plate that was hanging and supported from above.

Protein Oxidation and Plasma Leptin Levels

In the study by Ordonez et al. (2010) [44], a Sport Tester PE3000 telemetric heart rate monitor and blood samples were used to measure carbonyl and protein content in the blood. In studies by Ordonez et al. (2013) [46], Ordonez et al. (2014) [48], and Rosety-Rodriguez et al. (2014) [51], a wireless heart rate monitor, bioelectrical impedance analysis, anthropometric tapes, and blood samples were applied to measure fat mass percentage and distribution, leptin and adipokine plasma levels, maximum oxygen uptake, waist circumference, and waist-to-hip ratio; the experiment by Rosety-Rodriguez et al. (2014) [51] additionally used plasma levels of IL-6 and high-sensitivity CRP; and the work by Ordonez et al. (2014) [48] additionally used fibrinogen and a1-antitrypsin.

Executive Function, Cognitive Performance and Verbal Fluency

In studies by Chen et al. (2015) [1], (2016) [2], and (2019) [3], protocols such as the Peabody Picture Vocabulary Test, Physical Activity Readiness Questionnaire, and vision and hearing tests were carried out. The time between the presence of a possible stimulus until the initiation of movement, the Knock-Tap test, and the Dimensional Change Card Sort Test were used in studies by Chen et al. (2015) [1] and (2016) [2] to assess executive function. Chen et al. (2016) [2] performed an experiment to evaluate cognitive performance, where participants were asked to respond to a blue light by pressing a specific button with their right index finger, and to a white light by pressing a different button with their left index finger. Chen et al. (2019) [3] conducted a study in which participants were asked to complete a verbal test. The initial 2 parts of the test required them to list as many words as they could think of that were associated with specific categories such as animals or food and drink, within a 60 s time limit. The final 2 parts of the test required them to list words beginning with the letters S and F, also within a 60 s time limit.

3.2. Outcomes

Two of the studies did not provide sufficient data for calculating mean differences (MD); therefore, they are only presented in narrative form. Ulrich et al. (1992) [31] tested alternating stepping patterns of 11-month-old infants with Down syndrome. The infants produced alternating steps on a moving treadmill, contrary to on a steady treadmill on which they produced no steps. There was no statistically significant difference in the number of steps taken at each of the three speeds of the treadmill. In the study it was found that the participants took a higher number of alternating steps (241 steps) than single (101), parallel (22), or double (4). In the study by Ulrich et al. (1995) [32], correlation was found between the development and shifting between stepping patterns produced by treadmill stimulation in infants with DS. The authors found that infants produce more alternating steps while growing up and a there is a shift from different step types to consistent and dominant alternation in stepping. The study conducted by Ulrich et al. (2001) [33] aimed to investigate whether infants with DS undergoing treadmill training achieved earlier development of three specific locomotor behaviors when compared to a control group. The mean advantages of those who participated in the treadmill training compared to the control group are that they were able to raise themselves to stand in 60 days, walk with help in 73 days, and walk independently after 101 days. However, in the comparison with the control group, only results referring to the walking with help reached statistical significance. Carmeli et al. (2002) [34] performed a study, the aim of which was to evaluate how treadmill exercise impacts leg strength and dynamic balance in older individuals diagnosed with DS. In the experimental group, peak torque %BW, and average power %BW of quadriceps and hamstrings of aged individuals with DS significantly increased after completing the treadmill protocol. Dynamic balance performance was also significantly improved in the walking group. The study by Carmeli et al. (2004) [35] focused on walking performance of elderly people with DS and intermittent claudication. Upon completion of the training program, all participants exhibited notable enhancements in their walking speed, distance, and duration. Individuals experiencing intermittent claudication also reported a reduction in pain levels. Additionally, significant improvements were observed in the participants' blood hemodynamic parameters. Wu et al. (2007) [36] measured basic gait parameters and walking onset. There were three groups, which differed in walking experience (which was adjusted for in data analysis): control, low-, and high-intensity treadmill training. The high-intensity training group started walking at a younger age in comparison to the control. In terms of walking onset, there were no differences between low- and high-intensity training groups. An analysis conducted after the study showed that, aside from stride length, there were no significant differences in gait parameters between the groups. The high-intensity (HI) group had a significantly greater stride length compared to the control (C) group, while there was no significant difference between the low-intensity-generalized (LG) and HI groups. Stride length was the gait parameter that mainly contributed to the difference between the groups. The HI group showed a significantly longer stride length than the C group. Overall, the results suggest that the HI treadmill intervention led to earlier walking onset and advanced gait patterns, particularly in terms of stride length, in infants with DS. Wu et al. (2008) [37] measured the percentage of locomotor strategy and adjustment in obstacle clearance in a group of new walking toddlers with DS. The study participants trained in two different protocols: high and low intensity. The high-intensity group (84.3%) used walking strategies to avoid the obstacle more often than the low-intensity group (67.8%). There were no significant differences between the groups in terms of gait parameters in the five steps before approaching the obstacle. The study by Ulrich et al. (2008) [38] focused on stepping and motor development while comparing two groups: one with high-intensity and second with low-intensity treadmill training. There was no notable distinction between the groups in terms of the quantity of alternating steps taken at the start of the study. The group that engaged in high-intensity activities achieved all of the motor milestones more quickly than the group with low-intensity activities. However, only the accomplishment of item 52, which involved raising oneself to a standing position, was deemed statistically significant. The study by Angulo-Barroso et al. (2008) [39] compared physical activity in individuals with DS performing two treadmill training protocols: high intensity and low intensity, respectively. Infants receiving the high-intensity training had higher levels of High-act than infants in the low-intensity training group. Infants in the low-intensity group had higher duration of Low-act. The study by Angulo-Barroso et al. (2008) [40] explored the long-term effects of high-intensity and low-intensity treadmill training on gait development in walkers with DS. Six basic gait parameters were measured: normalized velocity, cadence, step length, step width, double support percentage, and dynamic base. On average, the high-intensity group performed better, producing higher normalized velocity and cadence, and lower double support percentage. Moreover, both groups significantly reduced foot rotation asymmetry over time, but no difference was observed between the groups. The study by Mendonca et al. (2009) [41] had a single group design, which measured the effect of treadmill training on submaximal and peak aerobic capacity. After the intervention, researchers observed decreased fat mass, increased absolute fat-free mass, and improved peak exercise capacity. There were no significant differences in participants' body weight, BMI, resting VO2, or heart rate. In the study by Looper et al. (2010) [42], researchers compared the effect of treadmill training combined with orthoses (experimental group) and treadmill training alone (control group). In both groups, all scores in the Gross Motor Function Measure were increased, although the control group had higher scores in the standing and walking, running, and jumping subscales. In their research, Wu et al. (2010) [43] conducted a comparison between low-intensity and high-intensity treadmill training groups in terms of joint kinematics. Results showed that both groups made significant progress in terms of joint kinematics during gait follow-up. However, in the high-intensity training group, peak ankle plantar flexion occurred at or before toe-off, and there was an increase in the forward thigh swing after toe-off time. The study by Ordonez et al. (2012) [44] explored the correspondence with treadmill training and protein oxidation based on the carbonyl content in individuals with DS. In the intervention group, plasmatic carbonyl content significantly decreased, in contrast to the control group in which the researchers observed no significant changes. El-Meniawy et al. (2012) [45] compared the effects of suspension therapy and treadmill training on balance in children with DS. The mean differences in overall, anterior-posterior, and medio-lateral stability indexes suggest that both interventions are beneficial for balance; however, the Student *t*-test showed no significant differences between the groups. The study by Lin et al. (2012) [47] measured the strength of lower-extremity muscle and agility in two groups: the first experimental, which performed treadmill training, and the second, control group. Significant differences in muscle strength were observed for hip flexors, extensors and abductors, knee flexors and extensors, and ankle plantar flexors in the experimental group. Treadmill training also increased participants' agility scores. The study by Ordonez et al. (2013) [46] explored the anti-inflammatory effect of treadmill training by measuring leptin levels in obese women with DS. Participants were divided into control and experimental groups. There were no significant changes in adiponectin plasma levels between the groups, but they differed in leptin plasmatic levels, which decreased in the training group. Fat mass percentage and waist-to-hip ratio were also reduced in the training group. Moreover, researchers found a positive association between leptin and WHR and a negative association between adiponectin and waist circumference. The study by Ordonez et al. (2014) [48] assessed the impact of treadmill training on plasmatic levels of pro-inflammatory cytokines and body composition in obese women with DS. In the intervention group, plasmatic levels of TNF- α, IL-6, high-sensitivity C-reactive protein, and fibrinogen were significantly decreased. Similarly, both waist-to-hip ratio and fat mass percentage decreased in the training group. No significant changes were observed in the control group. The experiment by Rodenbusch et al. (2013) [49] was a single group design study that aimed to explore the effect of treadmill inclination on gait of children with DS. In spatio-temporal variables, a reduction in cadence and an increase in cycle and swing time were observed, after the upward treadmill inclination was increased. In angular variables, researchers observed an increased angle at initial contact in hip, knee, and ankle joints, increased maximum flexion angle in the hip joint, maximum plantarflexion at pre-swing, and an increase in maximum dorsiflexion in stance, while walking on an inclined surface. The study by Chen et al. (2014) [50] explored the effect of treadmill training on grip strength in individuals with DS. Both study groups' post-test scores were significantly different, although the intervention group had higher grip strength scores. In the study by Rosety-Rodriguez et al. (2014) [51], researchers analyzed the longitudinal effect of treadmill training on pro-inflammatory cytokines and body mass composition. There were no significant changes over time in the control group. The intervention group showed a significant increase in plasma levels of IL-6 and hs-CRP (P = 0.026) at 3 months following the completion of the training program, compared to the post-test results. Moreover, there was a significant increase in both IL-6 and hs-CRP at 6 months after training, compared to the measurements taken just 1 month after the end of the intervention program. The study by Chen et al. (2015) [52] focused on the effect of treadmill training on choice-response time, attention shifting, and inhibition. The ANCOVA showed that both choice-response time and attention shifting did not significantly differ between the study groups. A statistically significant difference was present only in inhibition favoring the exercise group. Chen et al. (2016) [53] explored the dose–response relationship between exercise and cognitive performance represented by information processing, attentional switching, and inhibitory control. The ANCOVA and polynomial contrast showed no significant changes nor any trend in terms of attentional shifting. Inhibitory control improved linearly with intensity of training. Regarding information processing speed, the low-intensity group performed better than both the control group and the high-intensity group. The purpose of the study by Chen et al. (2019) [54] was to assess the relationship between exercise and cognitive functions, and especially their verbal fluency aspects. The ANCOVA, Levene's test, and normality check showed only a significant quadratic trend for semantic fluency. The study by Alsakhawi et al. (2019) [8] compared treadmill training and core stability exercises to the effects of treatment in the control group (receiving standard physiotherapy) in order to determine which intervention is more effective in improving balance in children with DS. Functional balance and the overall stability index were significantly higher in stability and treadmill training groups MD, respectively, although there was no significant difference between intervention groups.

Overall, the studies suggest that treadmill training can have positive effects on various aspects of motor development, such as increasing the number of alternating steps, improving leg strength and dynamic balance, and enhancing walking performance in individuals with DS. Additionally, treadmill training has been found to have anti-inflammatory effects and improve body composition in obese individuals with DS. The studies also suggest that treadmill training can improve cognitive functions, particularly inhibitory control and information processing speed, in individuals with DS.

3.3. Ethical Issues Concerning Studies Involving Infants and Children

In the analysis of ethical approvals and informed consent across the studies involving infants and children, there is a broad yet uneven adherence to the established ethical guidelines of biomedical research. Studies involving infants, including those by Ulrich et al. (1992) [31], Ulrich et al. (1995) [32], Ulrich et al. (2001) [33], Wu et al. (2007) [36], Wu et al. (2008) [37], Ulrich et al. (2008) [38], Angulo-Barroso et al. (2008) [40], Looper et al. (2010) [42], and Wu et al. (2010) [43], demonstrate a general compliance with the principle of informed consent. They reported obtaining consent from parents or legal guardians, thereby respecting their autonomy and their right to decide on their infants' participation in their respective studies. However, the reporting on the acquisition of ethical approval from relevant institutional review boards or ethics committees in these studies was inconsistent. The studies by Ulrich et al. (2001) [33], Wu et al. (2007) [36], Wu et al. (2008) [37], Ulrich et al. (2008) [38], Looper et al. (2010) [42], Angulo-Barroso et al. (2008) [40], and Wu et al. (2010) [43] explicitly stated that they received ethical approval, indicating their commitment to the principles of beneficence and non-maleficence. These principles require researchers to ensure that the potential benefits of their research outweigh any potential harm to the participants. On the other hand, the studies by Ulrich et al. (1992) [31] and Ulrich et al. (1995) [32], while acknowledging the acquisition of informed consent, did not explicitly mention obtaining ethical approval.

When it comes to studies involving children, such as those by Ordonez et al. (2012) [44], El-Meniawy et al. (2012) [45], Lin et al. (2012) [47], Rodenbusch et al. (2013) [49], and Chen et al. (2019) [54], the adherence to the ethical principles of informed consent and ethical approval appears to be more consistent. All these studies reported obtaining informed consent from parents or legal guardians, respecting their autonomy and their right to decide on their children's participation in the research. Furthermore, the studies by Ordonez et al. (2012) [44], Rodenbusch et al. (2013) [49], and Chen et al. (2019) [54] explicitly stated that they received ethical approval from the respective institutional ethics committees or review boards. These reports confirm their commitment to ensuring the safety and well-being of the child participants in their research. However, the studies by El-Meniawy et al. (2012) [45] and Lin et al. (2012) [47] reported obtaining informed consent but did not specify the acquisition of ethical approval.

In conclusion, while most studies demonstrate a broad adherence to the principles of informed consent and ethical approval, there is a need for more consistent and explicit reporting of ethical aspects in research focusing on infants and children with DS, particularly regarding ethical approval. This is crucial not only to uphold the ethical integrity of research, but also to ensure the trust of the public and the scientific community in the research process and its outcomes.

3.4. Risk of Bias

In this section we present the results of risk of bias assessment for 15 studies: 9 randomized controlled trials and 6 parallel randomized trials with no control group. A summary of the analysis across all domains for each individual study can be seen in Table 2 and the proportion of studies in each domain is shown below. Only three studies met the criteria for an overall low risk of bias, six studies were identified as having some concerns, and six were identified as having a high risk of bias.

- Domain 1: Risk of bias arising from randomization process

Six studies had a well-described randomization process [8,36,40,43,47,48]; therefore, we identified them as having low risk of bias in this domain. We assessed seven studies [8,31,37–39,46,52] as having some concerns, because of evasive descriptions of randomization and concealment and lack of differences between study groups. Two studies had high risk of bias in this domain due to no allocation concealment and baseline differences between study groups [42,50].

- Domain 2: Risk of bias due to deviations from the intended interventions

Seven studies had no dropouts and no deviations from the outcome; on that account we judged them as having low risk of bias [8,33,37,45–48]. Seven studies were marked as having some concerns because of a lack of intent to treat analysis when dropouts occurred. [36,38–40,43,50,52]. The study by Looper et al. (2010) had a high risk of bias in this domain since the lack of concealment of caregivers led to deviations from the protocol [42].

- Domain 3: Risk of bias due to missing outcome data

Nine studies were assessed as having low risk of bias because there were no missing data [1,2,6–9,11–13]. Five studies had a high risk of bias because of missing data and "per protocol" or "as treated" analysis model; also, no sensitivity analysis was conducted [5,6,9,10,14]. In the study by Wu et al. (2010) [12], the participants also dropped out during the trial, but the recruited group was 10% bigger as a precaution of missing data.

- Domain 4: Risk of bias in measurement of the outcome

We found no potential misconduct in this domain; in most studies, researchers used adequate scales and tools for an outcome measure. Furthermore, in most cases the outcome assessor was unaware of the group assignment or this knowledge was unlikely to affect the assessment. Therefore, none of the included studies had a high risk of bias in this domain and only one was judged as having some concerns, because of the authors' statement that the assessor was not blinded and this is likely to interfere with the results [12].

- Domain 5: Risk of bias in selection of the reported results

All of the included reports had a low risk of bias in domain 5, because we found no evidence that authors used multiple outcome measurements or analysis.

Table 2. Risk of bias: authors' judgment about risk of bias for each domain for each included study eligible for assessment. Colors used in the table represent—green: low risk of bias; yellow: some concerns about bias; red: high risk of bias.

Randomized Controlled Trials	**Domain 1**	**Domain 2**	**Domain 3**	**Domain 4**	**Domain 5**	**Overall**
Ulrich et al. (2001) [33]	Some concerns	Low risk	Low risk	Low risk	Low risk	Some concerns
Wu et al. (2007) [36]	Low risk	Some concerns	High risk	Low risk	Low risk	High risk
Looper et al. (2010) [42]	HIgh risk	High risk	High risk	Some concerns	Low risk	High risk
Lin et al. (2012) [47]	Low risk	Low risk	Low risk	Low risk	Low risk	Low risk
Ordonez et al. (2013) [46]	Some concerns	Low risk	Low risk	Low risk	Low risk	Some concerns
Ordonez et al. (2014) [48]	Low risk	Low risk	Low risk	Low risk	Low risk	Low risk
Chen et al. (2014) [50]	HIgh risk	Some concerns	Low risk	Low risk	Low risk	High risk
Chen et al. (2015) [52]	Some concerns	Some concerns	High risk	Low risk	Low risk	High risk
Alsakhawi et al. (2019) [8]	Low risk	Low risk	Low risk	Low risk	Low risk	Low risk
Randomized Trials with no Control	**Domiain 1**	**Domain 2**	**Domain 3**	**Domain 4**	**Domain 5**	**Overall**
Wu et al. (2008) [37]	Some concerns	Low risk	Low risk	Low risk	Low risk	Some concerns
Ulrich et al. (2008) [38]	Some concerns	Some concerns	High risk	Low risk	Low risk	High risk
Angulo-Barroso et al. (2008) [39]	Some concerns	Some concerns	High risk	Low risk	Low risk	High risk
Angulo-Barroso, Wu et al. (2008) [40]	Low risk	Some concerns	Low risk	Low risk	Low risk	Some concerns

Table 2. *Cont.*

Randomized Trials with no Control	Domaiin 1	Domain 2	Domain 3	Domain 4	Domain 5	Overall
Wu et al. (2010) [43]	Low risk	Some concerns	Some concerns	Low risk	Low risk	Some concerns
EL-Meniawy et al. (2012) [45]	Some concerns	Low risk	Low risk	Low risk	Low risk	Some concerns

4. Discussion

Down syndrome is a genetic disorder impacting physical and cognitive development. Exercise, particularly treadmill training, offers benefits such as enhanced strength, balance, and physical functioning. This review examined treadmill training's advantages for individuals with DS by evaluating 25 studies with 687 participants, assessing its effectiveness across all ages [8,31–54]. Because we were unable to conduct a meta-analysis, we narratively reported results. Unlike other articles, we considered all benefits of treadmill use for individuals with Down syndrome across various domains [21–27]. Compared to other physical activities such as aerobic exercise, swimming, and cycling, treadmill training has proven to be one of the more effective, simpler, and safer methods of helping people with DS with their various health problems [55–57]. The review of the studies faced several challenges. One of them was a wide spectrum of age groups in the examined populations. The manuscripts we analyzed referred to infants [31–33,36–40,42,43], children and adolescents [44,45,47,49,54], and adult DS patients [34,35,41,46,48,50–54]. The number of participants in the study group also varied among the papers, with two cases having very small sample sizes of less than ten [31,32]. Most of the studies had an intermediate size, between 10 and 30 participants [34,35,41,46,48,50–54]. Only three papers had larger study groups with more than thirty participants [8,44,47]. All the studies applied treadmill training as the form of intervention; however, there were some differences concerning speed, duration, and inclination of the treadmill exercise. In the studies involving infants [31–33,42], the participants were supported by other individuals, trainings were repeated up to five times a week, and the whole program lasted from eight repetitions of training to a longer period of several months. In the studies involving adults [34–37,40,41,43–49,51], the trainings lasted from 6 to 40 min and were repeated from 3 to 5 times a week. The training program lasted from 3 months to 14 weeks. Various assessment tools were used to evaluate outcomes such as motor skills, cardiovascular function, cognitive performance, and verbal fluency. These tools included scales for infant development, camera recordings, balance and motion capture systems, heart rate monitors, and tests for cognitive and verbal abilities. Effects measures in mean differences (MDs) of the analyzed studies could be presented for eighteen studies, and two of the studies did not provide sufficient data for calculating MD. The most commonly observed outcomes of the trainings were significant improvements in walking speed, distance, and duration, and better motor development in children.

This systematic review examining the benefits of treadmill training for individuals with Down syndrome (DS) unveils several important implications for practice. Firstly, treadmill training can accelerate walking onset and enhance gait development in children with DS, helping them walk with improved patterns at an earlier age. Secondly, adults with DS can experience anti-inflammatory advantages and cognitive function improvements through treadmill training.

As a form of exercise that is both safe and easy to perform for people of all ages, treadmill training presents a practical, accessible solution for those with DS. In essence, these findings emphasize the significance of treadmill training as a safe, effective method for boosting physical and cognitive functions in people with DS.

To provide a more comprehensive understanding of the effects of treadmill training, future studies should concentrate on several key areas: assessing cognitive function, including memory and attention; determining the optimal duration and frequency of treadmill training for maximum benefits; investigating the potential of treadmill training as a preventative measure against Alzheimer's disease in people with DS; and employing an appropriate control group, such as individuals with DS who do not partake in treadmill training, to accurately gauge the intervention's effects.

Despite the comprehensive approach and extensive analysis, this manuscript has several limitations that should be acknowledged. Firstly, the studies included in this review were conducted at different times and locations, which could have resulted in variations in methodology and results. Secondly, the majority of studies included were conducted on small sample sizes, which may limit the generalizability of the findings. Thirdly, the studies varied in terms of the age ranges of participants, which may have impacted the effectiveness of the interventions. Fourthly, the studies also varied in terms of the intervention duration and frequency, which makes it difficult to draw conclusions about the optimal parameters for effective intervention. Lastly, the risk of bias assessment revealed that some studies had high risk of bias, particularly in the domains of randomization and missing outcome data, which may have influenced the results. These limitations suggest a need for further well-designed studies with larger sample sizes and consistent intervention protocols to fully explore the effects of treadmill training on individuals with Down syndrome.

5. Conclusions

Our systematic review found that treadmill training can help children with Down syndrome develop walking and gait patterns at an earlier age, and can provide anti-inflammatory and cognitive benefits for adults with DS. Treadmill training is safe and easy to implement, and has the potential to improve physical and cognitive functions. Future studies should focus on assessing cognitive function, determining the optimal duration and frequency of training, exploring the potential of treadmill training as a preventative intervention for Alzheimer's disease, and using appropriate control groups.

Author Contributions: Conceptualization, K.K. (Krzysztof Krysta), K.K. (Karolina Kamińska) and M.C.; methodology, M.C.; validation, K.K. (Krzysztof Krysta) and M.K.; formal analysis, M.C.; investigation, K.K. (Karolina Kamińska) and M.C.; writing—original draft preparation, K.K. (Karolina Kamińska) and M.C.; writing—review and editing, K.K. (Krzysztof Krysta) and M.K.; visualization, M.C. and K.K. (Karolina Kamińska); Supervision, K.K. (Krzysztof Krysta) and M.K.; project administration, M.C. All authors have read and agreed to the published version of the manuscript.

Funding: The cost of publication was covered by Medical University of Silesia. Funding reference number KDP-2/06/2023.

Institutional Review Board Statement: Not applicable.

Informed Consent Statement: Not applicable.

Data Availability Statement: No new data were created.

Conflicts of Interest: The authors declare no conflict of interest.

References

1. Lukowski, A.F.; Milojevich, H.M.; Eales, L. Chapter Eight—Cognitive Functioning in Children with Down Syndrome: Current Knowledge and Future Directions. In *Advances in Child Development and Behavior*; Benson, J.B., Ed.; JAI: Stamford, CT, USA, 2019; Volume 56, pp. 257–289.
2. Diamandopoulos, K.; Green, J. Down syndrome: An integrative review. *J. Neonatal Nurs.* **2018**, *24*, 235–241. [CrossRef]
3. Asim, A.; Kumar, A.; Muthuswamy, S.; Jain, S.; Agarwal, S. "Down syndrome: An insight of the disease". *J. Biomed. Sci.* **2015**, *22*, 41. [CrossRef]
4. Antonarakis, S.E.; Skotko, B.G.; Rafii, M.S.; Strydom, A.; Pape, S.E.; Bianchi, D.W.; Sherman, S.L.; Reeves, R.H. Down syndrome. *Nat. Rev. Dis. Prim.* **2020**, *6*, 9. [CrossRef]
5. Bayley, N. Manual for the Bayley Scales of Infant Development (No Title). 1969. Available online: https://cir.nii.ac.jp/crid/1130000795357934080 (accessed on 21 April 2023).
6. Henderson, S. Some aspects of the development of motor control in Down's syndrome. *Themes Mot. Dev.* **1986**, 69–92.
7. Pitetti, K.H.; Climstein, M.; Mays, M.J.; Barrett, P.J. Isokinetic arm and leg strength of adults with Down syndrome: A comparative study. *Arch. Phys. Med. Rehabil.* **1992**, *73*, 847–850.

8. Alsakhawi, R.S.; Elshafey, M.A. Effect of Core Stability Exercises and Treadmill Training on Balance in Children with Down Syndrome: Randomized Controlled Trial. *Adv. Ther.* **2019**, *36*, 2364–2373. [CrossRef]
9. Gustafson, G.E. Effects of the ability to locomote on infants' social and exploratory behaviors: An experimental study. *Dev. Psychol.* **1984**, *20*, 397. [CrossRef]
10. Inadera, H. The usefulness of circulating adipokine levels for the assessment of obesity-related health problems. *Int. J. Med. Sci.* **2008**, *5*, 248. [CrossRef]
11. Malamud, N. Neuropathology of Organic Brain Syndromes Associated with Aging. In *Aging and the Brain: The Proceedings of the Fifth Annual Symposium Held at the Texas Research Institute of Mental Sciences in Houston, October 1971*; Springer: Berlin/Heidelberg, Germany, 1995; pp. 63–87.
12. Leverenz, J.B.; Raskind, M.A. Early amyloid deposition in the medial temporal lobe of young Down syndrome patients: A regional quantitative analysis. *Exp. Neurol.* **1998**, *150*, 296–304. [CrossRef]
13. Adams, D.; Oliver, C. The relationship between acquired impairments of executive function and behaviour change in adults with Down syndrome. *J. Intellect. Disabil. Res.* **2010**, *54*, 393–405. [CrossRef]
14. Daunhauer, L.A.; Fidler, D.J.; Will, E. School function in students with Down syndrome. *Am. J. Occup. Ther.* **2014**, *68*, 167–176. [CrossRef]
15. Nash, H.M.; Snowling, M.J. Semantic and phonological fluency in children with Down syndrome: Atypical organization of language or less efficient retrieval strategies? *Cogn. Neuropsychol.* **2008**, *25*, 690–703. [CrossRef]
16. Mehrholz, J.; Kugler, J.; Storch, A.; Pohl, M.; Hirsch, K.; Elsner, B. Treadmill training for patients with Parkinson Disease. An abridged version of a Cochrane Review. *Eur. J. Phys. Rehabil. Med.* **2016**, *52*, 704–713.
17. Tally, Z.; Boetefuer, L.; Kauk, C.; Perez, G.; Schrand, L.; Hoder, J. The efficacy of treadmill training on balance dysfunction in individuals with chronic stroke: A systematic review. *Top. Stroke Rehabil.* **2017**, *24*, 539–546. [CrossRef]
18. Han, Y.-G.; Yun, C.-K. Effectiveness of treadmill training on gait function in children with cerebral palsy: Meta-analysis. *J. Exerc. Rehabil.* **2020**, *16*, 10. [CrossRef]
19. Kleim, J.A.; Jones, T.A. Principles of experience-dependent neural plasticity: Implications for rehabilitation after brain damage. *J. Speech Lang. Hear. Res.* **2008**, *51*, S225–S239. [CrossRef]
20. Ringenbach, S.D.; Albert, A.R.; Chen, C.-C.; Alberts, J.L. Acute bouts of assisted cycling improves cognitive and upper extremity movement functions in adolescents with Down syndrome. *Ment. Retard.* **2014**, *52*, 124–135. [CrossRef]
21. Damiano, D.L.; DeJong, S.L. A systematic review of the effectiveness of treadmill training and body weight support in pediatric rehabilitation. *J. Neurol. Phys. Ther.* **2009**, *33*, 27–44. [CrossRef]
22. Valentín-Gudiol, M.; Mattern-Baxter, K.; Girabent-Farrés, M.; Bagur-Calafat, C.; Hadders-Algra, M.; Angulo-Barroso, R.M. Treadmill interventions in children under six years of age at risk of neuromotor delay. *Cochrane Database Syst. Rev.* **2017**, *2017*, CD009242. [CrossRef]
23. Paleg, G.; Romness, M.; Livingstone, R. Interventions to improve sensory and motor outcomes for young children with central hypotonia: A systematic review. *J. Pediatr. Rehabil. Med.* **2018**, *11*, 57–70. [CrossRef]
24. Valentin-Gudiol, M.; Bagur-Calafat, C.; Girabent-Farrés, M.; Hadders-Algra, M.; Mattern-Baxter, K.; Angulo-Barroso, R. Treadmill interventions with partial body weight support in children under six years of age at risk of neuromotor delay: A report of a Cochrane systematic review and meta-analysis. *Eur. J. Phys. Rehabil. Med.* **2013**, *49*, 67–91.
25. Pino-Ramos, G.-D.; Romero-Galisteo, R.P.; Pinero-Pinto, E.; Lirio-Romero, C.; Palomo-Carrión, R. Eficacia de la marcha en cinta rodante sobre el desarrollo motor de niños con parálisis cerebral y síndrome de down. *Medicina* **2021**, *81*, 367–374.
26. Andriolo, R.B.; El Dib, R.; Ramos, L.; Atallah, Á.N.; da Silva, E.M. Aerobic exercise training programmes for improving physical and psychosocial health in adults with Down syndrome. *Cochrane Database Syst. Rev.* **2005**.
27. Zwicker, J.G.; Mayson, T.A. Effectiveness of treadmill training in children with motor impairments: An overview of systematic reviews. *Pediatr. Phys. Ther.* **2010**, *22*, 361–377. [CrossRef]
28. Page, M.J.; McKenzie, J.E.; Bossuyt, P.M.; Boutron, I.; Hoffmann, T.C.; Mulrow, C.D.; Shamseer, L.; Tetzlaff, J.M.; Akl, E.A.; Brennan, S.E. The PRISMA 2020 statement: An updated guideline for reporting systematic reviews. *Int. J. Surg.* **2021**, *88*, 105906. [CrossRef]
29. Sterne, J.A.; Savović, J.; Page, M.J.; Elbers, R.G.; Blencowe, N.S.; Boutron, I.; Cates, C.J.; Cheng, H.-Y.; Corbett, M.S.; Eldridge, S.M. RoB 2: A revised tool for assessing risk of bias in randomised trials. *BMJ* **2019**, *366*, l4898. [CrossRef]
30. Higgins, J.P.; Thomas, J.; Chandler, J.; Cumpston, M.; Li, T.; Page, M.J.; Welch, V.A. *Cochrane Handbook for Systematic Reviews of Interventions*; John Wiley & Sons: Hoboken, NJ, USA, 2019.
31. Ulrich, B.D.; Ulrich, D.A.; Collier, D.H. Alternating stepping patterns: Hidden abilities of 11-month-old infants with Down syndrome. *Dev. Med. Child Neurol.* **1992**, *34*, 233–239. [CrossRef]
32. Ulrich, B.D.; Ulrich, D.A.; Collier, D.H.; Cole, E.L. Developmental shifts in the ability of infants with Down syndrome to produce treadmill steps. *Phys. Ther.* **1995**, *75*, 14–23. [CrossRef]
33. Ulrich, D.A.; Ulrich, B.D.; Angulo-Kinzler, R.M.; Yun, J. Treadmill training of infants with Down syndrome: Evidence-based developmental outcomes. *Pediatrics* **2001**, *108*, E84. [CrossRef]
34. Carmeli, E.; Kessel, S.; Coleman, R.; Ayalon, M. Effects of a treadmill walking program on muscle strength and balance in elderly people with Down syndrome. *J. Gerontol. A Biol. Sci. Med. Sci.* **2002**, *57*, M106–M110. [CrossRef]

35. Carmeli, E.; Barchad, S.; Masharawi, Y.; Coleman, R. Impact of a walking program in people with down syndrome. *J. Strength Cond. Res.* **2004**, *18*, 180–184.
36. Wu, J.; Looper, J.; Ulrich, B.D.; Ulrich, D.A.; Angulo-Barroso, R.M. Exploring effects of different treadmill interventions on walking onset and gait patterns in infants with Down syndrome. *Dev. Med. Child Neurol.* **2007**, *49*, 839–845. [CrossRef]
37. Wu, J.; Ulrich, D.A.; Looper, J.; Tiernan, C.W.; Angulo-Barroso, R.M. Strategy adoption and locomotor adjustment in obstacle clearance of newly walking toddlers with Down syndrome after different treadmill interventions. *Exp. Brain Res.* **2008**, *186*, 261–272. [CrossRef]
38. Ulrich, D.A.; Lloyd, M.C.; Tiernan, C.W.; Looper, J.E.; Angulo-Barroso, R.M. Effects of intensity of treadmill training on developmental outcomes and stepping in infants with Down syndrome: A randomized trial. *Phys. Ther.* **2008**, *88*, 114–122. [CrossRef]
39. Angulo-Barroso, R.; Burghardt, A.R.; Lloyd, M.; Ulrich, D.A. Physical activity in infants with Down syndrome receiving a treadmill intervention. *Infant Behav. Dev.* **2008**, *31*, 255–269. [CrossRef]
40. Angulo-Barroso, R.M.; Wu, J.; Ulrich, D.A. Long-term effect of different treadmill interventions on gait development in new walkers with Down syndrome. *Gait Posture* **2008**, *27*, 231–238. [CrossRef]
41. Mendonca, G.V.; Pereira, F.D. Influence of long-term exercise training on submaximal and peak aerobic capacity and locomotor economy in adult males with Down's syndrome. *Med. Sci. Monit.* **2009**, *15*, Cr33–Cr39.
42. Looper, J.; Ulrich, D.A. Effect of treadmill training and supramalleolar orthosis use on motor skill development in infants with Down syndrome: A randomized clinical trial. *Phys. Ther.* **2010**, *90*, 382–390. [CrossRef]
43. Wu, J.; Looper, J.; Ulrich, D.A.; Angulo-Barroso, R.M. Effects of various treadmill interventions on the development of joint kinematics in infants with Down syndrome. *Phys. Ther.* **2010**, *90*, 1265–1276. [CrossRef]
44. Ordonez, F.J.; Rosety, I.; Rosety, M.A.; Camacho-Molina, A.; Fornieles, G.; Rosety, M.; Rosety-Rodriguez, M. Aerobic training at moderate intensity reduced protein oxidation in adolescents with Down syndrome. *Scand. J. Med. Sci. Sports* **2012**, *22*, 91–94. [CrossRef]
45. El-Meniawy, G.H.; Kamal, H.M.; Elshemy, S.A. Role of treadmill training versus suspension therapy on balance in children with Down syndrome. *Egypt. J. Med. Hum. Genet.* **2012**, *13*, 37–43. [CrossRef]
46. Ordoñez, F.J.; Fornieles-Gonzalez, G.; Camacho, A.; Rosety, M.A.; Rosety, I.; Diaz, A.J.; Rosety-Rodriguez, M. Anti-inflammatory effect of exercise, via reduced leptin levels, in obese women with Down syndrome. *Int. J. Sport Nutr. Exerc. Metab.* **2013**, *23*, 239–244. [CrossRef]
47. Lin, H.C.; Wuang, Y.P. Strength and agility training in adolescents with Down syndrome: A randomized controlled trial. *Res. Dev. Disabil.* **2012**, *33*, 2236–2244. [CrossRef]
48. Ordonez, F.J.; Rosety, M.A.; Camacho, A.; Rosety, I.; Diaz, A.J.; Fornieles, G.; Garcia, N.; Rosety-Rodriguez, M. Aerobic training improved low-grade inflammation in obese women with intellectual disability. *J. Intellect. Disabil. Res.* **2014**, *58*, 583–590. [CrossRef]
49. Rodenbusch, T.L.; Ribeiro, T.S.; Simão, C.R.; Britto, H.M.; Tudella, E.; Lindquist, A.R. Effects of treadmill inclination on the gait of children with Down syndrome. *Res. Dev. Disabil.* **2013**, *34*, 2185–2190. [CrossRef]
50. Chen, C.C.; Ringenbach, D.R.; Snow, M. Treadmill walking effects on grip strength in young men with Down syndrome. *Res. Dev. Disabil.* **2014**, *35*, 288–293. [CrossRef]
51. Rosety-Rodriguez, M.; Diaz, A.J.; Rosety, I.; Rosety, M.A.; Camacho, A.; Fornieles, G.; Rosety, M.; Ordonez, F.J. Exercise reduced inflammation: But for how long after training? *J. Intellect. Disabil. Res.* **2014**, *58*, 874–879. [CrossRef]
52. Chen, C.C.; Ringenbach, S.D.; Crews, D.; Kulinna, P.H.; Amazeen, E.L. The association between a single bout of moderate physical activity and executive function in young adults with Down syndrome: A preliminary study. *J. Intellect. Disabil. Res.* **2015**, *59*, 589–598. [CrossRef]
53. Chen, C.C.; Ringenbach, S.D. Dose-response relationship between intensity of exercise and cognitive performance in individuals with Down syndrome: A preliminary study. *J. Intellect. Disabil. Res.* **2016**, *60*, 606–614. [CrossRef]
54. Chen, C.J.J.; Ringenbach, S.D.R. The effect of acute exercise on the performance of verbal fluency in adolescents and young adults with Down syndrome: A pilot study. *J. Intellect. Disabil. Res.* **2019**, *63*, 614–623. [CrossRef]
55. Paul, Y.; Ellapen, T.J.; Barnard, M.; Hammill, H.V.; Swanepoel, M. The health benefits of exercise therapy for patients with Down syndrome: A systematic review. *Afr. J. Disabil.* **2019**, *8*, 576. [CrossRef]
56. Naczk, A.; Gajewska, E.; Naczk, M. Effectiveness of Swimming Program in Adolescents with Down Syndrome. *Int. J. Environ. Res. Public Health* **2021**, *18*, 7441. [CrossRef]
57. Ringenbach, S.D.; Holzapfel, S.D.; Mulvey, G.M.; Jimenez, A.; Benson, A.; Richter, M. The effects of assisted cycling therapy (ACT) and voluntary cycling on reaction time and measures of executive function in adolescents with Down syndrome. *J. Intellect. Disabil. Res.* **2016**, *60*, 1073–1085. [CrossRef]

brain sciences

MDPI

Review

The Impact of Motor-Cognitive Dual-Task Training on Physical and Cognitive Functions in Parkinson's Disease

Yi Xiao, Tianmi Yang and Huifang Shang *

Department of Neurology, Laboratory of Neurodegenerative Disorders, National Clinical Research Center for Geriatric, West China Hospital, Sichuan University, Chengdu 610017, China
* Correspondence: hfshang2002@126.com; Tel.: +86-189-8060-2127

Abstract: Rehabilitation is a high-potential approach to improving physical and cognitive functions in Parkinson's disease (PD). Dual-task training innovatively combines motor and cognitive rehabilitation in a comprehensive module. Patients perform motor and cognitive tasks at the same time in dual-task training. The previous studies of dual-task training in PD had high heterogeneity and achieved controversial results. In the current review, we aim to summarize the current evidence of the effect of dual-task training on motor and cognitive functions in PD patients to support the clinical practice of dual-task training. In addition, we also discuss the current opinions regarding the mechanism underlying the interaction between motor and cognitive training. In conclusion, dual-task training is suitable for PD patients with varied disease duration to improve their motor function. Dual-task training can improve motor symptoms, single-task gait speed, single-task steep length, balance, and objective experience of freezing of gait in PD. The improvement in cognitive function after dual-task training is mild.

Keywords: dual-task training; Parkinson's disease; cognition; motor

check for updates

Citation: Xiao, Y.; Yang, T.; Shang, H. The Impact of Motor-Cognitive Dual-Task Training on Physical and Cognitive Functions in Parkinson's Disease. *Brain Sci.* **2023**, *13*, 437. https://doi.org/10.3390/brainsci13030437

Academic Editors: Daniele Corbo and Andrea Loftus

Received: 29 October 2022
Revised: 15 February 2023
Accepted: 20 February 2023
Published: 3 March 2023

1. Introduction

Parkinson's disease (PD) is a neurodegenerative disease characterized by cardinal motor dysfunction (bradykinesia, resting tremor, rigidity, and postural instability) and non-motor symptoms such as cognitive impairments. Patients with PD have difficulty in daily activities and the difficulty increases with the disease progression. Although anti-Parkinson drugs can relieve motor symptoms, patients with PD suffer from gait and balance dysfunction in their daily lives during the disease's course [1]. No disease-modified treatment is available. In addition to the above-mentioned motor symptoms, there are specific motor dysfunctions that add to the burden of disease on patients. Freezing of gait (FoG), defined as sudden and usually brief episodes of inability to produce effective stepping, is a commonly reported motor dysfunction in PD. FoG is related to a higher risk of falls and significantly decreased patients' quality of life. As FoG shows a poor response to pharmacotherapy, other treatments such as rehabilitation have been proposed [2,3].

The prevalence of cognitive impairment is also high and increases with disease progression in patients with PD. At diagnosis, from 10% to 38% of de novo PD patients have mild cognitive impairment. After 5 years, around 14% to 28% of PD patients reached the point of dementia. Unfortunately, only symptomatic treatment is currently available for cognitive impairment and no treatment can modify the progression of cognitive decline. Preventing cognitive decline before the incidence of dementia has received increased attention in recent years [4].

A motor-cognitive task involves performing two independent tasks at the same time, for example, answering arithmetic questions while walking. The conduction of motor-cognitive tasks requires the involvement of both the motor and the cognitive systems. In patients with PD, the performance of the motor-cognitive task is decreased compared to age-, sex-, and education-matched healthy controls [5]. The dual-task cost is the deficit in task performance in the dual-task condition compared to that in the single-task condition. The dual-task cost is higher in patients with FoG compared to that in patients without FoG [6].

Rehabilitation is a potential way to improve physical and cognitive function in patients with PD [7]. Motor-cognitive dual-task training is a comprehensive rehabilitation method combining motor training and cognitive training [8]. The interaction between motor and cognitive tasks may have an extra benefit compared to single-task training [9]. A large number of studies researched the effect of dual-task training on improving gait, balance, motor symptoms, and cognition in patients with PD [10–12]. Several meta-analyses were conducted in the past few years but achieved controversial results [13–15]. In the current review, we summarize the current evidence of the physical and cognitive effect of dual-task training on patients with PD.

We searched the following online databases: the Cochrane Database of Systematic Reviews, MEDLINE, Scopus, EMBASE, and PROSPERO. The research was performed in September 2022. We used the following keywords and searched for them in the title and abstract: (dual task or dual-task or motor cognitive or motor-cognitive or cognitive-motor), and (Parkinson or Parkinson's disease). We also searched through the references of related articles to capture other studies that met the eligibility criteria. The PICO strategy was used to conduct the study selection. The eligibility criteria were: (1) included subjects were PD patients; (2) the intervention was dual-task training, and the control was other training or single-task training or no intervention; (3) outcomes were cognitive or physical functions and performances. The exclusion criteria were: (1) articles not written in English; (2) the full article was not available.

2. Construction of Motor-Cognitive Dual-Task Training

The basic method to construct motor-cognitive dual-task training is to add a cognitive task to a motor task [8]. The cognitive task can be simple (arithmetic problems) or complex (combining verbal fluency tasks and reciting switching and working memory tasks). The level of difficulty of motor tasks also varied from gait training to a highly intensive HiBalance program [16]. The difficulty of the training can be unstable and increase with time as the participants adapt to a lower level of difficulty and reach a previously settled aim. In addition, the use of electronic devices adds numerous advantages to training. The motor task and cognitive task can be combined through video-based or virtual-reality-based exergames, adding fun and convenience to the training [17]. Portable exercise instruments, such as stationary bikes and sensors, move the scenario of rehabilitation from the hospital or rehabilitation center to home.

3. Effects of Dual-Task Training on Cognitive and Physical Function

3.1. Motor Symptom

Previous studies found that motor symptoms were improved after dual-task training [18–20]. Twenty patients were randomized into a 24-session single-modal training group (performing gait and cognitive training sequentially) or a dual-task training group (performing gait and cognitive training simultaneously). Both groups had a significantly decreased Movement Disorder Society-Unified Parkinson's disease Rating Scale motor subscale score (a higher score indicating a more severe symptom) after 8 weeks of training. The dual-task training group had a larger decrease in scale score compared to the single-modal training group (15% vs. 8%). The improvement in motor symptoms could be preserved for at least 4 weeks [18]. Another study found that the shorter session of cognitive-cycling training (16 sessions) also showed benefits [20]. The Hoehn–Yahr stage of the included patients ranged from two to three, indicating that the dual-task training was suitable for mild to moderate PD [18–20]. Some studies assessed motor function outcomes using other scales. Previous studies found that performances of the Timed Up-and-Go Test and the 6-Minute Walk Test were improved after dual-task training [21].

3.2. Single Task Gait

Studies showed that some gait parameters in the single-task condition improved after dual-task training [16,22,23]. After 30 sessions of the HiBalance program, the training group had a significant improvement in speed and step length compared to the no-intervention group [16]. One study included seventeen patients and found that achieved walking distance increased after four weeks of training. The gait variability in the step length in the single-task condition decreased but did not reach statistical significance [23]. The cadence of the dual-task training group was not improved compared to that of the no-intervention group [16,24].

3.3. Dual-Task Gait

In daily life situations, the motor task is always accompanied by cognitive tasks, for instance, walking while reading signs. Dual-task walking can simulate the complex walking scenario in reality. Previous findings of the dual-task performance after dual-task training were controversial [16,18]. A RCT with one hundred participants found that the gait speed and step length in the dual-task condition was not improved [16]. However, another study included twenty patients and found increasing velocity and step length in dual-task conditions after training. The improvement in the step width and arm swing varied in different cognitive tasks [18]. The different cognitive tasks may contribute to inconsistent results.

3.4. Balance and Fall

Dual-task training can lead to an improvement in balance [10,11,16]. Large-sample-size RCTs found an improvement in balance assessed by the Mini-BESTest or Berg Balance Scale in the dual-task training group compared to the no-intervention group [10,16]. The results of previous studies regarding decreasing fall risk were controversial. An RCT with one hundred and twenty-one participants found that the 24-week frequency of falls did not decrease after dual-task training [25]. However, another RCT with a total of twenty-one patients found that 8 weeks of dual-task training decreased the 30-day frequency of the falls by 60% [19]. The dual-task training may decrease the risk of all groups for a short duration after training but the benefit is not persistent.

3.5. FoG

Several primary studies indicated a potential improvement in FoG after dual-task training [23,26,27]. A randomized, single-blinded, cross-over design trial included forty-six patients with PD and FoG. Patients received the cognitively challenging Agility Boot Camp Program (dual-task training) and education (control), sequentially. Compared to education, dual-task training showed a significant improvement in the subjective experience of FoG assessed by the Freezing of Gait Questionnaire. But there was no significant improvement in the objective measurement of FoG [27]. An improvement in the Freezing of Gait Questionnaire score was also found after dual-task training in an RCT that included sixty participants [21]. Another study included seventeen patients with PD and also found a decrease in scores on the Freezing of Gait Questionnaire after dual-task training, but this decrease did not reach statistical significance [23]. Killane et al. included thirteen patients with PD and FoG and found the mean number of FOG episodes per trial for the dual tasks decreased from three to one after dual-task training [26].

3.6. Cognitive Function

The effects of dual-task training on cognition in PD have been rarely researched [28,29]. An RCT found that the dual-task training group had an increased performance of the cognitive task in dual-task walking compared to the no-intervention group [16]. A single-blind RCT included forty patients with PD and found that the dual-task group showed a trend but no significant improvement in the executive function test (Frontal Assessment Battery and Trail Making Test) after training compared to the single-task training group (gait training) [28]. The trend of improvement in the cognitive test (Stroop) after dual-task training was also found in another study, but this improvement failed to reach statistical significance [29].

4. Advantages and Enhancement of Dual-Task Training

4.1. Dual-Task vs. Single-Task Training

Dual-task training may have some advantages over single-motor training but further studies are needed because the current evidence is not adequate. An RCT found that the dual-task training group had better gait velocity and stride length compared to the single-motor training group [28]. However, another small-sample-size RCT found that there were no differences in gait and motor symptoms between the dual-task training group and the single-task training group [18]. The improvement in balance was also comparable between the two training groups [10]. One primary study indicated that the dual-task group performed better in mediolateral balance with an eye closed but performed worse in anteroposterior balance compared to the single-motor-training group [11]. However, a previous study found that dual-task training was more effective in the improvement of motor symptoms and decreased the frequency of falls in a 30-day period compared to the separate motor and cognitive training [19].

4.2. Enhancement of Dual-Task Training

Several studies researched combining dual-task training and other rehabilitation to boost their effectiveness. Dual-task training plus action observation and motor imagery had a better effect on motor and cognitive function [30]. The intervention group was asked to observe a task before conducting the task while the control group only conducted the task. The intervention group had greater improvements in gait, balance, and executive function compared to the control group [30]. An RCT found that transcranial direct current stimulation enhanced the effect of dual-task training. The transcranial direct current stimulation device was placed in a small bag positioned around the participant's hips, and was active (intervention group) or sham (control group) during the gait training. The cognitive performance during the dual-task and Timed Up-and-Go Test was improved. However, the gait and bradykinesia were not enhanced by the transcranial direct current stimulation [31]. The enhancement depends on the specific type of supplementary training. A previous study that included sixty participants showed that the multi-intervention group (combining dual-task training and aquatic therapy) did not improve motor function (FoG, balance, gait, and motor symptoms) compared to the single-intervention group (dual-task training) in PD [21].

5. Mechanism of Dual-Task Training

5.1. Motor Automaticity

One of the potential mechanisms of the physical benefit of dual-task training is the improvement in motor automaticity after training. Motor automaticity is the ability to conduct a skilled motor task without conscious attention or executive control, and plays an important role in performing dual tasks [32,33]. Functional magnetic resonance imaging has been used to assess the brain activity associated with motor automaticity in patients with PD. Participants practiced dual tasks until they could conduct the tasks automatically and then a visual letter-counting task was added to the dual tasks. Compared to healthy controls, fewer patients with PD automatically performed the combined tasks. The neuroimaging showed that only the bilateral superior parietal lobes and left insular cortex were less activated in PD when the proficiency of the task reached the automatic stage, while activities in more areas decreased in healthy controls [34]. In the automatic stage, the parietal cortex, premotor area, cerebellum, precuneus, and prefrontal cortex were more active in patients with PD compared to healthy controls [34]. The impaired motor automaticity in PD contributes to automaticity-associated motor deficits, such as akinesia, slowness of simple repetitive movements, shorter stride length, and FoG [35].

The striatum is the main pathological region in PD and is also related to the maintenance of the motor automaticity in PD [35,36]. Increased activity of the anterior putamen was found in patients with PD in the automatic stage compared to the controls [35]. Participants were asked to direct their attention back to the previously practiced automated tasks. Patients with PD had a decreased connectivity from the putamen to the motor cortex in the controlling attention stage compared to the automatic stage. However, in healthy controls, the activity and connectivity of the striatum at the controlling attention stage were comparable to those at the automatic stage [36]. Impaired sensorimotor striatum in PD results in the loss of previously learned automatic skills and the deficit in learning new skills [35].

Previous studies indicated that exercise can improve the function of the striatum. Positron emission tomography showed that expression of the striatal dopamine D2 receptor increased after high-intensity treadmill exercise in MPTP mouse models of PD compared to that in controls [37]. The improvements in striatum function were replicated in vivo in patients with PD. Four patients with PD were randomized into 24 sessions of an intensive treadmill exercise group or a control group. Increased dopamine D2 receptor binding potential and improved postural control were found in the exercise group after training, while there were no differences found in the control group [38]. Fifty-six patients with PD were randomized into an aerobic exercise group or stretching group and received six months of intervention. The performance of the oculomotor cognitive control task was improved in the exercise group compared to that of the stretching group. Functional magnetic resonance imaging found that the anterior putamen had increased functional connectivity with the sensorimotor cortex compared to the posterior putamen in the exercise group, while this phenomenon was absent in the stretching group [39]. Another study included ninety-one patients with PD and randomized them into the HiBalance group or the active control group (speech training). After 20 sessions of training, the HiBalance group had increased left putamen volumes and stronger thalamic–cerebellar connectivity in structural covariance networks, while the active control group showed no changes post-training [40].

5.2. Dual-Task Practice Advantage

In addition to the improvements achieved by motor training, there are additional benefits from combining motor and cognitive training, named the dual-task practice advantage. The dual-task practice advantage is described as an advantage in dual-task performance after dual-task practice compared to single-task practice (where tasks are completed sequentially) [41]. The dual-task practice advantage was found in a previous RCT of older adults. Older adults with balance problems were assigned to a single-task training group or dual-task training group (fixed-priority or variable-priority instructions). After four weeks of training, all the groups showed improvements in balance and single-task walking speed. However, only the dual-task training group had improved dual-task walking speed. In addition, the dual-task training group with variable-priority instructions was the only group that quickly showed improvement in the second week of training and maintained the training effect in the twelfth week after training [42].

The classical theoretical mechanism of the dual-task practice advantage is the allocation and scheduling hypothesis [41]. This hypothesis assumes that dual-task training improves the cognitive resource for allocation and scheduling tasks during comprehensive tasks. In a previous study, participants were arranged to conduct dual tasks, such as copying different sentences while reading stories, and then were tested for the meaning of the sentences. As they understood the meaning of sentences, the research considered that the completion of the two tasks was not automatically performed as one task but by switching between different tasks. In addition, they found that the skill to switch attention during dual tasks could be improved with practice [43]. Through dual-task training, patients can improve their ability to manage multiple tasks and the achievements can be transferred to other untrained dual tasks shared with the same neural circuit. A previous study divided dual-task performance-matched participants into hybrid training (combine single-task and dual-task training) or single-task training groups. A visual-manual task and an auditory-vocal task were trained. After training, the hybrid training group was found to have better performance in practiced dual tasks and new dual tasks compared to the single-task training group [44]. Another study explored the synergistic effect of cognitive training and motor training. A total of one hundred and thirty-six older adults were randomized into four intervention groups: cognitive training plus motor training, cognitive control plus motor training, cognitive training plus motor control, and cognitive control plus motor control. The task-set cost is the ratio between the trial mixed with two different single tasks and the trial composed of pure single tasks, and reflects the ability to memorize specific requirements of different single tasks and respond to distinct single tasks. The cognitive training was associated with a reduction in the dual-task cost. The dual-task training group predicted a decreased task-set cost, which indicated that dual-task training improved the ability to manage and switch between different tasks [45].

The neuroimaging studies found increased connectivity in older adults after dual-task training, which supported the theory. The right cerebellar vermis, left lobule V of the cerebellar anterior lobe, and precuneus were activated in the dual-task processing. The dual-task training increased the functional connectivity between these cerebellar regions and cerebellar areas related to motor and cognitive control during dual tasks [46]. However, the existence of dual-task practice advantages in patients with PD remains unclear. One study found that compared to the sequential gait and cognitive training group, the simultaneous training group had better upper extremity performance [18]. However, another study evaluated the dual-task practice advantages in PD and found that the improvements in gait parameters were comparable between simultaneous motor-cognitive training and sequential motor-cognitive training [47]. Different domains of cognitive training may contribute to the reverse results of the previous studies, as the former study trained verbal fluency, working memory, discrimination and decision making, mental tracking, and reaction time, while the latter study trained language, memory, executive function, and attention. More studies are needed in the future to establish the theory of additional improvements resulting from dual-task training in PD.

5.3. Guided Plasticity Facilitation Hypothesis

The guided plasticity facilitation hypothesis may contribute to the improvement in cognition by motor-cognitive dual-task training. In this hypothesis, motor training facilitates the neuroplasticity and cognitive training guides the neuroplasticity. An experiment in mice found that exercise can stimulate hippocampal neurogenesis. Mouse undergoing exercise and cognitive stimuli (environmental enrichment) recorded a greater increase in new neurons in the dentate gyrus compared to that in mice with single exercise or cognitive stimuli [48].

Brain-derived neurotrophic factor (BDNF) is a potential factor that mediates the neurogenesis induced by exercise. BDNF is known as a synapse regulator and plays an important role in the underlying mechanism of learning and memory, and hippocampal long-term potentiation (synaptic efficacy enhancement) [49]. In patients with PD, the level of serum BDNF was found to be decreased compared to that of the healthy controls, and was related to the binding ratio of the presynaptic dopamine transporter in caudate and putamen [50]. Previous animal experiments found that the level of BDNF in the circulation increases after intensive exercise [51]. A meta-analysis confirmed that the level of serum BDNF significantly increased after exercise in patients with PD [52]. In addition, the volume of the left hippocampal subfields CA1, CA4/dentate gyrus, and subiculum showed a group-dependent increase in PD [53]. When BDNF function was inhibited using a specific immune-adhesin chimera that selectively binds BDNF, the benefit of exercise on cognition vanished. The learning and recall abilities of animals that received exercise and the BDNF blocker were reduced to those of the control level [54]. However, the increase in BDNF induced by exercise is temporary, which indicates that the interaction of the motor and cognitive training is influenced by the execution arrangement and interval of the two tasks. The level of serum BDNF was elevated at day seven of the motor training program compared to baseline in patients with PD, but showed no changes compared to baseline in the following days of training and at the sixty-day follow-up after the end of the training [55].

The two tasks induce neurogenesis through distinct but complementary mechanisms as exercise stimulates the proliferation of precursor cells, whereas cognitive stimuli promote the survival of the newborn cells [56]. The synergic effect of motor-cognitive tasks may be based on neurotrophic factors. Ninety-five healthy young participants were divided into exercise, dual-task training, and control groups. Participants who had a greater cognitive improvement after training had a greater increase in BDNF and insulin-like growth factor-1. These participants who had greater increases in neurotrophic factors after exercise also showed a greater improvement after dual-task training compared to single-exercise training [57]. In patients with PD with mild cognitive impairment, the level of serum BDNF increased after the training of executive cognition compared to the control group [58]. Further research is needed to establish the role of cognitive training in dual-task training and the neural mechanism behind the synergic effect of motor-cognitive tasks.

6. Future Studies

There were some limitations of the previous studies. First, the selections of motor and cognitive tasks to build dual-task training were different between studies and this may influence the results [15]. In addition, the settings of the control group may be another source of heterogeneity. There were three main types of control group. The first was no control group. Researchers evaluated the effect of dual-task training by comparing the performance of patients before and after training [59]. The second was that the control group received no intervention or maintained daily activity [60,61]. The third was that the control group had a different activity, such as a single motor or cognitive task [62,63], or performed motor and cognitive tasks separately [47]. The third kind of setting was more likely to achieve a negative result compared to the others because the positive effect of exercise on physical function has been proved by previous studies [14]. However, the results of the three kinds of studies were combined in the previous meta-analysis. This partly explained the controversial results of the previous meta-analysis [13–15]. The settings of control groups in future studies should be chosen carefully and more attention should be paid to the separate training of motor and cognitive tasks.

There were some gaps in the specific outcomes of dual-task training. First, the effect on cognition in PD deserves more attention. The positive effects of dual-task training on cognition were found in older adults and patients with Alzheimer's disease [64]. To date, few studies have researched the cognitive benefit of dual-task training in PD, and further studies are needed to determine whether dual-task training can kill two birds with one stone. [12,28,29] Second, because of the prosperity and diversity of rehabilitation methods in neurodegenerative disease, there is a trend to build comprehensive rehabilitation modules by combining different interventions to achieve better effectiveness [65]. Although the standardized mean difference of the different modules can be calculated and compared in a meta-analysis [14], the best way to establish evidence for a comparison between them is to use a randomized controlled trial. In addition, a training program with high-intensity and high-diversity tasks seems to result in a greater improvement in motor and cognitive function, which has guided the direction of the design of the training program [66]. However, the safety of the comprehensive module with challenging tasks should be tested because heavy training may be beyond the physical and cognitive endurance range of patients, resulting in negative results.

Individualized rehabilitation is also important, as a previous study found that the effects of dual-task training were different among patients with different characteristics. A worse cognitive status was associated with higher dual-task costs in cognition and gait in PD. The existence of other PD non-motor symptoms was related to a higher dual-task cost in cognition [67]. A large-sample-size RCT found that a lower dual-task gait velocity and a higher cognitive performance at baseline were related to a larger improvement in the dual-task velocity after training [68]. Different strategies used in dual tasks and different learning abilities may contribute to this phenomenon. Patients with PD with mild cognitive impairment used a posture-first strategy while patients with PD with normal cognition used a posture-second strategy [67]. Evidence showed that high physical activity was related to a slower apolipoprotein ε4-related cognitive decline in PD [69]. To date, the classification of PD patients in rehabilitation clinical trials has been based on motor symptoms. Future studies should consider more clinical and genetic characteristics as stratification factors and design individualized training programs for PD patients.

There are also some limitations of the current review. First, we did not pool the results of studies because the control intervention and the construction of dual-task training varied between studies. Second, we did not include studies not written in English. Third, we included RCT studies and non-RCT studies, so the quality of studies varied.

7. Conclusions

In the current review, we summarize the present evidence of the impact of dual-task training on motor and cognitive function in patients with PD. The summary of the current review please check Figure 1. Significant improvement has been observed in motor symptoms, single-task gait speed, single-task step length, and balance, but the effects of dual-task training on FoG, dual-task gait, and cognition are still being researched. Trends of improvement in these fields were observed in previous studies. Large-sample-size randomized control tails of dual-task training are needed to establish the effectiveness of this training. In addition, the settings for intervention and control groups should be more specific, and more attention should be paid to the additional benefit resulting from the interaction between motor and cognitive training and the mechanism of the synergic effect.

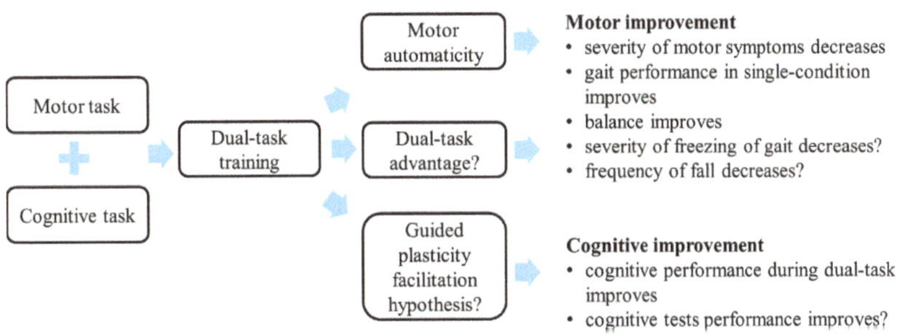

Figure 1. Summary of the motor and cognitive effect of the dual-task training.

Author Contributions: Drafting of the manuscript: Y.X. Review and modify the manuscript: Y.X., T.Y., and H.S. All authors have read and agreed to the published version of the manuscript.

Funding: This work was funded by the National Key Research and Development Program of China (Grant No. 2021YFC2501200) and the Sichuan Science and Technology Program (Grant No. 2022ZDZX0023).

Institutional Review Board Statement: Not applicable.

Informed Consent Statement: Not applicable.

Data Availability Statement: Not applicable.

Conflicts of Interest: The authors declare that they have no competing interests.

Abbreviations

PD: Parkinson's disease; FoG: freezing of gait.

References

1. Samii, A.; Nutt, J.G.; Ransom, B.R. Parkinson's disease. *Lancet* **2004**, *363*, 1783–1793. [CrossRef]
2. Zhang, W.S.; Gao, C.; Tan, Y.Y.; Chen, S.D. Prevalence of freezing of gait in Parkinson's disease: A systematic review and meta-analysis. *J. Neurol.* **2021**, *268*, 4138–4150. [CrossRef] [PubMed]
3. Okuma, Y. Freezing of gait and falls in Parkinson's disease. *J. Park. Dis.* **2014**, *4*, 255–260. [CrossRef]
4. Aarsland, D.; Batzu, L.; Halliday, G.M.; Geurtsen, G.J.; Ballard, C.; Ray Chaudhuri, K.; Weintraub, D. Parkinson disease-associated cognitive impairment. *Nat. Rev. Dis. Prim.* **2021**, *7*, 47. [CrossRef] [PubMed]
5. Rochester, L.; Hetherington, V.; Jones, D.; Nieuwboer, A.; Willems, A.M.; Kwakkel, G.; Van Wegen, E. Attending to the task: Interference effects of functional tasks on walking in Parkinson's disease and the roles of cognition, depression, fatigue, and balance. *Arch. Phys. Med. Rehabil.* **2004**, *85*, 1578–1585. [CrossRef]
6. de Souza Fortaleza, A.C.; Mancini, M.; Carlson-Kuhta, P.; King, L.A.; Nutt, J.G.; Chagas, E.F.; Freitas, I.F.J.; Horak, F.B. Dual task interference on postural sway, postural transitions and gait in people with Parkinson's disease and freezing of gait. *Gait Posture* **2017**, *56*, 76–81. [CrossRef] [PubMed]
7. Abbruzzese, G.; Marchese, R.; Avanzino, L.; Pelosin, E. Rehabilitation for Parkinson's disease: Current outlook and future challenges. *Park. Relat. Disord.* **2016**, *22* (Suppl. 1), S60–S64. [CrossRef]
8. Herold, F.; Hamacher, D.; Schega, L.; Müller, N.G. Thinking While Moving or Moving While Thinking—Concepts of Motor-Cognitive Training for Cognitive Performance Enhancement. *Front. Aging Neurosci.* **2018**, *10*, 228. [CrossRef]
9. Zhu, X.; Yin, S.; Lang, M.; He, R.; Li, J. The more the better? A meta-analysis on effects of combined cognitive and physical intervention on cognition in healthy older adults. *Ageing Res. Rev.* **2016**, *31*, 67–79. [CrossRef]
10. Perumal, V.; Melam, G.R.; Alhusaini, A.A.; Buragadda, S.; Sharma, N. Instruction prioritization in task-based balance training for individuals with idiopathic Parkinson's disease. *Somatosens. Mot. Res.* **2017**, *34*, 27–33. [CrossRef]
11. Fernandes, Â.; Rocha, N.; Santos, R.; Tavares, J.M. Effects of dual-task training on balance and executive functions in Parkinson's disease: A pilot study. *Somatosens. Mot. Res.* **2015**, *32*, 122–127. [CrossRef]
12. Salazar, R.D.; Ren, X.; Ellis, T.D.; Toraif, N.; Barthelemy, O.J.; Neargarder, S.; Cronin-Golomb, A. Dual tasking in Parkinson's disease: Cognitive consequences while walking. *Neuropsychology* **2017**, *31*, 613–623. [CrossRef] [PubMed]
13. Radder, D.L.M.; Lígia Silva de Lima, A.; Domingos, J.; Keus, S.H.J.; van Nimwegen, M.; Bloem, B.R.; de Vries, N.M. Physiotherapy in Parkinson's Disease: A Meta-Analysis of Present Treatment Modalities. *Neurorehabilit. Neural Repair* **2020**, *34*, 871–880. [CrossRef] [PubMed]
14. Hvingelby, V.S.; Glud, A.N.; Sørensen, J.C.H.; Tai, Y.; Andersen, A.S.M.; Johnsen, E.; Moro, E.; Pavese, N. Interventions to improve gait in Parkinson's disease: A systematic review of randomized controlled trials and network meta-analysis. *J. Neurol.* **2022**, *269*, 4068–4079. [CrossRef] [PubMed]
15. Li, Z.; Wang, T.; Liu, H.; Jiang, Y.; Wang, Z.; Zhuang, J. Dual-task training on gait, motor symptoms, and balance in patients with Parkinson's disease: A systematic review and meta-analysis. *Clin. Rehabil.* **2020**, *34*, 1355–1367. [CrossRef]
16. Conradsson, D.; Löfgren, N.; Nero, H.; Hagströmer, M.; Ståhle, A.; Lökk, J.; Franzén, E. The Effects of Highly Challenging Balance Training in Elderly With Parkinson's Disease: A Randomized Controlled Trial. *Neurorehabilit. Neural Repair* **2015**, *29*, 827–836. [CrossRef]
17. Chua, L.K.; Chung, Y.C.; Bellard, D.; Swan, L.; Gobreial, N.; Romano, A.; Glatt, R.; Bonaguidi, M.A.; Lee, D.J.; Jin, Y.; et al. Gamified Dual-Task Training for Individuals with Parkinson Disease: An Exploratory Study on Feasibility, Safety, and Efficacy. *Int. J. Environ. Res. Public Health* **2021**, *18*, 12384. [CrossRef]
18. Rosenfeldt, A.B.; Penko, A.L.; Streicher, M.C.; Zimmerman, N.M.; Koop, M.M.; Alberts, J.L. Improvements in temporal and postural aspects of gait vary following single- and multi-modal training in individuals with Parkinson's disease. *Park. Relat. Disord.* **2019**, *64*, 280–285. [CrossRef]
19. Penko, A.L.; Barkley, J.E.; Rosenfeldt, A.B.; Alberts, J.L. Multimodal Training Reduces Fall Frequency as Physical Activity Increases in Individuals With Parkinson's Disease. *J. Phys. Act. Health* **2019**, *16*, 1085–1091. [CrossRef]
20. Chang, H.C.; Chen, C.C.; Weng, Y.H.; Chiou, W.D.; Chang, Y.J.; Lu, C.S. The efficacy of cognitive-cycling dual-task training in patients with early-stage Parkinson's disease: A pilot study. *NeuroRehabilitation* **2020**, *47*, 415–426. [CrossRef]
21. Clerici, I.; Maestri, R.; Bonetti, F.; Ortelli, P.; Volpe, D.; Ferrazzoli, D.; Frazzitta, G. Land Plus Aquatic Therapy Versus Land-Based Rehabilitation Alone for the Treatment of Freezing of Gait in Parkinson Disease: A Randomized Controlled Trial. *Phys. Ther.* **2019**, *99*, 591–600. [CrossRef] [PubMed]
22. Fok, P.; Farrell, M.; McMeeken, J. The effect of dividing attention between walking and auxiliary tasks in people with Parkinson's disease. *Hum. Mov. Sci.* **2012**, *31*, 236–246. [CrossRef]
23. Wollesen, B.; Rudnik, S.; Gulberti, A.; Cordes, T.; Gerloff, C.; Poetter-Nerger, M. A feasibility study of dual-task strategy training to improve gait performance in patients with Parkinson's disease. *Sci. Rep.* **2021**, *11*, 12416. [CrossRef] [PubMed]
24. Canning, C.G.; Ada, L.; Woodhouse, E. Multiple-task walking training in people with mild to moderate Parkinson's disease: A pilot study. *Clin. Rehabil.* **2008**, *22*, 226–233. [CrossRef] [PubMed]

25. Strouwen, C.; Molenaar, E.; Münks, L.; Keus, S.H.J.; Zijlmans, J.C.M.; Vandenberghe, W.; Bloem, B.R.; Nieuwboer, A. Training dual tasks together or apart in Parkinson's disease: Results from the DUALITY trial. *Mov. Disord. Off. J. Mov. Disord. Soc.* **2017**, *32*, 1201–1210. [CrossRef]
26. Killane, I.; Fearon, C.; Newman, L.; McDonnell, C.; Waechter, S.M.; Sons, K.; Lynch, T.; Reilly, R.B. Dual Motor-Cognitive Virtual Reality Training Impacts Dual-Task Performance in Freezing of Gait. *IEEE J. Biomed. Health Inform.* **2015**, *19*, 1855–1861. [CrossRef] [PubMed]
27. King, L.A.; Mancini, M.; Smulders, K.; Harker, G.; Lapidus, J.A.; Ramsey, K.; Carlson-Kuhta, P.; Fling, B.W.; Nutt, J.G.; Peterson, D.S.; et al. Cognitively Challenging Agility Boot Camp Program for Freezing of Gait in Parkinson Disease. *Neurorehabilit. Neural Repair* **2020**, *34*, 417–427. [CrossRef]
28. San Martín Valenzuela, C.; Moscardó, L.D.; López-Pascual, J.; Serra-Añó, P.; Tomás, J.M. Effects of Dual-Task Group Training on Gait, Cognitive Executive Function, and Quality of Life in People With Parkinson Disease: Results of Randomized Controlled DUALGAIT Trial. *Arch. Phys. Med. Rehabil.* **2020**, *101*, 1849–1856.e1841. [CrossRef]
29. Pereira-Pedro, K.P.; de Oliveira, I.M.; Mollinedo-Cardalda, I.; Cancela-Carral, J.M. Effects of Cycling Dual-Task on Cognitive and Physical Function in Parkinson's Disease: A Randomized Double-Blind Pilot Study. *Int. J. Environ. Res. Public Health* **2022**, *19*, 7847. [CrossRef]
30. Sarasso, E.; Agosta, F.; Piramide, N.; Gardoni, A.; Canu, E.; Leocadi, M.; Castelnovo, V.; Basaia, S.; Tettamanti, A.; Volontè, M.A.; et al. Action Observation and Motor Imagery Improve Dual Task in Parkinson's Disease: A Clinical/fMRI Study. *Mov. Disord. Off. J. Mov. Disord. Soc.* **2021**, *36*, 2569–2582. [CrossRef]
31. Schabrun, S.M.; Lamont, R.M.; Brauer, S.G. Transcranial Direct Current Stimulation to Enhance Dual-Task Gait Training in Parkinson's Disease: A Pilot RCT. *PLoS ONE* **2016**, *11*, e0158497. [CrossRef] [PubMed]
32. Strouwen, C.; Molenaar, E.A.; Münks, L.; Keus, S.H.; Bloem, B.R.; Rochester, L.; Nieuwboer, A. Dual tasking in Parkinson's disease: Should we train hazardous behavior? *Expert Rev. Neurother.* **2015**, *15*, 1031–1039. [CrossRef] [PubMed]
33. Cameron, I.G.; Watanabe, M.; Pari, G.; Munoz, D.P. Executive impairment in Parkinson's disease: Response automaticity and task switching. *Neuropsychologia* **2010**, *48*, 1948–1957. [CrossRef]
34. Wu, T.; Hallett, M. A functional MRI study of automatic movements in patients with Parkinson's disease. *Brain A J. Neurol.* **2005**, *128*, 2250–2259. [CrossRef]
35. Wu, T.; Hallett, M.; Chan, P. Motor automaticity in Parkinson's disease. *Neurobiol. Dis.* **2015**, *82*, 226–234. [CrossRef] [PubMed]
36. Wu, T.; Liu, J.; Zhang, H.; Hallett, M.; Zheng, Z.; Chan, P. Attention to Automatic Movements in Parkinson's Disease: Modified Automatic Mode in the Striatum. *Cereb. Cortex* **2015**, *25*, 3330–3342. [CrossRef] [PubMed]
37. Vuckovic, M.G.; Li, Q.; Fisher, B.; Nacca, A.; Leahy, R.M.; Walsh, J.P.; Mukherjee, J.; Williams, C.; Jakowec, M.W.; Petzinger, G.M. Exercise elevates dopamine D2 receptor in a mouse model of Parkinson's disease: In vivo imaging with [(1)(8)F] fallypride. *Mov. Disord. Off. J. Mov. Disord. Soc.* **2010**, *25*, 2777–2784. [CrossRef]
38. Fisher, B.E.; Li, Q.; Nacca, A.; Salem, G.J.; Song, J.; Yip, J.; Hui, J.S.; Jakowec, M.W.; Petzinger, G.M. Treadmill exercise elevates striatal dopamine D2 receptor binding potential in patients with early Parkinson's disease. *Neuroreport* **2013**, *24*, 509–514. [CrossRef]
39. Johansson, M.E.; Cameron, I.G.M.; Van der Kolk, N.M.; de Vries, N.M.; Klimars, E.; Toni, I.; Bloem, B.R.; Helmich, R.C. Aerobic Exercise Alters Brain Function and Structure in Parkinson's Disease: A Randomized Controlled Trial. *Ann. Neurol.* **2022**, *91*, 203–216. [CrossRef]
40. Albrecht, F.; Pereira, J.B.; Mijalkov, M.; Freidle, M.; Johansson, H.; Ekman, U.; Westman, E.; Franzen, E. Effects of a Highly Challenging Balance Training Program on Motor Function and Brain Structure in Parkinson's Disease. *J. Park. Dis.* **2021**, *11*, 2057–2071. [CrossRef]
41. Strobach, T. The dual-task practice advantage: Empirical evidence and cognitive mechanisms. *Psychon. Bull. Rev.* **2020**, *27*, 3–14. [CrossRef] [PubMed]
42. Silsupadol, P.; Shumway-Cook, A.; Lugade, V.; van Donkelaar, P.; Chou, L.S.; Mayr, U.; Woollacott, M.H. Effects of single-task versus dual-task training on balance performance in older adults: A double-blind, randomized controlled trial. *Arch. Phys. Med. Rehabil.* **2009**, *90*, 381–387. [CrossRef]
43. Hirst, W.; Spelke, E.S.; Reaves, C.C.; Caharack, G.; Neisser, U. Dividing attention without alternation or automaticity. *J. Exp. Psychol. Gen.* **1980**, *109*, 98–117. [CrossRef]
44. Liepelt, R.; Strobach, T.; Frensch, P.; Schubert, T. Improved intertask coordination after extensive dual-task practice. *Q. J. Exp. Psychol.* **2011**, *64*, 1251–1272. [CrossRef] [PubMed]
45. Bherer, L.; Gagnon, C.; Langeard, A.; Lussier, M.; Desjardins-Crepeau, L.; Berryman, N.; Bosquet, L.; Vu, T.T.M.; Fraser, S.; Li, K.Z.H.; et al. Synergistic Effects of Cognitive Training and Physical Exercise on Dual-Task Performance in Older Adults. *J Gerontol. B Psychol. Sci. Soc. Sci.* **2021**, *76*, 1533–1541. [CrossRef] [PubMed]
46. Wu, T.; Liu, J.; Hallett, M.; Zheng, Z.; Chan, P. Cerebellum and integration of neural networks in dual-task processing. *Neuroimage* **2013**, *65*, 466–475. [CrossRef]
47. Geroin, C.; Nonnekes, J.; de Vries, N.M.; Strouwen, C.; Smania, N.; Tinazzi, M.; Nieuwboer, A.; Bloem, B.R. Does dual-task training improve spatiotemporal gait parameters in Parkinson's disease? *Park. Relat. Disord.* **2018**, *55*, 86–91. [CrossRef]
48. Fabel, K.; Wolf, S.A.; Ehninger, D.; Babu, H.; Leal-Galicia, P.; Kempermann, G. Additive effects of physical exercise and environmental enrichment on adult hippocampal neurogenesis in mice. *Front. Neurosci.* **2009**, *3*, 50. [CrossRef]

49. Lu, B.; Nagappan, G.; Lu, Y. BDNF and synaptic plasticity, cognitive function, and dysfunction. *Handb. Exp. Pharmacol.* **2014**, *220*, 223–250. [CrossRef]
50. Hernandez-Vara, J.; Saez-Francas, N.; Lorenzo-Bosquet, C.; Corominas-Roso, M.; Cuberas-Borros, G.; Lucas-Del Pozo, S.; Carter, S.; Armengol-Bellapart, M.; Castell-Conesa, J. BDNF levels and nigrostriatal degeneration in "drug naive" Parkinson's disease patients. An "in vivo" study using I-123-FP-CIT SPECT. *Park. Relat. Disord.* **2020**, *78*, 31–35. [CrossRef]
51. Walsh, J.J.; Tschakovsky, M.E. Exercise and circulating BDNF: Mechanisms of release and implications for the design of exercise interventions. *Appl. Physiol. Nutr. Metab. Physiol. Appl. Nutr. Metab.* **2018**, *43*, 1095–1104. [CrossRef] [PubMed]
52. Ruiz-Gonzalez, D.; Hernandez-Martinez, A.; Valenzuela, P.L.; Morales, J.S.; Soriano-Maldonado, A. Effects of physical exercise on plasma brain-derived neurotrophic factor in neurodegenerative disorders: A systematic review and meta-analysis of randomized controlled trials. *Neurosci. Biobehav. Rev.* **2021**, *128*, 394–405. [CrossRef] [PubMed]
53. Schaeffer, E.; Roeben, B.; Granert, O.; Hanert, A.; Liepelt-Scarfone, I.; Leks, E.; Otterbein, S.; Saraykin, P.; Busch, J.H.; Synofzik, M.; et al. Effects of exergaming on hippocampal volume and brain-derived neurotrophic factor levels in Parkinson's disease. *Eur. J. Neurol.* **2022**, *29*, 441–449. [CrossRef] [PubMed]
54. Vaynman, S.; Ying, Z.; Gomez-Pinilla, F. Hippocampal BDNF mediates the efficacy of exercise on synaptic plasticity and cognition. *Eur. J. Neurosci.* **2004**, *20*, 2580–2590. [CrossRef]
55. Angelucci, F.; Piermaria, J.; Gelfo, F.; Shofany, J.; Tramontano, M.; Fiore, M.; Caltagirone, C.; Peppe, A. The effects of motor rehabilitation training on clinical symptoms and serum BDNF levels in Parkinson's disease subjects. *Can. J. Physiol. Pharm.* **2016**, *94*, 455–461. [CrossRef]
56. Kempermann, G.; Gast, D.; Gage, F.H. Neuroplasticity in old age: Sustained fivefold induction of hippocampal neurogenesis by long-term environmental enrichment. *Ann. Neurol.* **2002**, *52*, 135–143. [CrossRef]
57. Heisz, J.J.; Clark, I.B.; Bonin, K.; Paolucci, E.M.; Michalski, B.; Becker, S.; Fahnestock, M. The Effects of Physical Exercise and Cognitive Training on Memory and Neurotrophic Factors. *J. Cogn. Neurosci.* **2017**, *29*, 1895–1907. [CrossRef]
58. Angelucci, F.; Peppe, A.; Carlesimo, G.A.; Serafini, F.; Zabberoni, S.; Barban, F.; Shofany, J.; Caltagirone, C.; Costa, A. A pilot study on the effect of cognitive training on BDNF serum levels in individuals with Parkinson's disease. *Front. Hum. Neurosci.* **2015**, *9*, 130. [CrossRef]
59. Brauer, S.G.; Morris, M.E. Can people with Parkinson's disease improve dual tasking when walking? *Gait Posture* **2010**, *31*, 229–233. [CrossRef]
60. Duncan, R.P.; Earhart, G.M. Randomized controlled trial of community-based dancing to modify disease progression in Parkinson disease. *Neurorehabilit. Neural Repair* **2012**, *26*, 132–143. [CrossRef] [PubMed]
61. de Bruin, N.; Doan, J.B.; Turnbull, G.; Suchowersky, O.; Bonfield, S.; Hu, B.; Brown, L.A. Walking with music is a safe and viable tool for gait training in Parkinson's disease: The effect of a 13-week feasibility study on single and dual task walking. *Park. Dis.* **2010**, *2010*, 483530. [CrossRef] [PubMed]
62. Costa-Ribeiro, A.; Andrade, S.; Ferrer, M.L.V.; Silva, O.; Salvador, M.L.S.; Smaili, S.; Lindquist, A.R.R. Can Task Specificity Impact tDCS-Linked to Dual Task Training Gains in Parkinson's Disease? A Protocol for a Randomized Controlled Trial. *Front. Aging Neurosci.* **2021**, *13*, 684689. [CrossRef] [PubMed]
63. Wong-Yu, I.S.; Mak, M.K. Multi-dimensional balance training programme improves balance and gait performance in people with Parkinson's disease: A pragmatic randomized controlled trial with 12-month follow-up. *Park. Relat. Disord.* **2015**, *21*, 615–621. [CrossRef]
64. Ali, N.; Tian, H.; Thabane, L.; Ma, J.; Wu, H.; Zhong, Q.; Gao, Y.; Sun, C.; Zhu, Y.; Wang, T. The Effects of Dual-Task Training on Cognitive and Physical Functions in Older Adults with Cognitive Impairment; A Systematic Review and Meta-Analysis. *J. Prev. Alzheimer's Dis.* **2022**, *9*, 359–370. [CrossRef]
65. Feng, Y.S.; Yang, S.D.; Tan, Z.X.; Wang, M.M.; Xing, Y.; Dong, F.; Zhang, F. The benefits and mechanisms of exercise training for Parkinson's disease. *Life Sci.* **2020**, *245*, 117345. [CrossRef] [PubMed]
66. Machado, S.; Teixeira, D.; Monteiro, D.; Imperatori, C.; Murillo-Rodriguez, E.; da Silva Rocha, F.P.; Yamamoto, T.; Amatriain-Fernández, S.; Budde, H.; Carta, M.G.; et al. Clinical applications of exercise in Parkinson's disease: What we need to know? *Expert Rev. Neurother.* **2022**, *22*, 771–780. [CrossRef] [PubMed]
67. Johansson, H.; Ekman, U.; Rennie, L.; Peterson, D.S.; Leavy, B.; Franzen, E. Dual-Task Effects During a Motor-Cognitive Task in Parkinson's Disease: Patterns of Prioritization and the Influence of Cognitive Status. *Neurorehabilit. Neural Repair* **2021**, *35*, 356–366. [CrossRef] [PubMed]
68. Strouwen, C.; Molenaar, E.; Münks, L.; Broeder, S.; Ginis, P.; Bloem, B.R.; Nieuwboer, A.; Heremans, E. Determinants of Dual-Task Training Effect Size in Parkinson Disease: Who Will Benefit Most? *J. Neurol. Phys.* **2019**, *43*, 3–11. [CrossRef]
69. Kim, R.; Park, S.; Yoo, D.; Jun, J.S.; Jeon, B. Association of Physical Activity and APOE Genotype With Longitudinal Cognitive Change in Early Parkinson Disease. *Neurology* **2021**, *96*, e2429–e2437. [CrossRef]

brain sciences

MDPI

Review

Impact of Physical Exercise Alone or in Combination with Cognitive Remediation on Cognitive Functions in People with Schizophrenia: A Qualitative Critical Review

Giacomo Deste [1,2], Daniele Corbo [3], Gabriele Nibbio [1], Mauro Italia [1], Dario Dell'Ovo [1], Irene Calzavara-Pinton [1], Jacopo Lisoni [2], Stefano Barlati [1,2], Roberto Gasparotti [3,4] and Antonio Vita [1,2,*]

1 Department of Clinical and Experimental Sciences, University of Brescia, 25123 Brescia, Italy
2 Department of Mental Health and Addiction Services, ASST Spedali Civili of Brescia, 25123 Brescia, Italy
3 Department of Medical and Surgical Specialties, Radiological Sciences and Public Health, University of Brescia, 25123 Brescia, Italy
4 Neuroradiology Unit, ASST Spedali Civili of Brescia, 25123 Brescia, Italy
* Correspondence: antonio.vita@unibs.it

Abstract: Physical exercise and cognitive remediation represent the psychosocial interventions with the largest basis of evidence attesting their effectiveness in improving cognitive performance in people living with schizophrenia according to recent international guidance. The aims of this review are to provide an overview of the literature on physical exercise as a treatment for cognitive impairment in schizophrenia and of the studies that have combined physical exercise and cognitive remediation as an integrated rehabilitation intervention. Nine meta-analyses and systematic reviews on physical exercise alone and seven studies on interventions combining physical exercise and cognitive remediation are discussed. The efficacy of physical exercise in improving cognitive performance in people living with schizophrenia is well documented, but more research focused on identifying moderators of participants response and optimal modalities of delivery is required. Studies investigating the effectiveness of integrated interventions report that combining physical exercise and cognitive remediation provides superior benefits and quicker improvements compared to cognitive remediation alone, but most studies included small samples and did not explore long-term effects. While physical exercise and its combination with cognitive remediation appear to represent effective treatments for cognitive impairment in people living with schizophrenia, more evidence is currently needed to better understand how to implement these treatments in psychiatric rehabilitation practice.

Keywords: cognitive remediation; integrated interventions; physical exercise; psychiatric rehabilitation; psychosocial interventions; schizophrenia; cognitive functioning; recovery

check for updates

Citation: Deste, G.; Corbo, D.; Nibbio, G.; Italia, M.; Dell'Ovo, D.; Calzavara-Pinton, I.; Lisoni, J.; Barlati, S.; Gasparotti, R.; Vita, A. Impact of Physical Exercise Alone or in Combination with Cognitive Remediation on Cognitive Functions in People with Schizophrenia: A Qualitative Critical Review. *Brain Sci.* **2023**, *13*, 320. https://doi.org/10.3390/brainsci13020320

Academic Editor: Fiorenzo Moscatelli

Received: 11 January 2023
Revised: 31 January 2023
Accepted: 11 February 2023
Published: 14 February 2023

1. Introduction

Schizophrenia (SCZ) represents a severe mental disorder which is often associated with poor functional outcomes [1–3]. Cognitive impairment is one of the core features of the disorder and has been considered of great relevance since its earliest conceptualizations [4,5]. In fact, deficits in both neurocognitive performance [6] and social cognition abilities [7] can be frequently observed in people living with SCZ. Cognitive deficits are one of the core determinants of functional impairment, showing an important negative impact on both functional capacity and real-world functioning [8–10], and they also represent one of the main limiting factors for the process of recovery in the context of psychiatric rehabilitation [11–14].

Pharmacological treatment is effective in reducing the core symptoms of SCZ [15,16], but its impact on cognition appears to be limited, as available medications provide minimal effects on cognitive functioning [17,18]. This has led to a considerable scientific and clinical interest regarding the potential usefulness of non-pharmacological interventions [3,19].

According to the most recent European Psychiatric Association guidance, the psychosocial interventions that have the most solid body of evidence confirming their effectiveness in improving cognitive performance in people living with SCZ are physical exercise (PE) and cognitive remediation (CR) [18].

PE can be considered an evidence-based intervention, as it provides consistent benefits for people living with mental disorders, and in people living with SCZ, physical activity is currently recommended as an adjunctive treatment in rehabilitation programs to improve symptoms severity and also overall quality of life [20,21]. While PE could also provide substantial benefits to cognitive performance, its moderators of effectiveness and the optimal modalities to deliver it in clinical practice to specifically target cognitive impairment in SCZ currently require better understanding.

CR is a behavioral training-based intervention that specifically targets cognitive processes and that has proven to be effective also in improving functional outcomes [22–25], and which appears to be more effective in the real-world context when implemented into structured rehabilitation programs or alongside other evidence-based treatments [23,26–28].

As both these interventions appear to be characterized by good practical feasibility and relatively low resource cost for implementation in rehabilitation practice, better understanding the extent of their effectiveness and their potential as a combined treatment program could represent a topic of considerable clinical interest.

The aims of the present narrative review are to provide a comprehensive overview of the literature attesting the effectiveness of PE as a treatment for cognitive impairment in people living with SCZ and of the studies that have combined PE and CR as an integrated rehabilitation intervention.

2. Methods

To ensure a comprehensive assessment of the relevant literature, works included in the present narrative review were drawn from a search conducted in open databases (PubMed, Scopus, and Google Scholar) with combinations of the following research terms: [physical AND (exercise OR activity)], [cognitive AND (remediation OR rehabilitation OR training)] and (schizophrenia OR psychosis). Given the substantial number on studies of the effects on PE on cognition in people living with SCZ, only systematic reviews and meta-analyses were considered for the first part of the present work, while individual studies were considered for interventions combining PE and cognitive remediation.

Results of this search were also checked against the results of a recent European Psychiatric Association guidance document on treatment of cognitive impairment [18] and from the systematic search conducted for a recent and comprehensive systematic review and meta-analysis on CR trials [29].

3. Results

3.1. Effectiveness of Physical Exercise on Cognition in Schizophrenia

In recent years, an ever-growing body of literature exploring the relationship between PE and cognition in people living with SCZ has been published, posing an interesting background for future research (Table 1).

Exploring the literature in chronological order, we found that in a 2012 systematic review including 10 randomized clinical trials (RCT) (for a total of 322 SCZ participants), it is stated that aerobic and strength exercises have been proven to be effective in improving short-term memory, a result that was correlated to an increase in the volume of the hippocampus [30].

Table 1. Studies on the effect of physical exercise on cognition in schizophrenia.

Year of Publication and Authors	Type of Study	Number of Included Studies (Total Number of Participants)	Summary
Vancampfort et al., 2012 [30]	Systematic Review	10 (322)	In one study, STM (correlated to an increased hippocampal volume) is improved by both AE and SE.
Dauwan et al., 2016 [31]	Meta-analysis	17 (659)	In one study, exercise improved verbal STM by 34% ($p < 0.05$) and brain volume was increased by 12%.
Firth et al., 2015 [32]	Meta-analysis	29 (1109)	Only yoga seems to influence the subdomain of LTM ($p < 0.05$), and to have a trend towards significance for AT and EF ($p = 0.08$).
Firth et al., 2017a [33]	Systematic Review	16 (423)	Improvements in brain structure and connectivity are associated to exercise's beneficial effects on cognition in SCZ. Some studies also found a positive correlation between BDNF and cognitive enhancements.
Firth et al., 2017b [34]	Meta-analysis	10 (385)	PE significantly improves cognition and "higher dosages" of PE and supervised PE are associated with larger improvements in global cognition. Moreover, PE is particularly beneficial for SC, WM, and AT. Finally, AE seems more effective than yoga.
Li et al., 2018 [35]	Systematic Review	7 (679)	In one RCT (n = 194, low quality), mindful exercise is significantly more effective than other forms of exercise in WM subtests (MD 0.39).
Van der Stouwe et al., 2018 [36]	Systematic Review	29 (4448)	The protective effects of exercise on the reduction of hippocampal volume and white matter tracts were similar for both SSD and HC and they showed a dose–response relationship, but the effects in people with SSD appear to be of reduced entity.
Dauwan et al., 2021 [37]	Meta-analysis	122 (7231)	PE was superior to TAU in improving AT and WM ($p < 0.009$), EF ($p = 0.013$), memory ($p = 0.038$), and PS ($p = 0.003$). Results were proven not to be age-dependent.
Fernández-Abascal et al., 2021 [38]	Meta-analysis	59 (4202)	Post-intervention benefit was found for many cognitive domains: STM and AT measured through the WAIS' forward and backward DSTs showed a significant effect at the end of the intervention ($p = 0.014$ for the forward test and $p < 0.001$ for the backward test), although the effects did not persist at follow-up ($p = 0.528$ and $p = 0.291$, respectively). Furthermore, different studies explored other outcomes through different measures: through the RAVLT test (significant only for the STM); through the Corsi direct block span for visual STM (not significant); through the Penn computerized neurocognitive battery for attention, SM, and WM (several domains showed a significant improvement for both yoga and other types of exercise, some persisting at follow-up as well); through the WCST for EF (significant improvement); through the MCCB for AT, memory, and SC (improvements seen in the scores correlated to improvement in physical performances); through the NES test for motor and sequencing skills (showing improvement but not maintained at follow-up); and other miscellaneous tests for SP, WM, VeL, and ViL (only WM was significantly improved). Only two studies found no post-intervention effects on cognition.

AE (aerobic exercise), AT (attention), DST (digit span test), EF (executive functioning), HC (healthy controls), LTM (long-term memory), MCCB (MATRICS Consensus Cognitive Battery), NES (Neurological Evaluation Scale), PE (physical exercise), PS (psychomotor speed), RCT (randomized controlled trial), RAVLT (Rey Auditory Verbal Learning Test), SC (social cognition), SCZ (schizophrenia), SE (strength exercise), SM (spatial memory), SP (speed of processing), SSD (schizophrenia spectrum disorders), STM (short-term memory), TAU (treatment as usual), VeL (verbal learning), ViL (visual learning), WAIS (Weschler Adult Intelligence Scale), WCST (Wisconsin Card Sorting Test), WM (working memory).

Then, in 2015, Firth et al. analyzed 17 trials (for a total of 659 non-affective psychotic patients), finding a positive effect of exercise on neurocognition and, in particular, in one study it was observed that exercise was associated to improvements in short-term verbal memory of 34% ($p < 0.05$) and to increase in brain volume of 12% (significantly more than that in one of the control groups) [32].

The next year, a Dutch review with broader inclusion criteria selected 29 studies (for a total of 1109 SCZ patients) and found that exercise in general had no demonstrated effect on cognition, with the exception of yoga which was seen to improve the cognitive subdomain of long-term memory (n = 184: Hedges' g = 0.32, $p < 0.05$), and with an additional trend towards significance for the subdomains of attention and executive functioning (n = 184: Hedges' g = 0.38, $p = 0.08$) [31].

In 2017, the group of Firth et al. published a review including 16 trials (for a total of 423 SCZ patients) of which 7 focused on neuroimaging: preliminary evidence indicated that cognitive improvements provided by PE for people living with SCZ are related to improvements in brain structure and connectivity, and some of these studies found a positive correlation between brain-derived neurotrophic factor (BDNF) and cognitive enhancements, indicating that neurogenesis could be the mechanism underlying the cognitive benefits of exercise in SCZ (current evidence, however, is too limited to provide definitive conclusions) [33].

In the same year, Firth and collaborators also conducted a meta-analysis presenting evidence that PE significantly improves cognition and that interventions using "higher dosages" of PE are related to greater improvements in global cognition. Moreover, subgroup analyses suggested that interventions including direct supervision from a physical activity instructor resulted in larger cognitive improvement. Moreover, domain-specific analyses highlighted that attention/vigilance, working memory, and social cognition had substantial improvements: all these domains are among those most commonly impaired in people living with SCZ and represent well-known predictors of real-world outcomes. Finally, the authors report that aerobic exercise (AE) may be more effective than yoga for cognitive outcomes in SCZ, unlike what was observed in previous research [34].

The next year, in 2018, a systematic review by Li et al. explored the efficacy of mindful (i.e., yoga, different forms of Tai Chi, and Qigong) versus non-mindful exercise in subjects with SCZ: the former came out as being significantly more effective than the latter in the accuracy index for working memory, but the RCT (n = 194) coming to the conclusion is described by Li et al. as being low-quality (MD = 0.39), whereas in the domains of "attention" and "social functioning", no clear difference was found (MD = −0.48) [35].

In the same year, a Dutch systematic review comparing the effects of AE in people living with a SCZ spectrum disorder (168 subjects) and healthy participants (for a total of 201 subjects) within a similar age range was published. Most studies included in this work effects on hippocampal volume, reporting positive outcomes: PE could help in preventing the reduction of hippocampal volume over time in people living with SCZ; in healthy subjects, changes in connectivity of the dorsolateral prefrontal cortex related to better cognitive performance were reported. In the comparisons between participants with SCZ spectrum disorders and healthy subjects, similar exercise-mediated effects with a dose–response relationship on hippocampal volume as well as on white matter tracts were observed. However, these effects appear to be smaller in people living with SCZ spectrum disorders. Finally, this systematic review underlines the need for further studies to investigate additional neural correlates beside the hippocampus [36].

In 2021, then, a meta-analysis by Dauwan et al. (including studies on neurological conditions such as Alzheimer's disease, Parkinson's disease, Huntington's disease, and multiple sclerosis, as well as mental disorders such as SCZ and major depressive disorder) showed that exercise provided greater gains than treatment as usual (TAU) in several cognitive domains, including attention and working memory ($p < 0.009$), executive functioning ($p = 0.013$), memory ($p = 0.038$), and psychomotor speed ($p = 0.003$), and these results were proven not to be age-dependent [37].

In the same year, a Spanish research team published a meta-analysis of 59 RCTs analyzing interventions on diet and physical activity in people living with non-affective psychosis, including also participants with first-episode psychosis (FEP). Significant improvements were observed for many cognitive domains: two studies evaluated short-term memory performance and attention through the WAIS' forward and backward digit span tests, with positive post-treatment effects for both measures (g = 0.309, $p = 0.014$ for the forward test and g = 0.621, $p < 0.001$ for the backward test), although neither had an effect that persisted at follow-up (g = −0.132, $p = 0.528$ and g = −0.201, $p = 0.291$, respectively). Furthermore, different studies explored other outcomes such as short- and long-term memory measured through the Rey Auditory Verbal Learning Test (RAVLT) (significant only for the short-term memory component); short-term visual memory measured through the Corsi direct block span (not significant); attention, spatial memory, and working memory measured via the Penn computerized neurocognitive battery (several domains showed a significant improvement for both yoga and other types of exercise, some persisting at follow-up as well); executive functions measured through the Wisconsin Card Sorting Test (WCST) (showing a significant improvement); attention, memory, social cognition, and other domains measured through the MATRICS Consensus Cognitive Battery (showing improvement in the scores correlated to improvement in physical performances); motor and sequencing skills measured through the Neurological Evaluation Scale (showing improvements after exercise interventions that were not maintained at follow-up); and speed of processing, working memory, verbal learning, and visual learning measured through a variety of tests (only working memory was significantly improved after the intervention). Only two studies found no post-intervention effect on cognition [38].

3.2. Combination of Physical Exercise and Cognitive Remediation for Treating Cognition in Schizophrenia

Both CR and PE have been reported to be effective on improving cognition in SCZ: several works in the recent literature compare the efficacy of the two interventions and explore whether they have a synergistic effect. In the next paragraph, we present the relevant literature in a chronological order. A summary of the studies can be found in Table 2.

Table 2. Studies on the effect of physical exercise in combination with cognitive remediation on cognition in schizophrenia.

Year of Publication and Authors	Type of Study	Groups	Total Number of Participants (Diagnosis)	Summary
Tan and King, 2013 [39]	RCT	PE and CR	70 (SCZ)	The CR group kept a greater improvement in all of neurocognitive tests when compared to the PE group ($p < 0.05$ for all tests).
Oertel-Knöchel et al., 2014 [40]	RCT	CR+AE, CR+RT, WW	51 (29 SCZ + 22 MDD)	Cognitive performances were improved in both treatment groups (CT+AE group better than CT+RT), particularly in the cognitive subdomains of SP ($p < 0.001$), WM ($p = 0.01$), and ViL ($p = 0.004$). Additionally, the results show that the improvement is greater in SCZ compared to MDD, suggesting that the combined intervention is even more effective in SCZ ($p < 0.001$).
Malchow et al., 2015 [41]	Clinical Trial	PE, PE + CACR, TT + CACR	65 (43 SCZ + 22 HC)	The performance in VLMT and WCST significantly improved in the PE + CACR group. The positive effects were not seen in the TT + CACR group.
Nuechterlein et al., 2016 [42]	Pilot Study	CT&E, CT	18 (recent onset SSD)	The ES for the improvement in the MCCB score was larger for CT&E patients when compared to CT patients: the largest differential gains that were seen for CT&E were SC (f = 0.65), WM (f = 0.50), SP (f = 0.38), and attention (f = 0.33).

Table 2. *Cont.*

Year of Publication and Authors	Type of Study	Groups	Total Number of Participants (Diagnosis)	Summary
Choi et al., 2019 [43]	RCT	PE, CT, PE + CT	85 (SCZ)	WM was improved in all intervention groups (vs. the CT only group) and mostly in the PE only group ($p < 0.001$). At follow-up instead, the only group still showing significant improvements in WM was the CT + PE group ($p < 0.001$). Moreover, concerning PS, the only group showing a more significant improvement (vs. other groups) was the PE only group ($p < 0.001$), whereas at follow-up, only the PE + CT group ($p < 0.001$).
Nuechterlein et al., 2022 [44]	RCT	CT&E, CT	47 (FEP)	PE + CT (compared to CT alone) provided faster gains in cognition and greater improvements in psychosocial functioning. Strong correlations were observed between cognitive gains at six months and proportion of PE completed (r = 0.56) and the number of homework PE sessions (r = 0.61). The transfer of cognitive improvement to real-world functioning appears to be related to the quantity of completed PE (r = 0.51), and to participation in both CT sessions (r = 0.46) and bridging groups (r = 0.56).
Dai et al., 2022 [45]	RCT	CAE, AE, TAU	82 (SCZ)	The CAE group had a significantly better performance in PS and cognitive flexibility when compared to the TAU and the AE groups ($p < 0.05$).

AE (aerobic exercise), CACR (computer-assisted cognitive remediation), CAE (cognitive remediation + aerobic exercise), CR (cognitive remediation), CT (cognitive training), CT&E (Cognitive Training & Exercise), ES (effect size), FEP (first episode psychosis), HC (healthy controls), MCCB (MATRICS Consensus Cognitive Battery), PE (physical exercise), RCT (randomized controlled trial), RT (relaxation training), SC (social cognition), SCZ (schizophrenia), SP (speed of processing), SSD (schizophrenia spectrum disorders), TAU (treatment as usual), TT (table tennis), ViL (visual learning), VLMT (Verbal Learning Memory Test), WCST (Wisconsin Card Sorting Test), WM (working memory), WW (wait and watch).

In 2013, Tan and King published their RCT on the effects of CR on functional outcomes in people living with SCZ. A total of 70 included participants were allocated to 2 groups: 34 subjects were randomized to the PE group, whereas 36 were randomized to the CR group. Neurocognitive and functional outcomes were measured at baseline, after 3 months (end of treatment), after 9 months, and after 1 year and 3 months from baseline. Over time, the CR group kept a greater improvement on all measures of neurocognition when compared to the PE group ($p < 0.05$ for all tests: Comprehensive Trail Making Test, RAVLT, Wechsler Adult Intelligence Scale's Digit Span Forward and Backward tests) [39].

In 2014, Oertel-Knöchel et al. published a study focusing on the effects on cognition of a combination of PE and cognitive training (CT) in SCZ (SCZ, n = 29) and major depressive disorder (MDD, n = 22). Patients were randomly allocated in the following groups: CT + PE (n = 16, of which 8 diagnosed with SCZ); CT + relaxation exercises (n = 17, of which 11 diagnosed with SCZ); and "waiting control group" to check for potential biases (n = 18, of which 10 diagnosed with SCZ). Amelioration of the cognitive performances was observed in both treatment groups, although the improvement was greater in the group undergoing CT + AE. Specifically, the results showed a significant post-intervention increase in performances in the cognitive subdomains of speed of processing ($p < 0.001$), working memory ($p = 0.01$), and visual learning ($p = 0.004$). Additionally, the results showed that the improvement in cognition was greater in patients living with SCZ compared to people suffering from MDD, suggesting that the combined intervention is even more effective in SCZ ($p < 0.001$) [40].

Malchow et al. conducted a single-center trial, taking place between 2010 and 2013 and evaluating the effects of bicycle ergometer training and add-on computer-assisted remediation (CACR) training, compared to a control group playing table tennis + CACR. The study included 22 subjects with SCZ and 22 healthy controls matched for age and sex; all the subjects underwent 3 months of endurance training (30 min, three times/week) and CACR training (30 min, two times/week) was added from week 6. An additional group including 21 subjects living with SCZ played table soccer and additionally received the same CACR training. The performance in the Verbal Learning Memory Test (VLMT) and WCST improved significantly in subjects with SCZ belonging to the endurance training + CACR group. The positive effects were specific to the endurance training + CACR group and were not observed in patients playing table soccer + CACR [41].

A pilot study conducted in 2016 by Neuchterlein et al. was designed to test the efficacy of enhancing CT with AE in patients with a recent onset of psychotic illness. A total of 18 patients were enrolled into two different groups: 7 subjects were allocated to the Cognitive Training & Exercise (CT&E) group and 9 to CT alone group for a 10-week period. All patients received treatment with a second-generation antipsychotic medication and regular psychiatric visits in addition to the study interventions. The effect size for the improvement in the MCCB score was larger for CT&E patients when compared to CT patients: Cohen's f is 0.48 and the single cognitive domains with the largest differential gains for CT&E (when compared to CT alone) were social cognition (f = 0.65), working memory (f = 0.50), speed of processing (f = 0.38), and attention/vigilance (f = 0.33) [42].

In 2020, Choi et al. randomly allocated 85 outpatients diagnosed with SCZ into 2 groups of 18 hours of either PE (29 subjects), CT focused on processing speed and working memory (27 subjects), or a time-matched combination of the two (29 subjects). Working memory was seen to be improved post-intervention in all groups, with greater gains in the PE only group, and to a lesser extent in the PE + CT group, when compared to the CT only group ($p < 0.001$). After 3 months from the end of the treatment instead, the only group still showing significant improvements in working memory was the CT + PE group ($p < 0.001$). Moreover, concerning processing speed, after the intervention, all groups showed better scores, but the only group showing a significant improvement compared to the others was the PE only group ($p < 0.001$), whereas at follow-up, the only group showing a persistent improvement compared to the other groups was the PE + CT group ($p < 0.001$) [43].

A very recent work by Neuchterlein et al. on the same topic explored the effect of a PE + CT program lasting 6 months compared to CT in 47 outpatients with FEP. The combination of AE and CT led to faster cognitive improvement and to larger gains in psychosocial functioning compared to CT. Both groups improved in a six-month period of CT treatment, but AE appeared to provide an additional boost to CT in the first three months. BDNF increase appeared to be associated with the dimension of cognitive improvement, but this association was observed only at a statistical trend-level. The authors reflect upon the possibility that people with FEP could be more motivated to participate in PE and present fewer medical comorbidities, limiting the access to PE than people living with more advanced stages of SCZ; the randomized treatments were delivered in the context of a psychiatric rehabilitation program, and this could facilitate larger effects on cognitive performance and its transfer to real-world functioning. The view that the quantity of completed exercise represented an essential factor in the cognitive improvement observed in the PE + CT group is backed also by correlations observed between improvements in cognitive performance and proportion of completed exercise (r = 0.56) and the number of homework sessions (r = 0.61). The transfer of improvement in cognitive performance to real-world functioning appeared to be related to the amount of completed exercise (r = 0.51), but it also appeared to be associated with the amount of participation in CT sessions as well as in bridging groups. These findings support the hypothesis that exercise represents the core factor behind cognitive improvement, while participation in other components of the intervention is responsible for improving generalization to real-world functioning [44].

Very recently, Dai et al. published an RCT comparing 3 groups of individuals living with SCZ: 31 subjects in the treatment as usual group, 26 subjects in the computerized CR therapy + aerobic exercise (CAE) group, and 25 subjects in the AE group. From the results, it is reported that the CAE group had a significantly better performance in terms of change of processing speed and cognitive flexibility scores when compared to the control and the AE groups after treatment ($p < 0.05$) [45].

4. Discussion

Considering the emerging evidence, there is a clear rationale to propose PE intervention to people living with SCZ.

PE interventions delivered in trials recruiting people living with SCZ and considering cognitive performance among the study outcomes included a wide variety of different types of PE. Among these, the most represented were AE, anaerobic exercise (also defined as strength exercise), mind-body exercise (also defined as mindful exercise, encompassing exercise such as yoga, Tai Chi, and Qigong), and dance movement therapy. AE included both light exercise requiring low consumption of energy, such as walking or light intensity exercise in standing or sitting position, and moderate or vigorous exercise requiring high energy consumption for several minutes, such as high intensity cycling, dancing, jogging, and running; some studies combined both AE and anaerobic exercise. Frequency of administration varied considerably between studies, ranging from 60 min once per week to 45–60 min three times per week for AE and anaerobic exercise; mind-body exercise interventions were implemented with higher intensity, up to 90 min seven times per week. The duration of the program also varied considerably, ranging from 4 to 56 weeks.

According to the results of several meta-analysis and systematic reviews, PE-based interventions could not only provide substantial benefits for metabolic, endocrinological, and health-related reasons, but also provide reliable improvement in cognitive performance.

In fact, several cognitive domains that are frequently impaired in people living with SCZ, including attention, short- and long-term memory, executive functions, working memory, and social cognition abilities, showed consistent gains in systematic reviews and meta-analytic evidence investigating the effects of PE.

In this regard, PE can be fully considered an evidence-based intervention for treatment of cognitive impairment in SCZ [18,21].

However, despite the wealth of the literature attesting its efficacy, moderators of participants response and optimal modalities of delivery of psychical exercise-based interventions still require to be further explored. For instance, in one of the most recent and well-conducted meta-analyses [34], no significant moderator of response emerged besides a trend–level relationship between greater amounts of exercise in minutes of activity per week of treatment and greater cognitive gains. Another recent meta-analysis [38] reported similar results, as it found a small superior effect of moderate-to-vigorous AE compared to light AE or mind-body exercise.

While these observations further confirm the clinical effectiveness of PE, as they show a dose–response effect, they do not provide useful insight to improve its usefulness in a clinical context and to better implement it in rehabilitation settings. In this regard, future research should focus on better identifying ideal candidates for this treatment, the optimal duration and intensity of programs, as well as barriers and facilitators for its implementation in rehabilitation settings [46].

As CR represents the psychosocial intervention with the highest level of recommendation in recent international guidance [18], and it appears to provide greater benefits on both cognitive and functional outcomes when combined with other interventions [23,28], pairing it with psychical exercise as a combined treatment appears to be a potentially effective strategy.

To date, to the best of our knowledge, no systematic assessment and meta-analysis focusing on interventions combining CR and PE has been performed. However, several individual studies, independently conducted in different research and clinical centers, have assessed the efficacy of this combination.

Most of these studies report a superior effect on cognitive outcomes of combined treatment compared to CR alone or to non-specific control conditions; one study in particular reported that adding PE appears to provide a 'boosting effect' on cognitive performance, leading to faster gains [44].

While these results are very interesting, it should also be pointed out that all of these studies included small samples of participants, so their results must be considered preliminary and do not offer a definite proof of superior efficacy or effectiveness of this combined treatment.

Instead, these promising results should represent the conceptual basis for additional confirmatory studies, including larger samples and conducted with a multicentric approach.

Moreover, long-term effectiveness of PE and combined interventions requires further study: durability of positive effects is essential if cognitive gains are to be translated into real-world outcomes [26,47], and while the results of several studies including follow-up observations suggest that CR does indeed provide lasting benefits [48–51], the long-term effect of combined approaches remains to be better explored, and their long-term superiority to CR alone remains to be demonstrated.

Finally, moderators of response of this combined treatment also remain to be further explored: possible moderators include both participant-related characteristics, such as duration of illness, duration of untreated psychosis, age of onset of the disorder, baseline cognitive functioning, and baseline symptoms severity, and treatment-related features, such as duration and intensity of both CR and PE, the presence of an active and trained therapist, and an individual or group delivery of treatment. Several of these factors appear to affect participants' response to CR interventions [22,23], but their impact on a combined treatment requires further research.

The present review has some notable points of strength. It presents a solid element of novelty, as it provides both a comprehensive update on systematic works that investigate the effectiveness of PE-based interventions on cognition in people living with SCZ and is, to the best of our knowledge, the first review to provide a collection of all studies integrating CR and PE interventions. As cognitive performance represents one of the stronger predictors of real-world rehabilitation outcomes and is also one of the main treatment targets of importance in the patient's perspective [52,53], providing further insight on effective ways to improve it represents an issue of clear scientific and clinical interest.

However, some limitations have to be acknowledged. As a narrative review, the studies summarized and discussed were not identified through a systematic search strategy, which was beyond the scope of the present work and may represent an interesting perspective for future research. Moreover, no quantitative analysis was performed for combined treatments; however, considering the wide heterogeneity of designs and control conditions of reported studies, a pooled effect estimate might not be sufficiently accurate, and certainty of the provided evidence may not be sufficient.

Despite these limitations, the reviews and the studies offer valuable insight on a topic of considerable scientific interest, highlighting that steps are required in clinical research to further improve cognitive and rehabilitation outcomes of people living with SCZ.

5. Conclusions

Several systematic reviews and meta-analysis attest to the efficacy of PE in improving cognitive performance in people living with SCZ, but moderators of participants response and optimal modalities of delivery remain to be identified.

Integrated interventions combining CR and PE appear to provide superior benefits and quicker improvements compared to cognitive radiation alone, but larger studies are required to confirm these findings and longitudinal observations are necessary to attest the durability of positive effects.

Future research should also focus on better understanding the barriers and facilitators for the implementation in rehabilitation practice of both PE and combined interventions.

Author Contributions: Conceptualization, G.D., D.C., G.N., S.B.; investigation, M.I. and D.D. and I.C.-P.; data curation, G.N., I.C.-P. and J.L.; writing—original draft preparation, M.I., D.D. and I.C.-P.; writing—review and editing, G.D., D.C., G.N., J.L. and S.B; supervision, R.G. and A.V. All authors have read and agreed to the published version of the manuscript.

Funding: This research received no external funding.

Institutional Review Board Statement: Not applicable.

Informed Consent Statement: Not applicable.

Data Availability Statement: Not applicable.

Conflicts of Interest: The authors declare no conflict of interest.

References

1. Galderisi, S.; Rossi, A.; Rocca, P.; Bertolino, A.; Mucci, A.; Bucci, P.; Rucci, P.; Gibertoni, D.; Aguglia, E.; Amore, M.; et al. The influence of illness-related variables, personal resources and context-related factors on real-life functioning of people with schizophrenia. *World Psychiatry* **2014**, *13*, 275–287. [CrossRef] [PubMed]
2. Harvey, P.D.; Strassnig, M. Predicting the severity of everyday functional disability in people with schizophrenia: Cognitive deficits, functional capacity, symptoms, and health status. *World Psychiatry* **2012**, *11*, 73–79. [CrossRef] [PubMed]
3. Maj, M.; Van Os, J.; De Hert, M.; Gaebel, W.; Galderisi, S.; Green, M.F.; Guloksuz, S.; Harvey, P.D.; Jones, P.B.; Malaspina, D.; et al. The clinical characterization of the patient with primary psychosis aimed at personalization of management. *World Psychiatry* **2021**, *20*, 4–33. [CrossRef]
4. Bleuler, E. Die Prognose Der Dementia Praecox (Schizophreniegruppe). *Allg. Z. Psychiat* **1908**, *65*, 436–480.
5. Kraepelin, E.; Barclay, R. Dementia Praecox. In *The Clinical Roots of the Schizophrenia Concept: Translations of Seminal European Contributions on Schizophrenia*; Cambridge University Press: Cambridge, UK, 1919; pp. 13–24.
6. Green, M.F.; Horan, W.P.; Lee, J. Nonsocial and social cognition in schizophrenia: Current evidence and future directions. *World Psychiatry* **2019**, *18*, 146–161. [CrossRef]
7. Pinkham, A.E.; Harvey, P.D.; Penn, D.L. Social Cognition Psychometric Evaluation: Results of the Final Validation Study. *Schizophr. Bull.* **2018**, *44*, 737–748. [CrossRef] [PubMed]
8. Deste, G.; Vita, A.; Nibbio, G.; Penn, D.L.; Pinkham, A.E.; Harvey, P.D. Autistic Symptoms and Social Cognition Predict Real-World Outcomes in Patients With Schizophrenia. *Front. Psychiatry* **2020**, *11*, 524. [CrossRef] [PubMed]
9. Fett, A.-K.J.; Viechtbauer, W.; Dominguez, M.-D.; Penn, D.L.; van Os, J.; Krabbendam, L. The relationship between neurocognition and social cognition with functional outcomes in schizophrenia: A meta-analysis. *Neurosci. Biobehav. Rev.* **2010**, *35*, 573–588. [CrossRef]
10. Galderisi, S.; Rucci, P.; Kirkpatrick, B.; Mucci, A.; Gibertoni, D.; Rocca, P.; Rossi, A.; Bertolino, A.; Strauss, G.P.; Aguglia, E.; et al. Interplay Among Psychopathologic Variables, Personal Resources, Context-Related Factors, and Real-life Functioning in Individuals With Schizophrenia: A Network Analysis. *JAMA Psychiatry* **2018**, *75*, 396–404. [CrossRef]
11. Giordano, G.M.; Galderisi, S.; Pezzella, P.; Perrottelli, A.; Bucci, P. Determinants of Clinical Recovery in Schizophrenia. In *Recovery and Major Mental Disorders*; Carpiniello, B., Vita, A., Mencacci, C., Eds.; Comprehensive Approach to Psychiatry; Springer International Publishing: Cham, Switzerland, 2022; pp. 23–43, ISBN 978-3-030-98301-7.
12. Spaulding, W.D.; Sullivan, M.E. Treatment of Cognition in the Schizophrenia Spectrum: The Context of Psychiatric Rehabilitation. *Schizophr. Bull.* **2016**, *42*, S53–S61. [CrossRef] [PubMed]
13. Vita, A.; Barlati, S. Recovery from schizophrenia. *Curr. Opin. Psychiatry* **2018**, *31*, 246–255. [CrossRef] [PubMed]
14. Wykes, T.; Dunn, G. Cognitive deficit and the prediction of rehabilitation success in a chronic psychiatric group. *Psychol. Med.* **1992**, *22*, 389–398. [CrossRef] [PubMed]
15. Galderisi, S.; Mucci, A.; Dollfus, S.; Nordentoft, M.; Falkai, P.; Kaiser, S.; Giordano, G.M.; Vandevelde, A.; Nielsen, M.; Glenthøj, L.B.; et al. EPA guidance on assessment of negative symptoms in schizophrenia. *Eur. Psychiatry* **2021**, *64*, e23. [CrossRef] [PubMed]
16. Leucht, S.; Leucht, C.; Huhn, M.; Chaimani, A.; Mavridis, D.; Helfer, B.; Samara, M.; Rabaioli, M.; Bächer, S.; Cipriani, A.; et al. Sixty Years of Placebo-Controlled Antipsychotic Drug Trials in Acute Schizophrenia: Systematic Review, Bayesian Meta-Analysis, and Meta-Regression of Efficacy Predictors. *Am. J. Psychiatry* **2017**, *174*, 927–942. [CrossRef]
17. Veselinović, T.; Neuner, I. Progress and Pitfalls in Developing Agents to Treat Neurocognitive Deficits Associated with Schizophrenia. *CNS Drugs* **2022**, *36*, 819–858. [CrossRef]
18. Vita, A.; Gaebel, W.; Mucci, A.; Sachs, G.; Barlati, S.; Giordano, G.M.; Nibbio, G.; Nordentoft, M.; Wykes, T.; Galderisi, S. European Psychiatric Association guidance on treatment of cognitive impairment in schizophrenia. *Eur. Psychiatry* **2022**, *65*, e57. [CrossRef] [PubMed]
19. Solmi, M.; Croatto, G.; Piva, G.; Rosson, S.; Fusar-Poli, P.; Rubio, J.M.; Carvalho, A.F.; Vieta, E.; Arango, C.; DeTore, N.R.; et al. Efficacy and acceptability of psychosocial interventions in schizophrenia: Systematic overview and quality appraisal of the meta-analytic evidence. *Mol. Psychiatry* **2022**, *28*, 354–368. [CrossRef] [PubMed]
20. Ashdown-Franks, G.; Firth, J.; Carney, R.; Carvalho, A.F.; Hallgren, M.; Koyanagi, A.; Rosenbaum, S.; Schuch, F.B.; Smith, L.; Solmi, M.; et al. Exercise as Medicine for Mental and Substance Use Disorders: A Meta-review of the Benefits for Neuropsychiatric and Cognitive Outcomes. *Sports Med.* **2020**, *50*, 151–170. [CrossRef] [PubMed]
21. Stubbs, B.; Vancampfort, D.; Hallgren, M.; Firth, J.; Veronese, N.; Solmi, M.; Brand, S.; Cordes, J.; Malchow, B.; Gerber, M.; et al. EPA guidance on physical activity as a treatment for severe mental illness: A meta-review of the evidence and Position Statement from the European Psychiatric Association (EPA), supported by the International Organization of Physical Therapists in Mental Health (IOPTMH). *Eur. Psychiatry* **2018**, *54*, 124–144. [CrossRef]

22. Lejeune, J.A.; Northrop, A.; Kurtz, M.M. A Meta-analysis of Cognitive Remediation for Schizophrenia: Efficacy and the Role of Participant and Treatment Factors. *Schizophr. Bull.* **2021**, *47*, 997–1006. [CrossRef]
23. Vita, A.; Barlati, S.; Ceraso, A.; Nibbio, G.; Ariu, C.; Deste, G.; Wykes, T. Effectiveness, Core Elements, and Moderators of Response of Cognitive Remediation for Schizophrenia: A Systematic Review and Meta-Analysis of Randomized Clinical Trials. *JAMA Psychiatry* **2021**, *78*, 848–858. [CrossRef] [PubMed]
24. Wykes, T.; Huddy, V.; Cellard, C.; McGurk, S.R.; Czobor, P. A Meta-Analysis of Cognitive Remediation for Schizophrenia: Methodology and Effect Sizes. *Am. J. Psychiatry* **2011**, *168*, 472–485. [CrossRef] [PubMed]
25. Yeo, H.; Yoon, S.; Lee, J.; Kurtz, M.M.; Choi, K. A meta-analysis of the effects of social-cognitive training in schizophrenia: The role of treatment characteristics and study quality. *Br. J. Clin. Psychol.* **2021**, *61*, 37–57. [CrossRef] [PubMed]
26. Bowie, C.R.; Bell, M.D.; Fiszdon, J.M.; Johannesen, J.K.; Lindenmayer, J.-P.; McGurk, S.R.; Medalia, A.A.; Penadés, R.; Saperstein, A.M.; Twamley, E.W.; et al. Cognitive remediation for schizophrenia: An expert working group white paper on core techniques. *Schizophr. Res.* **2020**, *215*, 49–53. [CrossRef] [PubMed]
27. Montemagni, C.; Del Favero, E.; Riccardi, C.; Canta, L.; Toye, M.; Zanalda, E.; Rocca, P. Effects of Cognitive Remediation on Cognition, Metacognition, and Social Cognition in Patients With Schizophrenia. *Front. Psychiatry* **2021**, *12*, 649737. [CrossRef] [PubMed]
28. Nibbio, G.; Barlati, S.; Cacciani, P.; Corsini, P.; Mosca, A.; Ceraso, A.; Deste, G.; Vita, A. Evidence-Based Integrated Intervention in Patients with Schizophrenia: A Pilot Study of Feasibility and Effectiveness in a Real-World Rehabilitation Setting. *Int. J. Environ. Res. Public Health* **2020**, *17*, 3352. [CrossRef] [PubMed]
29. Vita, A.; Barlati, S.; Ceraso, A.; Deste, G.; Nibbio, G.; Wykes, T. Acceptability of cognitive remediation for schizophrenia: A systematic review and meta-analysis of randomized controlled trials. *Psychol. Med.* **2022**, 1–11. [CrossRef]
30. Vancampfort, D.; Probst, M.; Skjaerven, L.H.; Catalán-Matamoros, D.; Gyllensten, A.L.; Gómez-Conesa, A.; Ijntema, R.; De Hert, M. Systematic Review of the Benefits of Physical Therapy Within a Multidisciplinary Care Approach for People With Schizophrenia. *Phys. Ther.* **2012**, *92*, 11–23. [CrossRef]
31. Dauwan, M.; Begemann, M.J.H.; Heringa, S.M.; Sommer, I. Exercise Improves Clinical Symptoms, Quality of Life, Global Functioning, and Depression in Schizophrenia: A Systematic Review and Meta-analysis. *Schizophr. Bull.* **2015**, *42*, 588–599. [CrossRef]
32. Firth, J.; Cotter, J.; Elliott, R.; French, P.; Yung, A.R. A systematic review and meta-analysis of exercise interventions in schizophrenia patients. *Psychol. Med.* **2015**, *45*, 1343–1361. [CrossRef]
33. Firth, J.; Cotter, J.; Carney, R.; Yung, A.R. The pro-cognitive mechanisms of physical exercise in people with schizophrenia. *Br. J. Pharmacol.* **2017**, *174*, 3161–3172. [CrossRef]
34. Firth, J.; Stubbs, B.; Rosenbaum, S.; Vancampfort, D.; Malchow, B.; Schuch, F.; Elliott, R.; Nuechterlein, K.H.; Yung, A.R. Aerobic Exercise Improves Cognitive Functioning in People With Schizophrenia: A Systematic Review and Meta-Analysis. *Schizophr. Bull.* **2016**, *43*, 546–556. [CrossRef] [PubMed]
35. Li, J.; Shen, J.; Wu, G.; Tan, Y.; Sun, Y.; Keller, E.; Jiang, Y.; Wu, J. Mindful exercise versus non-mindful exercise for schizophrenia: A systematic review and meta-analysis of randomized controlled trials. *Complement. Ther. Clin. Pract.* **2018**, *32*, 17–24. [CrossRef] [PubMed]
36. van der Stouwe, E.; van Busschbach, J.; de Vries, B.; Cahn, W.; Aleman, A.; Pijnenborg, G. Neural correlates of exercise training in individuals with schizophrenia and in healthy individuals: A systematic review. *NeuroImage Clin.* **2018**, *19*, 287–301. [CrossRef] [PubMed]
37. Dauwan, M.; Begemann, M.J.H.; Slot, M.I.E.; Lee, E.H.M.; Scheltens, P.; Sommer, I.E.C. Physical exercise improves quality of life, depressive symptoms, and cognition across chronic brain disorders: A transdiagnostic systematic review and meta-analysis of randomized controlled trials. *J. Neurol.* **2019**, *268*, 1222–1246. [CrossRef] [PubMed]
38. Fernández-Abascal, B.; Suárez-Pinilla, P.; Cobo-Corrales, C.; Crespo-Facorro, B.; Suárez-Pinilla, M. In-and outpatient lifestyle interventions on diet and exercise and their effect on physical and psychological health: A systematic review and meta-analysis of randomised controlled trials in patients with schizophrenia spectrum disorders and first episode of psychosis. *Neurosci. Biobehav. Rev.* **2021**, *125*, 535–568. [CrossRef] [PubMed]
39. Tan, B.-L.; King, R. The effects of cognitive remediation on functional outcomes among people with schizophrenia: A randomised controlled study. *Aust. N. Z. J. Psychiatry* **2013**, *47*, 1068–1080. [CrossRef]
40. Oertel-Knöchel, V.; Mehler, P.; Thiel, C.; Steinbrecher, K.; Malchow, B.; Tesky, V.; Ademmer, K.; Prvulovic, D.; Banzer, W.; Zopf, Y.; et al. Effects of aerobic exercise on cognitive performance and individual psychopathology in depressive and schizophrenia patients. *Eur. Arch. Psychiatry Clin. Neurosci.* **2014**, *264*, 589–604. [CrossRef] [PubMed]
41. Malchow, B.; Keller, K.; Hasan, A.; Dörfler, S.; Schneider-Axmann, T.; Hillmer-Vogel, U.; Honer, W.G.; Schulze, T.G.; Niklas, A.; Wobrock, T.; et al. Effects of Endurance Training Combined With Cognitive Remediation on Everyday Functioning, Symptoms, and Cognition in Multiepisode Schizophrenia Patients. *Schizophr. Bull.* **2015**, *41*, 847–858. [CrossRef]
42. Nuechterlein, K.H.; Ventura, J.; McEwen, S.C.; Gretchen-Doorly, D.; Vinogradov, S.; Subotnik, K.L. Enhancing Cognitive Training Through Aerobic Exercise After a First Schizophrenia Episode: Theoretical Conception and Pilot Study. *Schizophr. Bull.* **2016**, *42* (Suppl. S1), S44–S52. [CrossRef] [PubMed]

43. Choi, J.; Taylor, B.; Fiszdon, J.M.; Kurtz, M.M.; Tek, C.; Dewberry, M.J.; Haber, L.C.; Shagan, D.; Assaf, M.; Pearlson, G.D. The synergistic benefits of physical and cognitive exercise in schizophrenia: Promoting motivation to enhance community effectiveness. *Schizophr. Res. Cogn.* **2019**, *19*, 100147. [CrossRef] [PubMed]
44. Nuechterlein, K.H.; McEwen, S.C.; Ventura, J.; Subotnik, K.L.; Turner, L.R.; Boucher, M.; Casaus, L.R.; Distler, M.G.; Hayata, J.N. Aerobic exercise enhances cognitive training effects in first-episode schizophrenia: Randomized clinical trial demonstrates cognitive and functional gains. *Psychol. Med.* **2022**, 1–11. [CrossRef] [PubMed]
45. Dai, Y.; Ding, H.; Lu, X.; Wu, X.; Xu, C.; Jiang, T.; Ming, L.; Xia, Z.; Song, C.; Shen, H.; et al. CCRT and aerobic exercise: A randomised controlled study of processing speed, cognitive flexibility, and serum BDNF expression in schizophrenia. *Schizophrenia* **2022**, *8*, 84. [CrossRef] [PubMed]
46. Vita, A.; Barlati, S. The Implementation of Evidence-Based Psychiatric Rehabilitation: Challenges and Opportunities for Mental Health Services. *Front. Psychiatry* **2019**, *10*. [CrossRef] [PubMed]
47. Wykes, T. Cognitive remediation—Where are we now and what should we do next? *Off. J. Ital. Soc. Psychopathol.* **2018**, *24*, 57–61.
48. Deste, G.; Barlati, S.; Cacciani, P.; DePeri, L.; Poli, R.; Sacchetti, E.; Vita, A. Persistence of effectiveness of cognitive remediation interventions in schizophrenia: A 1-year follow-up study. *Schizophr. Res.* **2015**, *161*, 403–406. [CrossRef] [PubMed]
49. Horan, W.P.; Dolinsky, M.; Lee, J.; Kern, R.S.; Hellemann, G.; Sugar, C.A.; Glynn, S.M.; Green, M.F. Social Cognitive Skills Training for Psychosis With Community-Based Training Exercises: A Randomized Controlled Trial. *Schizophr. Bull.* **2017**, *44*, 1254–1266. [CrossRef]
50. Tan, S.; Zhu, X.; Fan, H.; Tan, Y.; Yang, F.; Wang, Z.; Zhao, Y.; Fan, F.; Guo, J.; Li, Z.; et al. Who will benefit from computerized cognitive remediation therapy? Evidence from a multisite randomized controlled study in schizophrenia. *Psychol. Med.* **2019**, *50*, 1633–1643. [CrossRef]
51. Vaskinn, A.; Løvgren, A.; Egeland, M.K.; Feyer, F.K.; Østefjells, T.; Andreassen, O.A.; Melle, I.; Sundet, K. A randomized controlled trial of training of affect recognition (TAR) in schizophrenia shows lasting effects for theory of mind. *Eur. Arch. Psychiatry Clin. Neurosci.* **2019**, *269*, 611–620. [CrossRef]
52. Correll, C.U.; Ismail, Z.; McIntyre, R.S.; Rafeyan, R.; Thase, M.E. Patient Functioning, Life Engagement, and Treatment Goals in Schizophrenia. *J. Clin. Psychiatry* **2022**, *83*, 42345. [CrossRef]
53. Mohr, P.; Galderisi, S.; Boyer, P.; Wasserman, D.; Arteel, P.; Ieven, A.; Karkkainen, H.; Pereira, E.; Guldemond, N.; Winkler, P.; et al. Value of schizophrenia treatment I: The patient journey. *Eur. Psychiatry* **2018**, *53*, 107–115. [CrossRef] [PubMed]

brain sciences

MDPI

Article

A Large Video Set of Natural Human Actions for Visual and Cognitive Neuroscience Studies and Its Validation with fMRI

Burcu A. Urgen [1,2,3], Hilal Nizamoğlu [2,3,4], Aslı Eroğlu [2,3] and Guy A. Orban [5,*]

1 Department of Psychology, Bilkent University, 06800 Ankara, Türkiye
2 Interdisciplinary Neuroscience Graduate Program, Bilkent University, 06800 Ankara, Türkiye
3 Aysel Sabuncu Brain Research Center and National Magnetic Resonance Research Center (UMRAM), Bilkent University, 06800 Ankara, Türkiye
4 Department of Psychology, Justus Liebig University in Giessen, 35394 Giessen, Germany
5 Department of Medicine and Surgery, University of Parma, 43125 Parma, Italy
* Correspondence: guy.orban@kuleuven.be

Abstract: The investigation of the perception of others' actions and underlying neural mechanisms has been hampered by the lack of a comprehensive stimulus set covering the human behavioral repertoire. To fill this void, we present a video set showing 100 human actions recorded in natural settings, covering the human repertoire except for emotion-driven (e.g., sexual) actions and those involving implements (e.g., tools). We validated the set using fMRI and showed that observation of the 100 actions activated the well-established action observation network. We also quantified the videos' low-level visual features (luminance, optic flow, and edges). Thus, this comprehensive video set is a valuable resource for perceptual and neuronal studies.

Keywords: vision; action observation; fMRI; action videos; naturalistic stimuli; cognitive neuroscience; systems neuroscience

check for updates

Citation: Urgen, B.A.; Nizamoğlu, H.; Eroğlu, A.; Orban, G.A. A Large Video Set of Natural Human Actions for Visual and Cognitive Neuroscience Studies and Its Validation with fMRI. *Brain Sci.* **2023**, *13*, 61. https://doi.org/10.3390/brainsci13010061

Academic Editors: Daniele Corbo and Sien Hu

Received: 30 November 2022
Revised: 14 December 2022
Accepted: 22 December 2022
Published: 29 December 2022

1. Introduction

One of the most important skills social organisms possess is the ability to perceive the actions of conspecifics in their environment. This skill has a survival value since it allows the organisms to take the appropriate action based on what they perceive. For instance, if somebody smiles at you, you probably smile back but if that person is about to attack you, you will probably want to flee. Therefore, the perceptual and neural mechanisms of action perception have received great interest from psychologists, systems, and cognitive neuroscientists.

Although there is a growing body of research in observed action processing [1], the range of actions used in such empirical studies has been limited. Since the beginning of 1990s, the majority of neuroscience studies have used *grasping* as the exemplary action to study action observation [1]. More recent work has extended this work by introducing other action categories, such as locomotion actions (e.g., walking), communicative actions (e.g., gestures), self-directed actions (e.g., scratching one's own body), interaction actions (e.g., hugging), vocal actions (e.g., singing), or other manipulative actions (e.g., pushing an object) [2–5]. This body of work has demonstrated the significance of studying action categories other than grasping as the neural activations for different actions were localized in different brain regions especially at the level of parietal cortex. However, each individual study created its own limited set of action videos, and it has been difficult to make comparisons across studies due to variations in terms of actors, scenes, video durations, and video quality (e.g., frame per seconds or resolution). Given the richness of the human action repertoire, the aim of the current study is to introduce a large video set of actions performed in natural settings, covering the entire human repertoire and make it available to researchers working primarily in the fields of visual, cognitive neuroscience and psychology. We believe that such an effort is necessary and timely given recent work, indicated above, that shows differences in neural representations of different actions [2–5], the move to use more naturalistic stimuli in neuroscience studies [6,7], and the availability of multivariate pattern analysis techniques to analyze experimental data that has a rich set of conditions, i.e., a wide variety of actions [8], instead of a few of them, as is traditionally done in previous studies. It is important to note that in creating this database, we aim to address the shortcomings of earlier attempts, such as those that provided action stimuli in the form of point-light displays (See Table 1 in [9], as they do not have high ecological validity, and the ones that come from the computer vision community ([10,11]), as they are usually unconstrained in terms of actors, context, camera angle, and movements.

The action set presented in the current paper has two major features. It represents, as far as we know, the first attempt in visual and cognitive neuroscience to systematically encompasses the entire human behavioral repertoire, with the exception of emotion-driven actions (e.g., sexual) and those involving implements (e.g., using a tool or driving a car). Second, it aims to concentrate on actions that are evolutionarily old, i.e., actions to which the human brain would have adapted during evolution. Thus, all videos were recorded in natural settings (beaches, parks, riverbanks) avoiding artificial structures in the background, and using natural objects as targets of the actions (e.g., we used stones, fruits, and pieces of wood instead of man-made items). These specifications define a homogeneous group of action exemplars and distinguish our set from the large set of 80 atomic actions, collected by the Google research group from the internet [12].

Crucially, the action exemplars in this set are unrelated to any a-priori categories. Such a stimulus-driven approach, inspired by an fMRI study of voxel-level selectivity for the meanings of words [13], can be considered complementary to the earlier studies in which the action exemplars were selected a priori as part of a single class (e.g., dragging, grasping, dropping, and pushing considered to be manipulative actions in earlier studies) [3,5,14]. On the other hand, it is important to note that, unlike the studies that use continuous natural movies (e.g., [13,15]), our stimuli set includes human actions without the clutter of other movements such as those of objects or other actors. Therefore, it constitutes a more suitable dataset for researchers who would like to study visual action perception and processing in humans.

2. Materials and Methods

In this section, we describe the stimulus set and the post-processing of the videos. The stimulus set is freely available and can be downloaded from https://osf.io/u62bp/?view_only=393a2924aa05461394fe9f3171863b94 (accessed on 29 December 2022). We also carried out a validation study with fMRI to show that our stimuli drive the regions established to be associated with the processing of actions, also known as the Action Observation Network [1,7,8]. The fMRI data can also be downloaded from the same link above.

2.1. Stimulus Set

Actors: Four actors performed the actions (2 males, 2 females). One additional female actor accompanied some actors in videos portraying actions that involved two individuals. They were undergraduate students at the University of Parma. We did not choose professional actors for two reasons. First, we wanted to record actions that were as natural as possible, without stylized movements that could be introduced by professional actors. Second, we wanted to have variability in the body movements which reflected individual differences. Before the recording of each action video, all actors were directed concerning how to perform the action and instructed to perform the action as naturally as possible within 3 s. For each action, several recordings were taken one after the other to make sure that the action was performed as intended in terms of timing and naturalness, and the best recording was chosen during the post-processing of videos. The actors were paid for their participation in the recordings and gave informed consent for their videos to be published in scientific journals.

Actions: One hundred different actions were recorded for several seconds (at least 3 s) and each action was performed by 3 of the 4 actors. So, in total, we recorded 300 videos with a fixed camera. Actions involved various effectors including the fingers, hand, arm, foot, mouth, upper body, or full body. One or 2 actors were portrayed in each video. When two actors were present, one could target the other with his action, or the two could interact. The actions were recorded from a lateral viewpoint. A sample frame from each video is presented in Figure 1. Table 1 lists the actions shown in the 100 videos and the actors who perform each of them. The different actions are described in the Supplementary Information.

Figure 1. Sample frames from the stimulus set, one frame from each action exemplar. The first row corresponds to action exemplars 1–10, the second row corresponds to action exemplars 11–20, and so on. The action numbers refer to the numbers in Table 1.

Table 1. Number and name of action exemplars, with the actors performing them (M1: male actor 1, M2: male actor 2, F1: female actor 1, F2: female actor 2).

Action Exemplar No	Action Exemplar Name	Actors
1	Measuring with fingers	F1, F2, M1
2	Shouting	F1, M1, M2
3	Carrying with head and hands	F1, M1, M2
4	Caressing another person	F1, F2, M1
5	Free style swimming	F2, M1, M2

Table 1. *Cont.*

Action Exemplar No	Action Exemplar Name	Actors
6	Kicking wood with feet	F2, M1, M2
7	Dragging	F1, F2, M2
8	Reaching	F1, F2, M2
9	Measuring a long distance with feet	F1, F2, M1
10	Crushing a leaf with fingers	F1, F2, M2
11	Fanning with leaf	F1, F2, M2
12	Pushing a small stone	F1, F2, M2
13	Dropping a small stone	F1, F2, M2
14	Ridiculing another person	F1, F2, M1
15	Massaging own cheek	F1, F2, M2
16	Scratching own cheek	F1, F2, M2
17	Swallowing	F1, F2, M1
18	Yawning with hand	F1, F2, M2
19	Licking an orange	F1, F2, M2
20	Gazing at an object	F1, F2, M1
21	Peeling a fruit	F1, F2, M1
22	Filling a hole with hand	F1, F2, M1
23	Hitting own cheek	F1, F2, M1
24	Swimming back style	F1, M1, M2
25	Displacing wood	F1, F2, M2
26	Weighing an object with one hand	F1, F2, M2
27	Climbing down a tree	F1, F2, M1
28	Whistling	F2, M1, M2
29	Measuring with hands	F1, F2, M1
30	Picking a fruit from a tree	F2, M1, M2
31	Kicking horizontally	F2, M1, M2
32	Kicking vertically	F2, M1, M2
33	Carrying with head	F1, M1, M2
34	Blowing a leaf	F1, M1, M2
35	Chasing another person	F1, M1, M2
36	Struggling	F2, M1, M2
37	Waving goodbye	F1, M1, M2
38	Beating with a piece of wood	F1, M1, M2
39	Carrying with shoulder and hand	F1, F2, M2
40	Reprimanding a person	F2, M1, M2
41	Biting a banana	F2, M1, M2
42	Fighting with another person	F2, M1, M2
43	Washing own body	F2, M1, M2
44	Foraging	F1, M1, M2
45	Stretching own body	F1, M1, M2
46	Writing with fingers	F1, M1, M2

Table 1. *Cont.*

Action Exemplar No	Action Exemplar Name	Actors
47	Charging to attack	F1, F2, M1
48	Diving	F2, M1, M2
49	Pointing nearby	F2, M1, M2
50	Squeezing an orange	F1, F2, M2
51	Forbidding with fingers	F2, M1, M2
52	Wrapping a stone	F1, M1, M2
53	Grasping	F2, M1, M2
54	Carrying on shoulder	F1, F2, M1
55	Running	F1, F2, M2
56	Burying in the sand	F2, M1, M2
57	Stopping a person	F1, F2, M2
58	Pushing a person	F1, M1, M2
59	Kissing a person	F1, F2, M1
60	Throwing and catching a small piece of wood	F1, M1, M2
61	Caressing own cheek	F2, M1, M2
62	Overtaking an obstacle	F1, F2, M1
63	Touching another person on the shoulder	F1, F2, M2
64	Building pyramid from sand	F1, F2, M1
65	Hiding an object behind back	F2, M1, M2
66	Washing fruit	F1, F2, M1
67	Carrying with hands	F1, F2, M2
68	Walking	F1, F2, M2
69	Marching	F1, F2, M1
70	Rubbing own cheek	F1, M1, M2
71	Throwing nearby	F1, M1, M2
72	Masticating	F2, M1, M2
73	People meeting	F2, M1, M2
74	Climbing up a tree	F1, F2, M2
75	Dancing with another person	F2, M1, M2
76	Getting up	F1, M1, M2
77	Crawling	F1, M1, M2
78	Spitting a piece of banana	F1, F2, M1
79	Doing gymnastics with both feet and arms	F1, F2, M2
80	Pushing a large object	F2, M1, M2
81	Rolling body sidewise	F1, F2, M1
82	Walking on hand and knees	F1, M1, M2
83	Laughing together with another person	F2, M1, M2
84	Carrying with one hand	F1, F2, M1
85	Singing a song	F1, F2, M1
86	Weighing an object with two hands	F1, M1, M2
87	Pointing distantly	F1, F2, M2

Table 1. *Cont.*

Action Exemplar No	Action Exemplar Name	Actors
88	Drinking with hands	F2, M1, M2
89	Massaging another person	F1, F2, M1
90	Drinking with mouth	F1, F2, M2
91	Measuring height with own body	F1, M1, M2
92	Pinching off piece of banana	F1, F2, M1
93	Erasing	F1, F2, M2
94	Hugging a person (passive)	F1, M1, M2
95	Speaking with another person	F1, F2, M1
96	Rotating a stone	F2, M1, M2
97	Measuring a short distance with feet	F1, M1, M2
98	Hugging each other	F1, M1, M2
99	Digging a hole with a hand	F2, M1, M2
100	Throwing far	F1, F2, M1

2.2. *Post-Processing of the Videos*

The videos were recorded using Panasonic HCX 900 camcorders. After recording, the videos were edited using Final Cut Pro software and 3-s clips were made. The frame rate of these videos was 50 fps, so each video consisted of 150 frames. The size (height and width) of the frames was set to 314 × 410 pixels. The 3-s videos were then exported in .avi format and compressed using MPlayer's *mencoder* command (http://www.mplayerhq.hu/, accessed on 29 December 2022).

The videos portray different action exemplars in natural settings, which entails variations in low-level features such as luminance, motion, or edges. We quantified those variables for each video. Hence, they can be used as variables of no interest in the experimental designs to minimize the effects of such low-level factors.

2.3. *Data Validation*

We validated our stimuli set with an fMRI experiment. Four human subjects participated in our study (2 females and 2 males; Mean Age: 26.5). Ethical approval was received from the Human Research Ethics Committee of Bilkent University.

2.3.1. fMRI Experiment

Each participant underwent two fMRI sessions, each having 8 runs. In each session, the 100 action exemplar videos were split across odd and even runs. In each of the odd runs (i.e., runs one, three, five, and seven), the first 50 action exemplars were shown in a random order as mini-blocks of three video versions of the same action presented consecutively. Each video lasted 3 s, and thus, each mini-block was shown for a total of 9 s. In each of the even runs (i.e., runs two, four, six, and eight), the other 50 action exemplars were shown as randomly ordered mini-blocks of three video versions as in the odd runs. So, in total, each of the 100 action exemplars was presented 24 times across the two sessions. The order of the mini-blocks and video versions in each mini-block was randomized across different runs. An inter-stimulus interval ranging between 1–2 s was included in between the mini-blocks. The total duration of each run was 553.36 s.

In order to keep their attention throughout the runs, a question was asked in each repetition cycle about the video that was just presented (e.g., "Was it climbing up a tree?") with a simple yes or no button-press response time period of 3 s. The periods of question were not included in the analysis.

2.3.2. fMRI Data Acquisition

Participants were scanned at National Magnetic Resonance Research Center (UMRAM) in Bilkent University by using a 3T Siemens TimTrio MR scanner with a 32-channel phase array head coil. In order to minimize head movement, relevant foam paddings were put under their skull, around their neck, and under their legs. Stimuli were presented on an MR-compatible LCD screen (TELEMED, 60 Hz refresh rate, 800 $\times$ 600 pixel, 32 inches) and seen through a mirror system mounted on top of the head coil that is 168 cm away.

A high-resolution T1-weighted anatomical image covering the entire brain was acquired before the functional scans using the following acquisition parameters: TE = 2.92 ms, TR = 2.6 s, flip angle = 12°, Acceleration factor = 2, 176 sagittal slices with 1 mm $\times$ 1 mm $\times$ 1 mm resolution). Later on, for each of the eight experimental runs, functional images were acquired using echo-planar imaging (EPI) sequence (TR = 3 s, TE = 30 ms, flip angle = 90°, 96 $\times$ 96 matrix with FOV 240, 49 horizontal slices with 2.5 mm slice thickness). Each run started with the collection of 5 dummy scans to ensure that MR signal reached a steady state.

2.3.3. fMRI Data Preprocessing

Results included in this paper are based on fMRI data preprocessed using fMRIPrep 20.1.1 ([9]; RRID:SCR_016216), which is based on Nipype 1.5.0 ([10]; RRID:SCR_002502).

Anatomical Data Preprocessing

A total of 2 T1-weighted (T1w) images were found within the input BIDS dataset. All of them were corrected for intensity non-uniformity (INU) with N4BiasFieldCorrection [16], distributed with ANTs 2.2.0 ([17], RRID:SCR_004757). The T1w-reference was then skull-stripped with a Nipype implementation of the antsBrainExtraction.sh workflow (from ANTs), using OASIS30ANTs as target template. Brain tissue segmentation of cerebrospinal fluid (CSF), white-matter (WM), and gray-matter (GM) was performed on the brain-extracted T1w using fast (FSL 5.0.9, RRID:SCR_002823, [18]). A T1w-reference map was computed after registration of 2 T1w images (after INU-correction) using mri_robust_template (FreeSurfer 6.0.1, [19]). Brain surfaces were reconstructed using recon-all (FreeSurfer 6.0.1, RRID:SCR_001847, [20]), and the brain mask estimated previously was refined with a custom variation of the method to reconcile ANTs-derived and FreeSurfer-derived segmentations of the cortical gray-matter of Mindboggle (RRID:SCR_002438, [21]). Volume-based spatial normalization to one standard space (MNI152NLin2009cAsym) was performed through nonlinear registration with antsRegistration (ANTs 2.2.0), using brain-extracted versions of both T1w reference and the T1w template. The following template was selected for spatial normalization: ICBM 152 Nonlinear Asymmetrical template version 2009c ([22], RRID:SCR_008796; TemplateFlow ID: MNI152NLin2009cAsym).

Functional Data Preprocessing

For each of the 16 BOLD runs recorded per subject (across all sessions), the following preprocessing was performed. First, a reference volume and its skull-stripped version were generated using a custom methodology of *fMRIPrep*. The BOLD reference was then co-registered to the T1w reference using bbregister (FreeSurfer) which implements boundary-based registration [23]. Co-registration was configured with six degrees of freedom. Head-motion parameters with respect to the BOLD reference (transformation matrices, and six corresponding rotation and translation parameters) are estimated before any spatiotemporal filtering using mcflirt (FSL 5.0.9, [24]). BOLD runs were slice-time corrected using 3dTshift from AFNI 20160207 ([25], RRID:SCR_005927). The BOLD time-series were resampled to surfaces on the following spaces: *fsaverage5*. The BOLD time-series (including slice-timing correction when applied) were resampled onto their original, native space by applying a single, composite transform to correct for head-motion and susceptibility distortions. These resampled BOLD time-series will be referred to as *preprocessed BOLD in original space*, or just *preprocessed BOLD*. The BOLD time-series were resampled into standard space, generating a *preprocessed BOLD run in ['MNI152NLin2009cAsym'] space*. First, a reference volume and its skull-stripped version were generated using a custom methodology of *fMRIPrep*. Several confounding time-series were calculated based on the *preprocessed BOLD*: framewise displacement (FD), DVARS and three region-wise global signals. FD and DVARS are calculated for each functional run, both using their implementations in *Nipype* (following the definitions by [26]). The three global signals are extracted within the CSF, the WM, and the whole-brain masks. Additionally, a set of physiological regressors were extracted to allow for component-based noise correction (*CompCor*, [27]). Principal components are estimated after high-pass filtering the preprocessed BOLD time-series (using a discrete cosine filter with 128s cut-off) for the two *CompCor* variants: temporal (tCompCor) and anatomical (aCompCor). tCompCor components are then calculated from the top 5% variable voxels within a mask covering the subcortical regions. This subcortical mask is obtained by heavily eroding the brain mask, which ensures it does not include cortical GM regions. For aCompCor, components are calculated within the intersection of the aforementioned mask and the union of CSF and WM masks calculated in T1w space, after their projection to the native space of each functional run (using the inverse BOLD-to-T1w transformation). Components are also calculated separately within the WM and CSF masks. For each CompCor decomposition, the k components with the largest singular values are retained, such that the retained components' time series are sufficient to explain 50 percent of variance across the nuisance mask (CSF, WM, combined, or temporal). The remaining components are dropped from consideration. The head-motion estimates calculated in the correction step were also placed within the corresponding confounds file. The confound time series derived from head motion estimates and global signals were expanded with the inclusion of temporal derivatives and quadratic terms for each [28]. Frames that exceeded a threshold of 0.5 mm FD or 1.5 standardized DVARS were annotated as motion outliers. All resamplings can be performed with a *single interpolation step* by composing all the pertinent transformations (i.e., head-motion transform matrices, susceptibility distortion correction when available, and co-registrations to anatomical and output spaces). Gridded (volumetric) resamplings were performed using antsApplyTransforms (ANTs), configured with Lanczos interpolation to minimize the smoothing effects of other kernels [29]. Non-gridded (surface) resamplings were performed using mri_vol2surf (FreeSurfer).

Many internal operations of fMRIPrep use Nilearn 0.5.2 ([30], RRID:SCR_001362), mostly within the functional processing workflow. For more details of the pipeline, see the section corresponding to workflows in fMRIPrep's documentation.

2.3.4. Activation Maps for Observed Actions

An important aim of the present study is to show that our novel stimuli set could drive the action observation network as previously identified, and to investigate whether it could be extended given that previous work used a limited number of actions. This is achieved by univariate analysis, which reveals the activation map of actions.

Following pre-processing, we ran a general linear model (GLM) composed of 8 regressors, including 1 regressor for all the action videos, 6 motion regressors (3 translations and 3 rotations), and a constant factor. All regressors were convolved with the default canonical hemodynamic response function in SPM12. The activation map was generated by the beta value corresponding to the action videos.

2.3.5. Definition of ROIs

We defined three a priori ROIs that represent the three levels of the Action Observation Network based on previous work: Lateral occipito-temporal cortex (LOTC), posterior parietal cortex (PPC), and premotor cortex (PMC) (Figure 2). LOTC included the regions of the action observation network based on the activation map (threshold at $p < 0.001$) of [3]. These regions included (1) the MT cluster [31,32], (2) a portion extending dorsally from the MT cluster onto the middle temporal gyrus referred to as MTG, and (3) a portion extending ventrally from the MT cluster onto occipital temporal sulcus that we refer to as

OTS. The MTG and OTS correspond to the upper bank and lower bank of the macaque STS anterior to the MT cluster [33], the two regions which project to the parietal level of the action observation network in the macaque [34]. They are therefore considered to be input regions for the next level, PPC.

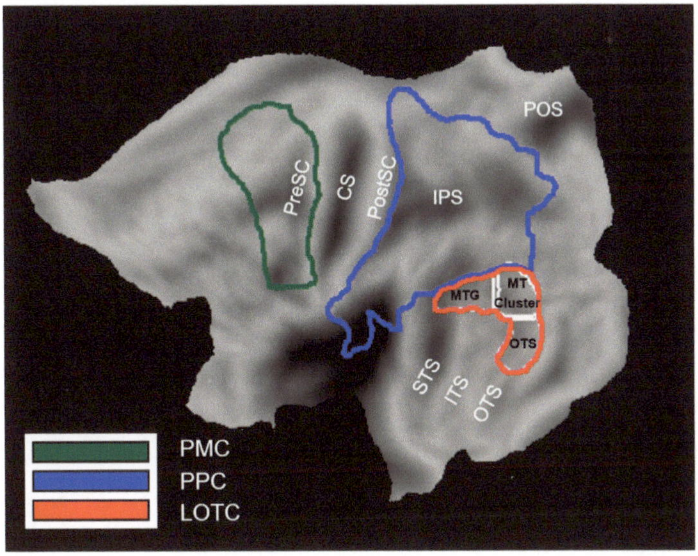

Figure 2. A priori ROIs of the Action Observation Network: LOTC (yellow line), PPC (blue line), PMC (green line). LOTC consists of three sub-regions including the MT cluster, MTG, and OTS. The anatomical landmarks are indicated in black: CS (Central sulcus), PreSC (Pre-central sulcus), PostCS (Post-central sulcus), IPS (intra-parietal sulcus), POS (parieto-occipital sulcus), STS (superior temporal sulcus), ITS (inferior temporal sulcus), OTS (occipito-temporal sulcus).

The PPC included the cyto-architectonic regions of SPL, IPS, IPL, and parietal operculum [35–38]. Its posterior boundary coursed between V6 and V6A [39] and extended between V3D and V7 [40].

The PMC included the cyto-architectonic supplementary, dorsal, and ventral premotor areas taken from the Anatomy software, but the ventral part was extended in the rostral direction to include regions that are responsive to observed actions according to [41,42].

All ROIs were defined on flat maps in Caret software [43]. Only ROIs in the left hemisphere were considered, because we now have considerable evidence that the position of the actor in the visual field affects the lateralization of PPC activation in the action observation paradigm [41], with the activation being contralateral to the hemifield in which the actor is shown. In almost all videos (with only a few exceptions represented by the 2 swimming exemplars, diving, and rolling side-way), the body of the actor was either motionless in the right visual field or remained in this field where actions (e.g., walking) implied horizontal motion to the left, as the camera partially followed the action.

3. Results

3.1. Post-Processing of Video Stimuli

We quantified low-level visual features for each video, including luminance, motion, and edges. These features can be used as variables of no interest in fMRI experiments to minimize the effects of those low-level factors. The MATLAB codes that generate these features as well as the output of these codes for each video can be downloaded from https://osf.io/u62bp/?view_only=393a2924aa05461394fe9f3171863b94 (accessed on 29 December 2022).

Luminance: For a given video, we first found the average of the RGB pixel values in all 150 frames (temporal averaging), and then calculated the average of the pixels in the averaged frame (spatial averaging). Thus, we obtained a single luminance value characterizing each video. Figure 3A shows the histogram of the average luminance values over the 100 action exemplars (further averaged over the three versions of an exemplar performed by different actors), and Figure 3B shows the values for the different actions.

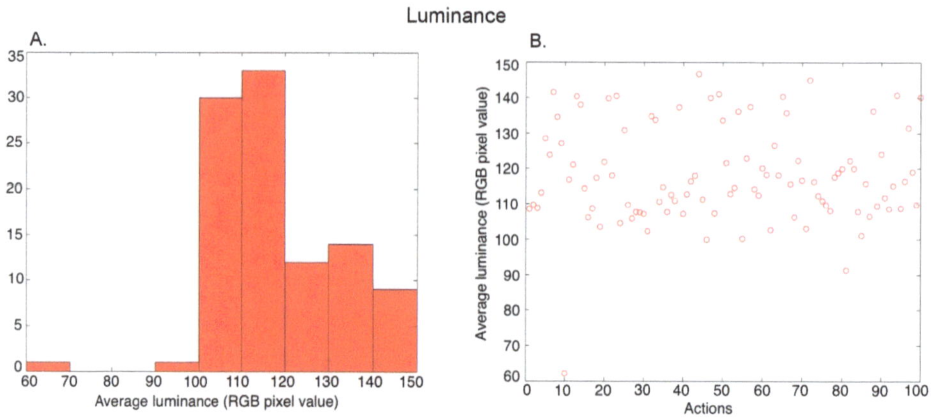

Figure 3. Characterization of the average luminance in the videos. (**A**) Histogram of the average luminance values across 100 action exemplars (averaged over the 3 versions of each action performed by different actors), (**B**) Average luminance of the 100 action exemplars.

Motion: We computed the mean speed in each video using an algorithm by [44]. The local motion vector was computed for each pixel in the image on a frame-by-frame basis. We performed temporal averaging (across frames) and spatial averaging (within a frame) (see Luminance above) to obtain one value for each video. Figure 4A shows the histogram of the average speed values covering the 100 action exemplars (further averaged over the three versions of each exemplar performed by different actors), and Figure 4B shows the values for the different actions.

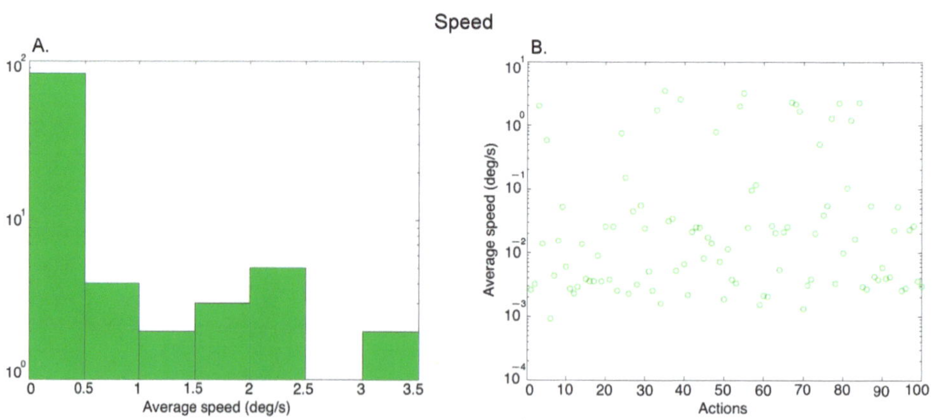

Figure 4. Characterization of the average speed in the videos. (**A**) Histogram of the average speed values across 100 action exemplars (averaged over the 3 versions of each action performed by different actors), (**B**) Average speed of the 100 action exemplars.

Edges (Form): We passed each video frame of each action exemplar through a set of Gabor filters [45] to extract the edge information. We used Gabors of 5 scales and 8 orientations using the Gabor filtering algorithm described in [46]. We performed temporal averaging (across frames) and spatial averaging (within a frame) (see Luminance above) to obtain one output value for each video. Figure 5A shows the histogram of the average edge information covering the 100 action exemplars (further averaged over the three versions of an exemplar performed by different actors), and Figure 5B shows the edge information for the different actions.

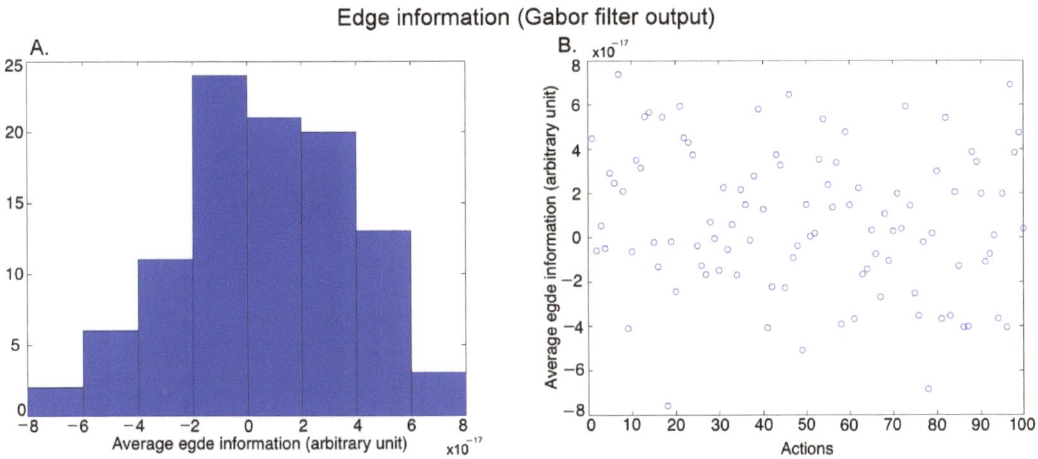

Figure 5. Characterization of the average edge information in the videos. (**A**) Histogram of the average edge information across 100 action exemplars (averaged over the 3 versions of each action performed by different actors), (**B**) Average edge information of the 100 action exemplars.

3.2. fMRI Activation Maps

Observation of the 100 actions used in the present study activated the three levels of the action observation network ($p < 0.001$ uncorrected): LOTC, PPC, and to a lesser degree PMC in the left hemisphere, as expected (Figure 6). The LOTC level included activations in the MT cluster as well as MTG and to some extent OTS, as previously defined (See Section 2.3.5). The PPC level included activations in functionally defined areas DIPSM in all participants and DIPSA in subjects 1, 2, and 4. Other PPC level activations in cyto-architectonic areas of the inferior parietal lobule (IPL) include PFcm, PGa, and PGp in subject 1; PFcm, PFm, PGa, and PGp in subject2; PFcm and PFt in subject 3, and finally PFop, PGa, and PGp in subject 4. In addition, superior parietal lobule areas in dorsal postcentral gyrus were activated in subjects 2, 3, and 4. The PMC level included areas in the anterior part of the dorsal premotor and ventral premotor cortex in Subject 2 and the posterior part of the ventral premotor cortex in Subject 3.

In addition to the ROIs of the action observation network, several other areas were activated by the observation of actions. These included mainly the early visual cortex, extending into the neighboring parieto-occipital sulcus, including V7 [47], in Subject 1. Additional activations included a medial frontal site in Subjects 1, 2, and 3 and another small cluster neighboring the parieto-occipital sulcus (POS) in Subjects 2 and 4.

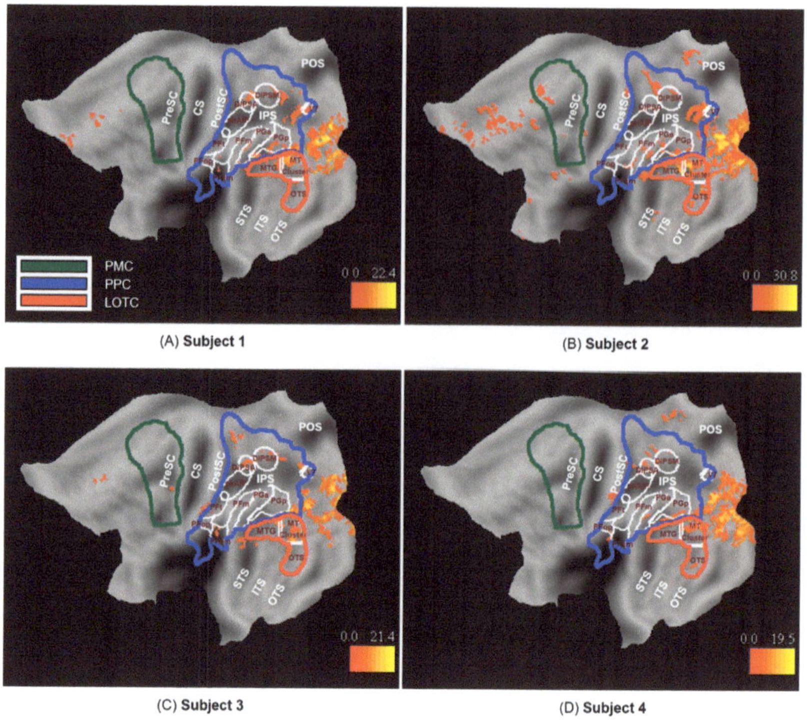

Figure 6. Activation map for observed actions ($p < 0.001$ uncorrected, k = 10) on the flattened left hemisphere of subjects (**A**) 1, (**B**) 2, (**C**) 3, and (**D**) 4.

4. Discussion

We describe a video stimulus set that consists of 100 different natural actions covering most of the human repertoire. Each action is performed by multiple actors. The low-level features of the videos were quantified, allowing them to be factored out in future experiments. Our fMRI validation study showed that the novel stimuli presented here drove the action observation network. The weak activation of the premotor level likely reflects a combination of two factors: a lower level of activation due to distance from the retina [34], as documented in many studies (e.g., [2]), and an increased selectivity of premotor voxels whereby these voxels are activated by only a small number of observed actions, typical for the present study.

To the best of our knowledge, this is the largest action database to be made available for use in psychology and cognitive neuroscience research. Earlier work used *grasping* as the exemplary action for a long-time in action observation research [1]. More recent work has introduced different action categories such as locomotion, communicative, self-directed, interaction, and vocal actions [2–5]. However, each study was constrained by a small set of action videos that was created for the purposes of that study and this made it difficult to compare the results across different studies due to variations in actors, scenes, video durations, and video quality. There are video databases that display actions in the form of point-light displays (See the list in [9]) to overcome the visual differences in the stimuli, but the shortcoming of these databases is that the stimuli are not naturalistic enough and lack ecological validity. There are some action databases with naturalistic actions, such as [48], but it focuses on actions that have emotional content. There are yet other naturalistic action databases, such as UCF50 [10] or HMDB51 [11], but their target is usually the computer vision community. Computer vision research has different constraints such as multiple cameras, camera motion, or clutter in the videos. Therefore, the databases have been created in accordance with the problems needing to be solved from a computer-vision perspective. Given that we are at the early stages of understanding the perceptual and neural mechanisms processing observed actions, it is necessary to initiate such studies using a stimulus set that is natural yet simple and sufficiently controlled to facilitate interpretation. In this respect, they differ from the set of 80 atomic actions [12].

We believe that our database will prove useful for researchers who intend to study the perceptual and neural differences between observing action exemplars such as locomotion or communicative actions, as well as interactions between two individuals, extending ongoing fMRI work in passive subjects with different action classes [2–5]. Our stimulus set can, however, be used in a much wider set of behavioral and neuroimaging studies, allowing some generic plausible models to be built for action perception. Indeed, our stimulus set can be used in an array of visual tasks. A first set are discrimination tasks, probing the identity of the observed actions, such as identification or same–different tasks [49,50]. If, as has been proposed [51], observed actions of different classes, such as manipulation or locomotion, are processed in different PPC regions, one would expect action discriminability, whether measured perceptually or in neural activity, to depend on the classes involved. A second set of tasks are classification studies probing the semantic categories of observed actions. The classification of static images has received a lot of attention comparing human and deep network performance, in an effort to model object processing in the ventral pathway [52]. This can be extended to the classification of videos to model observed action processing in the dorsal pathway. Yet another set of studies are similarity studies. Subjects have to rate how close two actions are, allowing to derive the distances between observed actions in perceptual space, which can be compared using RSA to the distance between these actions in a neural space derived from single cell recordings or fMRI activations. Such perceptual and neuronal studies would stimulate the computational modelling of observed actions. There have been a few modelling attempts to explain the neural mechanisms of observed actions [53–56], but so far these have been limited to only a few action exemplars such as locomotion or grasping and such modelling efforts would benefit from testing a larger set of actions. It is noteworthy that in the present stimulus set the actions were performed by several actors, which makes it easy to design control tasks, requiring subjects to discriminate or classify actors. Extending this further, the video set can also be used in studies for person or gender identification from body movements as we have multiple actors performing the same action with the same background.

A limitation of our stimuli set is that all actions take place in an outdoor scene. Therefore, researchers who are interested in contextual effects, such as the scene in which the action takes place, may not find sufficient variability in the stimulus set, although some actions were set on a beach, or in a lake, in addition to the grassy landscape. However, many of these actions can be performed indoors as well, and hence from an action-identity point of view, many of them can still be used in behavioral and neuroimaging experiments probing action observation.

Another limitation of the stimuli set is that we do not systematically control the emotional content of the actions. Most actions can be considered as neutral, such as locomotion, but some of them have an inherently positive valence, such as laughing, or negative valence, such as signing "no". Therefore, the set may not be optimal for studies aiming to systematically investigate the emotional content of the actions.

5. Conclusions

In summary, we believe that our stimuli set will be beneficial to the scientific community studying action perception from behavioral, neuro-scientific, and computational perspectives, particularly those who wish to move away from mere grasping and reaching as prototypical actions. As the stimulus set can be combined not only with fMRI, but also MEG, EEG, and stereo-EEG, it should find a wide range of applications from the bedside to the laboratory.

Supplementary Materials: The following supporting information can be downloaded at: https://www.mdpi.com/article/10.3390/brainsci13010061/s1.

Author Contributions: Conceptualization, B.A.U. and G.A.O.; methodology, B.A.U., H.N., A.E. and G.A.O.; software, B.A.U., H.N. and A.E.; validation, B.A.U., H.N. and A.E.; formal analysis, B.A.U., H.N. and A.E.; investigation, B.A.U., H.N., A.E. and G.A.O.; resources, B.A.U. and G.A.O.; data curation, B.A.U., H.N. and A.E.; writing—original draft preparation, B.A.U.; writing—review and editing, B.A.U., H.N., A.E. and G.A.O.; visualization, B.A.U., H.N. and A.E.; supervision, B.A.U. and G.A.O.; project administration, B.A.U. and G.A.O.; funding acquisition, B.A.U. and G.A.O. All authors have read and agreed to the published version of the manuscript.

Funding: This research was funded by ERC Advanced Grant, grant number ERC 2012-ADG 323606 to Guy A. Orban, and TUBITAK 3501 Career Grant, grant number 119K654 to Burcu A. Urgen.

Institutional Review Board Statement: The study was conducted in accordance with the Declaration of Helsinki and approved by the Human Subjects Research Ethics Committee of Bilkent University (approval number 2019_01_29_01; date 29 January 2019).

Informed Consent Statement: Informed consent was obtained from all subjects involved in the study. The data preprocessing section was used verbatim from the boilerplate text that was automatically generated by fMRIPrep with the express intention that users should copy and paste this text into their manuscripts unchanged. It is released under the CC0 license.

Data Availability Statement: The data and stimuli presented in this study are available on Open Science Framework platform: https://osf.io/u62bp/?view_only=393a2924aa05461394fe9f3171863b94 (accessed on 29 December 2022).

Acknowledgments: The authors would like to thank the actors who participated in the recording of the videos and Stefania Ferri for recording the videos.

Conflicts of Interest: The authors declare no conflict of interest.

References

1. Caspers, S.; Zilles, K.; Laird, A.R.; Eickhoff, S.B. ALE meta-analysis of action observation and imitation in the human brain. *Neuroimage* **2010**, *50*, 1148–1167. [CrossRef] [PubMed]
2. Abdollahi, R.O.; Jastorff, J.; Orban, G.A. Common and Segregated Processing of Observed Actions in Human SPL. *Cereb. Cortex* **2012**, *23*, 2734–2753. [CrossRef] [PubMed]
3. Ferri, S.; Rizzolatti, G.; Orban, G.A. The organization of the posterior parietal cortex devoted to upper limb actions: An fMRI study. *Hum. Brain Mapp.* **2015**, *36*, 3845–3866. [CrossRef] [PubMed]
4. Corbo, D.; Orban, G.A. Observing Others Speak or Sing Activates Spt and Neighboring Parietal Cortex. *J. Cogn. Neurosci.* **2017**, *29*, 1002–1021. [CrossRef]
5. Urgen, B.A.; Orban, G.A. The unique role of parietal cortex in action observation: Functional organization for communicative and manipulative actions. *NeuroImage* **2021**, *237*, 118220. [CrossRef]
6. Sonkusare, S.; Breakspear, M.; Guo, C. Naturalistic Stimuli in Neuroscience: Critically Acclaimed. *Trends Cogn. Sci.* **2019**, *23*, 699–714. [CrossRef]
7. Nastase, S.A.; Goldstein, A.; Hasson, U. Keep it real: Rethinking the primacy of experimental control in cognitive neuroscience. *Neuroimage* **2020**, *222*, 117254. [CrossRef]
8. Kriegeskorte, N.; Mur, M.; Bandettini, P.A. Representational similarity analysis—Connecting the branches of systems neuroscience. *Front. Syst. Neurosci.* **2008**, *2*, 4. [CrossRef]
9. Bidet-Ildei, C.; Francisco, V.; Decatoire, A.; Pylouster, J.; Blandin, Y. PLAViMoP database: A new continuously assessed and collaborative 3D point-light display dataset. *Behav. Res. Methods* **2022**, 1–22. [CrossRef]
10. Reddy, K.K.; Shah, M. Recognizing 50 human action categories of web videos. *Mach. Vis. Appl.* **2012**, *24*, 971–981. [CrossRef]
11. Kuehne, H.; Jhuang, H.; Garrote, E.; Poggio, T.; Serre, T. HMDB: A Large Video Database for Human Motion Recognition. In Proceedings of the 2011 International Conference on Computer Vision (ICCV), Barcelona, Spain, 6–13 November 2011.
12. Gu, C.; Sun, C.; Ross, D.A.; Vondrick, C.; Pantofaru, C.; Li, Y.; Vijayanarasimhan, S.; Toderici, G.; Ricco, S.; Sukthankar, R.; et al. AVA: A Video Dataset of Spatio-Temporally Localized Atomic Visual Actions. In Proceedings of the IEEE Conference on Computer Vision and Pattern Recognition (CVPR), Salt Lake City, UT, USA, 18–22 June 2018.
13. Huth, A.G.; Nishimoto, S.; Vu, A.T.; Gallant, J.L. A Continuous Semantic Space Describes the Representation of Thousands of Object and Action Categories across the Human Brain. *Neuron* **2012**, *76*, 1210–1224. [CrossRef] [PubMed]
14. Jastorff, J.; Begliomini, C.; Destro, M.F.; Rizzolatti, G.; Orban, G. Coding Observed Motor Acts: Different Organizational Principles in the Parietal and Premotor Cortex of Humans. *J. Neurophysiol.* **2010**, *104*, 128–140. [CrossRef] [PubMed]

15. Shahdloo, M.; Çelik, E.; Urgen, B.A.; Gallant, J.L.; Çukur, T. Task-Dependent Warping of Semantic Representations during Search for Visual Action Categories. *J. Neurosci.* **2022**, *42*, 6782–6799. [CrossRef] [PubMed]
16. Tustison, N.J.; Avants, B.B.; Cook, P.A.; Zheng, Y.; Egan, A.; Yushkevich, P.A.; Gee, J.C. N4ITK: Improved N3 bias correction. *IEEE Trans. Med. Imaging* **2010**, *29*, 1310–1320. [CrossRef]
17. Avants, B.; Epstein, C.; Grossman, M.; Gee, J. Symmetric diffeomorphic image registration with cross-correlation: Evaluating automated labeling of elderly and neurodegenerative brain. *Med. Image Anal.* **2008**, *12*, 26–41. [CrossRef]
18. Zhang, Y.; Brady, M.; Smith, S. Segmentation of brain MR images through a hidden Markov random field model and the expectation-maximization algorithm. *IEEE Trans. Med. Imaging* **2001**, *20*, 45–57. [CrossRef]
19. Reuter, M.; Rosas, H.D.; Fischl, B. Highly accurate inverse consistent registration: A robust approach. *Neuroimage* **2010**, *53*, 1181–1196. [CrossRef]
20. Dale, A.M.; Fischl, B.; Sereno, M.I. Cortical surface-based analysis. I. Segmentation and surface reconstruction. *NeuroImage* **1999**, *9*, 179–194. [CrossRef]
21. Klein, A.; Ghosh, S.S.; Bao, F.S.; Giard, J.; Häme, Y.; Stavsky, E.; Lee, N.; Rossa, B.; Reuter, M.; Neto, E.C.; et al. Mindboggling morphometry of human brains. *PLoS Comput. Biol.* **2017**, *13*, e1005350. [CrossRef]
22. Fonov, V.; Evans, A.; McKinstry, R.; Almli, C.; Collins, D. Unbiased nonlinear average age-appropriate brain templates from birth to adulthood. *Neuroimage* **2009**, *47*, S102. [CrossRef]
23. Greve, D.N.; Fischl, B. Accurate and robust brain image alignment using boundary-based registration. *Neuroimage* **2009**, *48*, 63–72. [CrossRef] [PubMed]
24. Jenkinson, M.; Bannister, P.; Brady, M.; Smith, S. Improved Optimization for the Robust and Accurate Linear Registration and Motion Correction of Brain Images. *Neuroimage* **2002**, *17*, 825–841. [CrossRef] [PubMed]
25. Cox, R.W.; Hyde, J.S. Software Tools for Analysis and Visualization of fMRI Data. *NMR Biomed. Int. J. Devoted Dev. Appl. Magn. Reson. In Vivo* **1997**, *10*, 171–178. [CrossRef]
26. Power, J.D.; Mitra, A.; Laumann, T.O.; Snyder, A.Z.; Schlaggar, B.L.; Petersen, S.E. Methods to detect, characterize, and remove motion artifact in resting state fMRI. *Neuroimage* **2014**, *84*, 320–341. [CrossRef]
27. Behzadi, Y.; Restom, K.; Liau, J.; Liu, T.T. A component based noise correction method (CompCor) for BOLD and perfusion based fMRI. *Neuroimage* **2007**, *37*, 90–101. [CrossRef]
28. Satterthwaite, T.D.; Wolf, D.H.; Ruparel, K.; Erus, G.; Elliott, M.A.; Eickhoff, S.B.; Gennatas, E.D.; Jackson, C.; Prabhakaran, K.; Smith, A.; et al. Heterogeneous impact of motion on fundamental patterns of developmental changes in functional connectivity during youth. *Neuroimage* **2013**, *83*, 45–57. [CrossRef]
29. Lanczos, C. Evaluation of noisy data. *J. Soc. Ind. Appl. Math. Ser. B Numer. Anal.* **1964**, *1*, 76–85. [CrossRef]
30. Abraham, A.; Pedregosa, F.; Eickenberg, M.; Gervais, P.; Mueller, A.; Kossaifi, J.; Gramfort, A.; Thirion, B.; Varoquaux, G. Machine learning for neuroimaging with scikit-learn. *Front. Neuroinform.* **2014**, *8*, 14. [CrossRef]
31. Tootell, R.; Reppas, J.; Kwong, K.; Malach, R.; Born, R.; Brady, T.; Rosen, B.; Belliveau, J. Functional analysis of human MT and related visual cortical areas using magnetic resonance imaging. *J. Neurosci.* **1995**, *15*, 3215–3230. [CrossRef]
32. Kolster, H.; Peeters, R.; Orban, G.A. The Retinotopic Organization of the Human Middle Temporal Area MT/V5 and Its Cortical Neighbors. *J. Neurosci.* **2010**, *30*, 9801–9820. [CrossRef]
33. Jastorff, J.; Popivanov, I.D.; Vogels, R.; Vanduffel, W.; Orban, G.A. Integration of shape and motion cues in biological motion processing in the monkey STS. *Neuroimage* **2012**, *60*, 911–921. [CrossRef] [PubMed]
34. Nelissen, K.; Borra, E.; Gerbella, M.; Rozzi, S.; Luppino, G.; Vanduffel, W.; Rizzolatti, G.; Orban, G. Action Observation Circuits in the Macaque Monkey Cortex. *J. Neurosci.* **2011**, *31*, 3743–3756. [CrossRef] [PubMed]
35. Caspers, S.; Geyer, S.; Schleicher, A.; Mohlberg, H.; Amunts, K.; Zilles, K. The human inferior parietal cortex: Cytoarchitectonic parcellation and interindividual variability. *Neuroimage* **2006**, *33*, 430–448. [CrossRef] [PubMed]
36. Eickhoff, S.B.; Grefkes, C.; Zilles, K.; Fink, G.R. The Somatotopic Organization of Cytoarchitectonic Areas on the Human Parietal Operculum. *Cereb. Cortex* **2006**, *17*, 1800–1811. [CrossRef] [PubMed]
37. Choi, H.-J.; Zilles, K.; Mohlberg, H.; Schleicher, A.; Fink, G.R.; Armstrong, E.; Amunts, K. Cytoarchitectonic identification and probabilistic mapping of two distinct areas within the anterior ventral bank of the human intraparietal sulcus. *J. Comp. Neurol.* **2006**, *495*, 53–69. [CrossRef]
38. Scheperjans, F.; Eickhoff, S.B.; Hömke, L.; Mohlberg, H.; Hermann, K.; Amunts, K.; Zilles, K. Probabilistic Maps, Morphometry, and Variability of Cytoarchitectonic Areas in the Human Superior Parietal Cortex. *Cereb. Cortex* **2008**, *18*, 2141–2157. [CrossRef]
39. Pitzalis, S.; Sereno, M.I.; Committeri, G.; Fattori, P.; Galati, G.; Patria, F.; Galletti, C. Human V6: The Medial Motion Area. *Cereb. Cortex* **2010**, *20*, 411–424. [CrossRef]
40. Abdollahi, R.O.; Kolster, H.; Glasser, M.F.; Robinson, E.C.; Coalson, T.S.; Dierker, D.; Jenkinson, M.; Van Essen, D.C.; Orban, G.A. Correspondences between retinotopic areas and myelin maps in human visual cortex. *NeuroImage* **2014**, *99*, 509–524. [CrossRef]
41. Ferri, S.; Pauwels, K.; Rizzolatti, G.; Orban, G.A. Stereoscopically Observing Manipulative Actions. *Cereb. Cortex* **2016**, *26*, 3591–3610. [CrossRef]
42. Jastorff, J.; Abdollahi, R.O.; Fasano, F.; Orban, G.A. Seeing biological actions in 3D: An fMRI study. *Hum. Brain Mapp.* **2016**, *37*, 213–219. [CrossRef]
43. Van Essen, D.C.; Drury, H.A.; Dickson, J.; Harwell, J.; Hanlon, D.; Anderson, C.H. An integrated software suite for surface-based analyses of cerebral cortex. *J. Am. Med. Inform. Assoc.* **2001**, *8*, 443–459. [CrossRef] [PubMed]

44. Pauwels, K.; Van Hulle, M.M. Optic flow from unstable sequences through local velocity constancy maximization. *Image Vis. Comput.* **2009**, *27*, 579–587. [CrossRef]
45. Olshausen, B.A.; Field, D.J. Emergence of simple-cell receptive field properties by learning a sparse code for natural images. *Nature* **1996**, *381*, 607–609. [CrossRef] [PubMed]
46. Haghighat, M.; Zonouz, S.; Abdel-Mottaleb, M. CloudID: Trustworthy cloud-based and cross-enterprise biometric identification. *Expert Syst. Appl.* **2015**, *42*, 7905–7916. [CrossRef]
47. Georgieva, S.; Peeters, R.; Kolster, H.; Todd, J.T.; Orban, G.A. The processing of three-dimensional shape from disparity in the human brain. *J. Neurosci.* **2009**, *29*, 727–742. [CrossRef]
48. Keefe, B.D.; Villing, M.; Racey, C.; Strong, S.L.; Wincenciak, J.; Barraclough, N.E. A database of whole-body action videos for the study of action, emotion, and untrustworthiness. *Behav. Res. Methods* **2014**, *46*, 1042–1051. [CrossRef]
49. Platonov, A.; Orban, G.A. Action observation: The less-explored part of higher-order vision. *Sci. Rep.* **2016**, *6*, 36742. [CrossRef]
50. Platonov, A.; Orban, G.A. Not all observed actions are perceived equally. *Sci. Rep.* **2017**, *7*, 17084. [CrossRef]
51. Orban, G.A.; Sepe, A.; Bonini, L. Parietal maps of visual signals for bodily action planning. *Anat. Embryol.* **2021**, *226*, 2967–2988. [CrossRef]
52. Zhuang, C.; Yan, S.; Nayebi, A.; Schrimpf, M.; Frank, M.C.; DiCarlo, J.J.; Yamins, D.L. Unsupervised neural network models of the ventral visual stream. *Proc. Natl. Acad. Sci. USA* **2021**, *118*, e2014196118. [CrossRef]
53. Giese, M.A.; Poggio, T. Neural mechanisms for the recognition of biological movements. *Nat. Rev. Neurosci.* **2003**, *4*, 179–192. [CrossRef] [PubMed]
54. Tan, C.; Singer, J.M.; Serre, T.; Sheinberg, D.; Poggio, T. Neural Representation of Action Sequences: How Far Can a Simple Snippet-Matching Model Take Us? In Proceedings of the Advances in Neural Information Processing Systems, Lake Tahoe, NV, USA, 5–8 December 2013; pp. 593–601.
55. Fleischer, F.; Caggiano, V.; Thier, P.; Giese, M.A. Physiologically Inspired Model for the Visual Recognition of Transitive Hand Actions. *J. Neurosci.* **2013**, *33*, 6563–6580. [CrossRef] [PubMed]
56. Theusner, S.; De Lussanet, M.; Lappe, M. Action Recognition by Motion Detection in Posture Space. *J. Neurosci.* **2014**, *34*, 909–921. [CrossRef] [PubMed]

brain sciences

MDPI

Article

Hand and Foot Selection in Mental Body Rotations Involves Motor-Cognitive Interactions

Stephan Frederic Dahm [1,2,*], **Emiko J. Muraki** [3] and **Penny M. Pexman** [3]

[1] Department of Psychology, Faculty of Psychology and Sport Science, University of Innsbruck, 6020 Innsbruck, Austria
[2] Institute of Psychology, Department of Psychology and Sports Medicine, UMIT TIROL—Private University of Health Sciences and Health Technology, 6060 Hall in Tirol, Austria
[3] Department of Psychology, Faculty of Arts, University of Calgary, Calgary, AB T2N 1N4, Canada
* Correspondence: stephan.dahm@uibk.ac.at

Abstract: Action imagery involves the mental representation of an action without overt execution, and can contribute to perspective taking, such as that required for left-right judgments in mental body rotation tasks. It has been shown that perspective (back view, front view), rotational angle (head-up, head-down), and abstractness (abstract, realistic) of the stimulus material influences speed and correctness of the judgement. The present studies investigated whether left-right judgements are more difficult on legs than on arms and whether the type of limb interacts with the other factors. Furthermore, a combined score for speed and accuracy was explored to eliminate possible tradeoffs and to obtain the best possible measure of subjects' individual ability. Study 1 revealed that the front view is more difficult than the back view because it involves a vertical rotation in perspective taking. Head-down rotations are more difficult than head-up rotations because they involve a horizontal rotation in perspective taking. Furthermore, leg stimuli are more difficult than hand stimuli, particularly in head-down rotations. In Study 2, these findings were replicated in abstract stimuli as well as in realistic stimuli. In addition, perspective taking for realistic stimuli in the back view is easier than realistic stimuli in the front view or abstract stimuli (in both perspectives). We conclude that realistic stimulus material facilitates task comprehension and amplifies the effects of perspective. By replicating previous findings, the linear speed-accuracy score was shown to be a valid measure to capture performance in mental body rotations.

Keywords: motor imagery; action imagery ability; mental action representations

check for updates

Citation: Dahm, S.F.; Muraki, E.J.; Pexman, P.M. Hand and Foot Selection in Mental Body Rotations Involves Motor-Cognitive Interactions. *Brain Sci.* **2022**, *12*, 1500. https://doi.org/10.3390/brainsci12111500

Academic Editors: Daniele Corbo and Matteo Bologna

Received: 27 July 2022
Accepted: 31 October 2022
Published: 4 November 2022

1. Introduction

Perspective taking occurs in many situations of daily life. For example, one may withdraw one's own hand when watching another person putting their hand into a hot flame. Another example is a soccer coach who synchronously imitates a goal-kick of his striker without himself being in position or having a ball. The mechanisms underlying this aspect of everyday behavior are still not fully understood. Therefore, the present studies aimed to replicate and extend investigations on whether perspective taking follows certain rules that are based on action imagery theories [1] and embodied cognition [2]

As a specific type of imagination [3], *action imagery* (also called motor imagery) refers to the imagination of one's own movement without actually executing the movement [1,4]. In contrast to visual imagery, action imagery may involve the imagination of all types of perception that include motor, kinesthetic, tactile, and visual elements of oneself, of the environment, as well as of the action consequences [5,6]. The visual content can be imagined either from a first-person perspective (i.e., through one's own eyes during the action) or a third-person perspective (i.e., seeing oneself doing an action through an observer's eyes) [7]. It is proposed that action imagery draws on mechanisms from action execution [1,5,8] and recruits neural regions similar to those used for action execution [1,9].

In this way action imagery is consistent with theories of embodied cognition, which propose that offline cognitive processes such as memory or language processing engage perceptual and motor neural systems [10,11]. For instance, proprioception of one's own body posture is more relevant when imagining in the first-person perspective than in a third-person perspective [12], which indicates embodied cognition, e.g., a simulation of one's own actions and body [8,13]. Similarly, neural activation in somatosensory areas is stronger in the first-person perspective than in the third-person perspective [12].

As in overt action execution, in action imagery, inverse models select the motor commands for the upcoming action which involve information about the corresponding muscle activations. An efference copy of the motor commands is then used by forward models to predict the action consequences [5,14]. In action imagery, inhibition processes [15,16] are assumed to prevent the motor commands from overtly activating the effectors. In the present studies, we investigated the involvement of action imagery in perspective taking in a *mental body rotation task* [17] by showing whole body figures to participants who judged whether the figure raised the left or right limb. Solving mental (body) rotation tasks involves executive functions such as perceptual attention, visual working memory [18], and decision making [19]. Subjects need to quickly select between relevant and irrelevant features of the visual stimuli, keep the updated image stored in working memory in order to mentally rotate it, and then decide about the raised limb [20].

Left-right judgements in a mental body rotation task are presumed to involve perspective taking, i.e., imagining oneself in the position of the human figures [21]. It is assumed that perspective taking involves the same mechanisms that are used in action imagery [4,17,22]. For instance, one may imagine rotating their own body towards a position that is congruent with the presented figure. Hence, this *perspective taking* involves egocentric perspective transformations if the depicted position of the human figure differs from the participant's position, because the task requires that participants adjust their own reference frame to that of the depicted figure [23–25]. The idea that perspective taking involves a covert action of oneself has been supported by corresponding neural activity in motor areas [26] and somatosensory areas [27] during hand rotations. However, an alternative explanation is that perspective taking involves mechanisms required for mental object transformation that do not involve action imagery [17]. For instance, one may imagine rotating the figure towards a position that is congruent with one's own body position [28].

It has been shown that participants who are experts in executing body rotations outperformed control participants in left-right judgments that involve perspective taking [29], but not in same-different judgements that do not involve perspective taking [17,30]. In addition, experts in executing body rotations outperformed control participants only when making judgements about upside-down body positions, but not when making judgements about head-up body positions [17,30]. Furthermore, in experts and novices, response times and error rates increase with angular disparity from head-up positions [17,31–33] or with abstractness of the stimuli [30]. For instance, when using photographs rather than line drawing figures and stick figures, response times (RTs) were shorter for the more realistic photograph stimuli than the more abstract stimuli, particularly for experts in executing body rotations [30].

It has previously been shown that efficiency scores combining speed (RTs) and accuracy (error rates: ER) are better suited to show individual differences in laterality judgements [34]. Previous studies have used an inverse efficiency score—$RT/(1 - ER)$—to combine these measures, however, this score does not take into account differences in variance between the two measures [35,36]. Therefore, *linear speed-accuracy scores* (LISAS) have been proposed as an alternative way to combine speed and accuracy, while also considering their respective variances. The LISA score is practically identical to the response time in case a participant made no errors. However, the LISA score increases with an increase of committed errors. Hence, good performance is indicated by low LISA scores which indicate fast and correct responses. In comparison to analyses of RTs alone [37] or to separate analyses of RT and errors [17,30–32], the LISA score additionally tracks performance increments that are caused by (lacking) accuracy, as with a speed-accuracy trade-off [17,30,31].

In the present studies, our goal was to replicate previous findings of perspective and rotation effects on mental body rotation task performance using LISA scores. Hence, by investigating healthy participants and using a counterbalanced within-group experimental design, we expected better performance (lower LISA scores) in head-up positions than in head-down positions [17,32,33], and in a back view/egocentric perspective rather than in a front view/allocentric perspective [17]. Such effects of positional (in)congruency [38] between one's own body position and the depicted figure could indicate the involvement of action imagery in mental body rotations.

In addition to the replications of these previously observed effects, we used stimuli that included arm and leg items. Leg items have not previously been investigated in whole body laterality judgement tasks. By using leg items, we intended to increase task difficulty of the laterality judgements. In daily situations, distinguishing between the left and right leg is less common than distinguishing between the left and right arm. Furthermore, the lower limbs may be more sensitive for some motor control functions than the upper limbs [39]. In accordance with this assumption, RTs have been observed to be shorter in hand items than in foot items [37]. Therefore, we expected better performance (lower LISA scores) for raised arms than for raised legs. Effects of the limbs would provide additional evidence for the use of action imagery in mental body rotation tasks.

In the second study, we compared realistic stimuli (human-like, gender-free avatars) with the abstract stimuli used in the first study (elliptical drawings of a human body). Realistic stimuli are assumed to make the task more intuitive [30]. Using LISAS, we expected to replicate better performance in realistic than in abstract stimuli [30] not only for arm, but also for leg items.

2. Study 1

For the first study, we created abstract stimulus material with elliptical drawings of a human body. The aim of Study 1 was to replicate findings of perspective and rotational angles on linear speed-accuracy scores. Further, we manipulated the raised limb, expecting higher scores on raised legs than on raised arms.

2.1. Methods of Study 1

2.1.1. Participants of Study 1

The link for participation was provided to interested students at the University of Calgary. Participants were not familiarized with the task prior to the study. Because the data collection was online where distractions cannot be controlled for, a rigorous outlier analysis was performed on RTs and error rates. Of 182 English-speaking participants, 27 participants were excluded from analysis because they were just clicking through ($n = 9$; error rates close to 50% and RTs below 500 ms), they were inattentive or distracted ($n = 7$; error rates close to 50%), they did not comply with the instructions ($n = 9$; error rates close to 100% in front view, but not in back view), due to missing values in RT of correct responses ($n = 2$) or due to LISA scores above 3 *SD* from the mean ($n = 1$). The remaining 153 participants (128 females, 24 males, 1 else) were on average 21 years old ($SD = 4.5$, range from 18 to 44 years). The laterality index (assessed with the Edinburgh Handedness Inventory, [40]) ranged from −100 to +100 with the mean ($M = 73.9$, $SD = 37$) indicating mainly right-handers. The Vividness of Movement Imagery Questionnaire 2 VMIQ-2; [7] indicated that participants had clear and vivid external visual imagery ($M \pm SD = 2.2 \pm 0.9$), internal visual imagery ($M \pm SD = 1.8 \pm 0.8$), and kinesthetic imagery ($M \pm SD = 2 \pm 0.9$). All participants gave informed consent. The study was in accordance with the Belmont Report [41] and was approved by the local ethics committee.

The data reported here were collected in a larger experiment involving language processing tasks that are not the focus of the present manuscript [42], and the sample size was selected based on having sufficient power to detect effects in the language tasks.

For the purposes of the present analysis, we conducted a post-hoc power analysis for an interaction between eight conditions (the combination of perspective, rotation and limb) using G*Power [43]. The LISA score was the primary outcome measure. We assumed an effect size of $f = 0.25$ and alpha was set at 0.05, which resulted in the power (1-beta) of 0.87.

2.1.2. The Stimulus Material of Study 1

We created gender-free items that showed an abstract depiction of a human figure (Figure 1) either in front view (i.e., the figure is facing the participant) or in back view (i.e., the figure is looking in the same direction as the participant). The eight rotational angles were 0, 45, 90, 135, 180, 225, 270, and 315. The figure raised one limb (left arm, right arm, left leg, right leg). This resulted in a total of 64 stimuli.

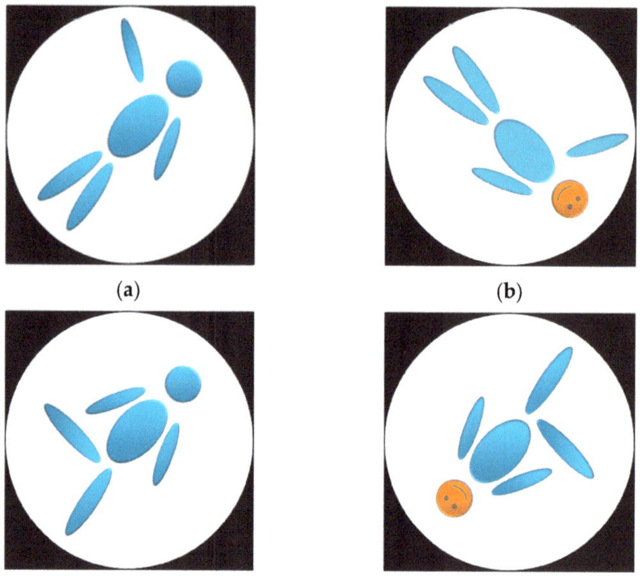

(a) (b) (c) (d)

Figure 1. Depiction of the abstract stimuli in Study 1. (**a**) Back view of the left arm raised. (**b**) Front view of the right arm raised. (**c**) Back view of the left leg raised. (**d**) Front view of the right foot raised.

2.1.3. Task and Procedure of Study 1

The experiment was created using the PsychoPy Builder interface [44] and run online using PsychoJS Version 3.2 [45]. Participants were told that the visual stimuli represent the figure of a human body which will be rotated. They were asked to press the 'D' key if the raised limb (foot or arm) is left and to press the 'K' key if the raised limb is right. Participants were told that in the front view they could see the face of the figure, whereas in the back view no face was visible. They performed eight familiarization trials with stimuli featuring every combination of perspective (i.e., front or back view), limb (i.e., arm or leg) and side (i.e., left or right), all presented at 0 rotation. This was followed by a block of 64 trials in which the stimuli (or conditions) were presented in random order. Participants' responses triggered the presentation of the next stimulus.

2.1.4. Data Analysis of Study 1

RT was defined as the interval between presentation of the stimulus and participants' response. The error rate indicates the percentage of incorrect responses. Median RTs and the percentage of errors were calculated for each perspective, rotation, and limb. Rotational angles were differentiated between head-up (−45, 0, +45) and head-down (−135, 180, 135), which have been shown to make the largest difference to behavioral responses [30]. Analysis of RTs and error rates revealed that the main effects could be observed in both measures (see Supplementary Materials). To take into account the speed-accuracy tradeoff in individuals, linear integrated speed-accuracy scores (LISAS) were calculated [35]. Because RTs are usually not normally distributed, we used the median and median absolute deviation (MAD; instead of the mean and standard deviation) when calculating these scores. Hence, either the median RT of correct responses (if the error rate is 0%) or median RT of correct responses + error rate × *MAD* (RT)/*SD* (ER) was used. For the repeated measures ANOVA, partial eta squared ($\eta_p{}^2$) is reported as effect size. Further comparisons were conducted using t-tests with Holm adjusted pairwise comparisons with Cohen's *d* as effect size. Where appropriate, minimum ($p_{\min}$) or maximum ($p_{\max}$) statistical values are reported. For all analyses, the probability of errors of the first kind was set at $\alpha = 0.05$. The tidyverse 1.3.1 [46] and rstatix 0.7.0 [47] packages were used for analysis using version 1.2.5033 of RStudio [48] and the ggplot2 package for the generation of graphs [49]. Data as well as the syntax for data analyses are available at https://osf.io/ymf8w.

2.2. Results of Study 1

A repeated measures ANOVA with the within-subject factors of perspective (front view, back view), rotation (head-up, head-down), and limb (arm, leg) was calculated on the LISAS. Means and standard errors of LISAS are shown in Figure 2.

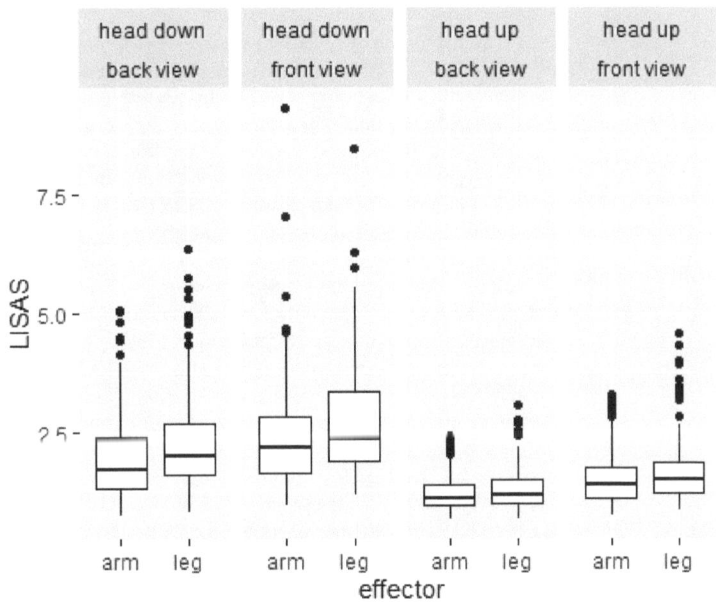

Figure 2. Boxplots of the linear integrated speed-accuracy scores (LISAS) depending on rotation (head-down, head-up), perspective (back view, front view), and limb (arm, leg) in Study 1.

The significant main effect of *perspective*, F (1, 152) = 121.7, $p < 0.001$, $\eta_p{}^2 = 0.45$, indicated significantly higher scores in the front view ($M \pm SD = 2.1 \pm 1.1$) than the back view ($M \pm SD = 1.7 \pm 0.9$). The significant main effect of *rotation*, F (1, 152) = 360.7, $p < 0.001$, $\eta_p{}^2 = 0.7$, indicated significantly higher scores in head-down rotations ($M \pm SD = 2.4 \pm 1.1$) than head-up rotations ($M \pm SD = 1.4 \pm 0.6$). The significant main effect of *limb*, F (1, 152) = 95, $p < 0.001$, $\eta_p{}^2 = 0.39$, indicated significantly higher scores for leg items ($M \pm SD = 2 \pm 1$) than arm items ($M \pm SD = 1.8 \pm 0.9$). The significant interaction between *rotation and limb*, F (1, 152) = 13.6, $p < 0.001$, $\eta_p{}^2 = 0.08$, indicated that the difference between limbs was significantly larger in head-down rotations ($\Delta M = 0.3$) than in head-up rotations ($\Delta M = 0.1$; $p < 0.001$, $d = 0.3$). All remaining interactions were not significant, $\eta_p^2 < 0.01$.

2.3. Discussion of Study 1

Visual inspection of the effect sizes of RT, ER (in the Supplementary Materials) and LISAS indicated that effect sizes in RT and ER were lower than in the LISA scores. Hence, LISA scores reflect the performance better than single analyses of RT or ER, as the effects in both measures and the relative values considering the variances of the two measures are taken into account in the LISAS. For instance, one participant may slow down in RTs to keep the ER low in more difficult items. In contrast, another participant may accept an increase in ER to keep RTs at the same level in more difficult items. In case of speed-accuracy tradeoffs in individual participants, the effects in RT and ER can also neutralize each other in the LISAS, making it a more appropriate measure, particularly to compare between individuals who differ in speed-accuracy preferences. This could also explain the small correlations between ER and RT (as shown in the Supplementary Materials).

As observed previously [17], performance was better in the back view than in the front view. It is assumed that this results from participants imagining a *perspective rotation* on the vertical axis, which is necessary in the front view, but not in the back view. In the back view, the participant's body is already congruent with the body position of the figure of the stimulus. Hence, when putting oneself in the perspective of the targeted stimulus figure to accurately make the left-right judgement, one uses imagery in a way that is consistent with the principles of embodied cognition [2,10].

Like the perspective factor, we also replicated the *rotation effects* observed in previous studies [17,32,33]. Performance was better in head-up rotations (rotations of maximal 45 degrees to the left or right) than in head-down rotations (rotations between 135 and 225 degrees). Similar to the perspective factor, it is assumed that this results from an imagined rotation. Such rotations on the horizontal axis are more time consuming and produce more errors proportional to the rotational angle [17,31–33]. In the present study, LISAS were higher in head-down rotations than in head-up rotations. Head-up stimuli do not involve such time-consuming and error-prone imagined rotations, which explains the better performance compared to head-down stimuli. This applies even for those head-up stimuli that involve slight rotational angles of 45 degrees.

As expected, the raised *limb* had an additional influence on mental body rotations. This was observed particularly in difficult rotational angles when the head is down. Performance was lower in leg stimuli than in arm stimuli, which may emerge due to various reasons. First, in daily life, we use our legs less often than our arms [39]. This applies particularly for students when they are sitting in the classroom or office. Second, distinguishing between left and right is often less important for legs than it is for arms. For instance, the right hand is typically used for a handshake, whereas there is no such greeting method or other everyday behavior for legs that favors left or right. Third, congruency with the response action [38] may have increased these effects, as participants responded with the fingers of the hands and not the feet. Hence, responding with the fingers (of the hand) to hand stimuli was more congruent than responding with the fingers (of the hand) to feet stimuli. In any event, differentiating between left and right legs increased task difficulty, which may be useful to measure individuals' action imagery ability more precisely [4].

3. Study 2

In Study 2, our goal was to replicate and extend the findings of Study 1. Therefore, we expected higher LISA scores in raised legs than in raised arms. Furthermore, we used more realistic stimuli (human-like avatars) to compare them with the abstract stimuli of Study 1. Using LISA scores, we expected better performance for realistic stimuli than for abstract stimuli [30].

As a secondary research aim, we assessed participants' general self-efficacy [50] and a German version [51] of vividness in action imagery (VMIQ-2, [7]) to test discriminative and convergent validity. As observed previously [52], we expected low to moderate ($0.2 < r < 0.5$) correlations between these self-assessment questionnaires (self-efficacy and VMIQ-2). Most importantly, we expected low correlations between self-efficacy and LISA scores of the mental body rotations, as they assess different constructs using different methods. In contrast, action imagery self-assessment questionnaires may measure the same construct as objective measures such as the mental body rotation task. Therefore, we expected low to moderate correlations between the VMIQ-2 ratings and LISA scores.

3.1. *Methods of Study 2*

3.1.1. Participants of Study 2

The link for participation was disseminated by student project members to their friends and to interested students at the UMIT—the Tyrolean Private University. As in Study 1, a rigorous outlier analysis was performed on RTs and error rates. Of 146 German-speaking participants, 24 participants were excluded from analysis because they were just clicking through ($n = 6$; error rates close to 50% and RTs below 500 ms), they were inattentive or distracted ($n = 2$; error rates close to 50% and large variance in RTs), they did not comply with the instructions in abstract stimuli ($n = 9$; error rates close to 100% in back view, but not in front view), or due to extreme outliers in RTs (above 5 s; $n = 3$) or error rates (above 50% in several but not all conditions; $n = 4$). The remaining 122 participants (72 females, 50 males) were on average 28.5 years old ($SD = 10.3$, range from 18 to 71 years) and mainly right-handed, as self-reported by the participants ($N = 109$). A German version [51,52] of the VMIQ-2 [7] indicated that participants had clear and vivid external visual imagery ($M \pm SD = 2.1 \pm 0.7$), internal visual imagery ($M \pm SD = 1.8 \pm 0.7$), and kinesthetic imagery ($M \pm SD = 1.9 \pm 0.7$). All participants gave informed consent. The study was in accordance with the Belmont Report [41] and was approved by the local ethics committee.

The required sample size for an interaction between 16 conditions (the combination of abstractness, perspective, rotation and limb) was estimated with G*Power [43]. The primary outcome measure used for sample size estimation was the LISA score. We assumed an effect size of $f = 0.25$. Alpha was set at 0.05 and the power (1-beta) at 0.8 which resulted in a minimum sample size of $N = 128$. Because the estimated sample size was not achieved in the final sample ($N = 122$), the power for medium effects ($f = 0.25$) was 0.78, which is only slightly below the recommended value of 0.8 [53].

3.1.2. The Stimulus Material of Study 2

In addition to the stimuli from Study 1, gender-free realistic avatars were created for purposes of Study 2, either in front view, i.e., the figure is facing the participant, or in the back view, i.e., the figure is looking in the same direction as the participant (Figure 3). The stimuli were created using makehuman in blender [54] by selecting 50% male and 50% female characteristics in body and face.

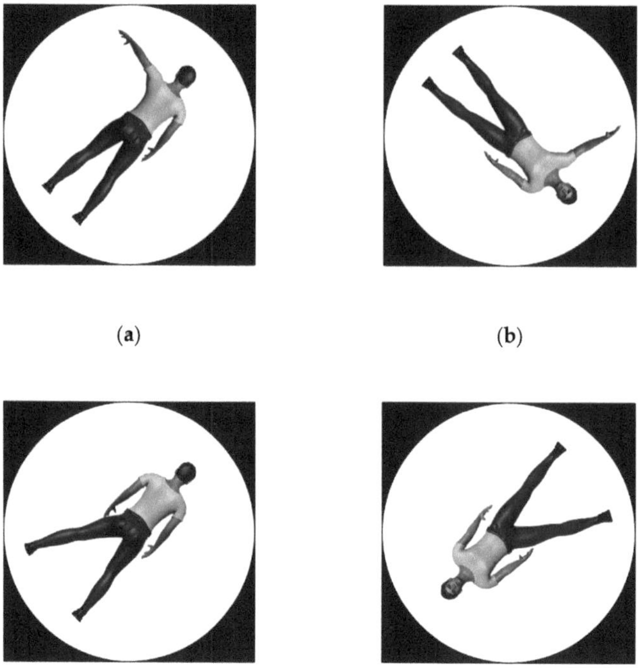

(a) (b) (c) (d)

Figure 3. Depiction of the realistic stimuli in Study 2. (**a**) Back view of the left arm raised. (**b**) Front view of the right arm raised. (**c**) Back view of the left leg raised. (**d**) Front view of the right foot raised.

3.1.3. Task and Procedure of Study 2

The experiment was run online using OpenSesameWeb Version 3.3.11 [55] and JATOS [56]. The experiment file is available at https://osf.io/ymf8w. As in Study 1, participants were instructed that the visual stimuli represent the figure of a human body that will be rotated. They were asked to press the 'X' key if the raised limb (foot or arm) is left and to press the 'Y' key if the raised limb is right. They were told that in the front view they could see the face of the figure, whereas in the back view no face was visible. To avoid learning effects during the assessment, participants started with four randomized familiarization blocks of 64 trials where either one of abstractness (abstract vs. realistic) or the limbs (arm vs. leg) was fixed for the block (e.g., a block of only abstract stimuli). This was followed by the assessment block of 128 trials in which all stimuli (or conditions) were presented in random order. A fixation dot was presented for 500 ms before each stimulus appeared on the screen. Participants' response triggered the presentation of the next fixation dot.

3.1.4. Data Analysis of Study 2

As in Study 1, linear integrated speed-accuracy scores (LISAS) were calculated and analyzed. Additionally, correlations between the vividness of movement imagery [51], self-efficacy [50], and the LISAS were calculated. Data as well as the syntax for data analyses are available at https://osf.io/ymf8w.

3.2. Results of Study 2

A repeated measures ANOVA with the within-subject factors of abstractness (abstract, realistic), perspective (front view, back view), rotation (head-up, head-down), and limb (arm, leg) was calculated on the LISAS. Means and standard errors of LISAS are shown in Figure 4. Results of the ANOVA are shown in Table 1.

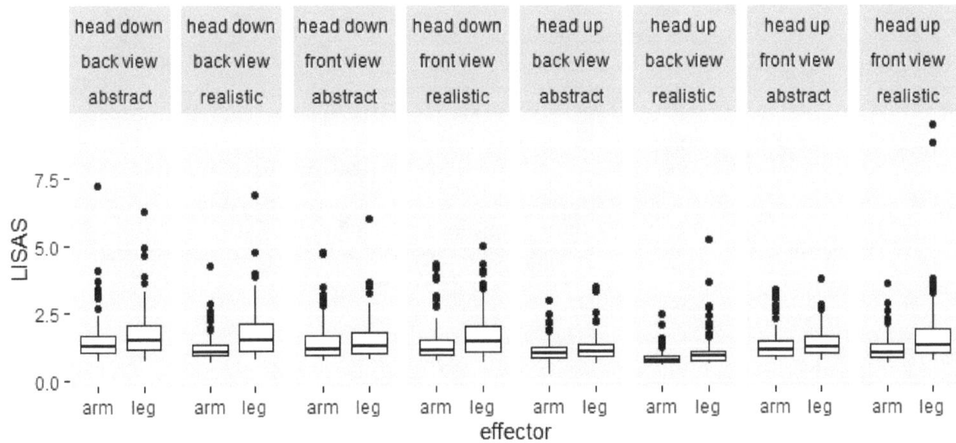

Figure 4. Boxplots of the linear integrated speed-accuracy scores (LISAS) depending on perspective (front view, back view), rotation (head-up, head-down), abstractness (realistic, abstract), and limb (arm, leg) in Study 2.

Table 1. Statistical values of the ANOVA ($df_1 = 1$, $df_2 = 121$) on LISAS in Study 2.

	F	p	η_p^2
Abstractness	5.2	0.025	0.41
Perspective	29.4	<0.001	0.20
Rotation	137.2	<0.001	0.53
Limb	93.0	<0.001	0.44
Abstractness × Perspective	42.0	<0.001	0.26
Abstractness × Rotation	0.1	0.803	<0.01
Abstractness × Limb	30.6	<0.001	0.20
Perspective × Rotation	67.3	<0.001	0.36
Perspective × Limb	0.2	0.643	<0.01
Rotation × Limb	2.9	0.094	0.02
Abstractness × Perspective × Rotation	0.3	0.565	<0.01
Abstractness × Perspective × Limb	5.8	0.018	0.05
Abstractness × Rotation × Limb	0.3	0.564	<0.01
Perspective × Rotation × Limb	22.3	<0.001	0.16
Abstractness × Perspective × Rotation × Limb	3.5	0.065	0.03

The significant main effect of *perspective* was modified by the significant interaction between abstractness and perspective. In realistic stimuli, the scores were significantly higher in the front view ($M \pm SD = 1.5 \pm 0.9$) than the back view ($M \pm SD = 1.2 \pm 0.7$, $p < 0.001$, $d = 0.71$). In abstract stimuli, the scores did not significantly differ between front view ($M \pm SD = 1.4 \pm 0.7$) and back view ($M \pm SD = 1.4 \pm 0.7$, $p = 0.287$, $d = 0.1$).

The significant main effect of *rotation* indicated significantly higher scores for head-down rotations ($M \pm SD = 1.5 \pm 0.8$) than head-up rotations ($M \pm SD = 1.2 \pm 0.7$). The significant interaction between rotation and perspective indicated that this difference was significantly larger in the back view ($\Delta M = 0.5$) than in the front view ($\Delta M = 1.3$, $p < 0.001$, $d = 0.74$).

The significant main effect of *limb* indicated significantly higher scores for leg stimuli ($M \pm SD = 14.7 \pm 0.8$) than arm stimuli ($M \pm SD = 1.2 \pm 0.6$). The significant interaction between abstractness and limb indicated that this difference was significantly larger for realistic stimuli ($\Delta M = 3.6$) than abstract stimuli ($\Delta M = 1.1$, $p < 0.001$, $d = 0.5$).

The significant main effect of *abstractness* was modified by the significant interaction between abstractness, perspective, and limb. For arm stimuli, the scores were significantly higher for abstract stimuli than realistic stimuli in the front view ($\Delta M = 0.11$, $p < 0.001$, $d = 0.32$) and back view ($\Delta M = 0.26$, $p < 0.001$, $d = 0.69$). In contrast, for leg stimuli, the scores were significantly lower for abstract stimuli than realistic stimuli in the front view ($\Delta M = -0.23$, $p < 0.001$, $d = -0.39$), but not in the back view ($\Delta M = 0.09$, $p = 0.067$, $d = 0.17$).

Pearson correlations between self-efficacy (SE, [49]) vividness of external visual (EVI), internal visual (IVI), and kinesthetic (KIN) imagery [7] and linear integrated speed-accuracy scores are shown in Figure 5. High interfactor-correlations were observed between the subdimensions of the VMIQ-2 [7]. Moderate correlations were observed between self-efficacy and vividness of action imagery. High correlations were observed between all LISA scores of the mental body rotations. Low correlations were observed between the questionnaires and the LISA scores.

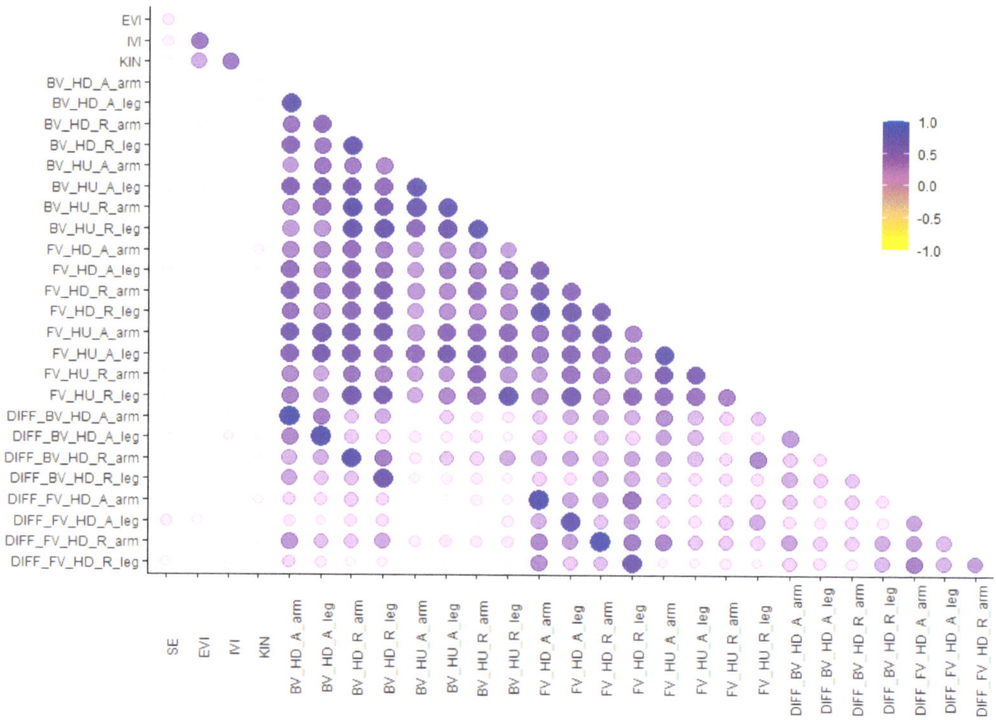

Figure 5. Pearson correlations between self-efficacy (SE; [49]) vividness of external visual (EVI), internal visual (IVI), and kinesthetic (KIN) imagery [7] and linear integrated speed-accuracy scores depending on perspective (front view: FV, back view: BV), rotation (head-up: HU, head-down: HD), abstractness (realistic: R, abstract: A), and limb (arm, leg). Additionally, correlations of the difference score (DIFF) of each condition with the baseline (back view head-up) are shown. Larger and darker circles indicate larger correlations.

In addition, we performed an *exploratory analysis* by calculating difference scores between the baseline condition without any rotational movement in the back view with heads up and the other conditions (e.g., front view head-down realistic legs—back view head-up realistic legs) to separate the unique variance of action imagery from other possible constructs in the mental body rotation task. Correlations of these difference scores are also shown in Figure 3.

3.3. Discussion of Study 2

As in Study 1, we observed higher scores for head-down rotations than head-up rotations, which is consistent with the findings of several previous studies [17,31–33]. In Study 2, this effect was even amplified in the back view, which allows a simple egocentric perspective without the need of a vertical rotation necessary for head-up rotations in the front view. Additionally, we observed higher scores for the front view than the back view [17]. However, this was observed for realistic but not abstract stimuli. The data pattern suggests that the egocentric perspective in realistic stimuli facilitates perspective taking, resulting in lower LISAS. An explanation for these findings may be that head-up figures in the back view do not require any mental rotations, only perspective taking, which is then facilitated if the stimuli are more realistic. Head-down figures in the back view require lateral horizontal mental rotations of one's own body to fully take the perspective and make a left/right judgement. Therefore, there was a large difference between the rotations in the back view. For the front view, however, both rotational angles (head-up and head-down) require a mental rotation. Head-up figures in the front view require a vertical mental rotation. One may assume that in head-down figures in the front view the vertical and lateral horizontal mental rotations are performed in a stepwise manner [33] which should have increased the scores considerably. However, this was not observed. Therefore, we assume that participants performed a frontal horizontal rotation (like a back-flip or front-flip), which was still slightly more difficult (or time-consuming) than a single vertical rotation, but easier (less time-consuming) than performing the rotations in a stepwise manner.

In accordance with Study 1, we observed higher scores for leg stimuli than arm stimuli. This was even amplified in realistic stimuli. One reason for this could be that the angle of the *limb* was more difficult to perceive in realistic stimuli than in abstract stimuli. Participants' verbal reports indicated that for realistic stimuli they needed to focus more strongly on both the foot the figure was standing on and the foot that was raised. It is possible that this was not the case in abstract stimuli where they focused on the raised limb only. From an action imagery point of view, this may indicate that for realistic stimuli, participants intended to replicate the balancing position on the standing foot to then imagine raising the other foot. This suggests that participants engaged in simulations of somatosensory experience to a greater degree for foot items, particularly for realistic stimuli. These assumptions are in line with the observation of lower scores in realistic stimuli than in abstract stimuli when arms were raised, but not when legs were raised. When legs were raised this was reversed in the front view. Hence, the allocentric perspective made the judgements on realistic leg figures more difficult.

The correlations in the present study showed that action imagery self-assessments via questionnaire and experimental assessments do not correlate. Similarly, such assessments do not correlate in visual imagery [3,57]. For instance, mental rotations of characters and numbers [58] did not correlate with the vividness of the visual imagery questionnaire [59]. In action imagery, it has been argued that the lack of correlations between experimental and self-report assessments [60] is due to the *methods of measurement* capturing different components of action imagery. However, such findings may also indicate that the different methods may capture constructs other than action imagery, because experimental methods always involve a combination of several abilities [3]. For instance, mental body rotations may not only require action imagery ability, but also independent constructs such as left-right disorientation, and closely related constructs such as fluid intelligence or working memory capacity. To rule out this argument, we calculated explorative difference scores that partialled out other constructs by using the position without any rotations (back view head-up) of the mental body rotation task as baseline. However, like the absolute scores, these difference scores did not correlate with the self-ratings of the VMIQ-2. Finally, the lack of correlation between self-ratings and objective measures in action imagery may be due to biases in self-reports [4,61] and a lack of variance in self-report questionnaires [52].

Interestingly, the difference scores were more strongly correlated with the absolute head-down scores than with the absolute head-up scores. This shows that the variance in the difference scores is mainly influenced by the rotation on the horizontal axis (which indicates the imagined action) and not as much by other constructs that may influence response times (e.g., general response times, working memory capacity). Furthermore, variance was also influenced by the perspective (rotations on the vertical axis), although not as strong as by rotations on the horizontal axis.

4. General Discussion

The results of both studies provide a replication of previous mental body rotation studies [17,31,32], showing that the rotational angle strongly affects participants' responses, not only in RTs but also in a combined measure of RT and ER. The LISA scores were higher (indicating worse performance) in head-down positions than in head-up positions. Furthermore, the scores were higher in the allocentric front view than in the egocentric back view [17], indicating that participants engaged in action imagery [1] for the task decisions, consistent with theories of embodied cognition [11]. Moreover, the scores were higher for abstract stimuli than realistic stimuli [30]. In addition to these replications, the present studies showed that leg judgments were more difficult to solve than arm judgements.

Regarding the *vertical and horizontal rotations*, it can be argued that this does not necessarily imply action imagery, i.e., a mental representation of one's own action in the absence of completing that action. Imagery may involve either a mental rotation of oneself or a mental rotation of the stimulus to gain congruency between one's own perspective and the depicted perspective [28]. However, it has been shown that hand judgments (as used in the present study) involve representations grounded in the motor system, while same-different judgements involve object-based representations grounded in the visual system [32].

Additionally, it can be argued that action imagery usually involves imagination of oneself, whereas the stimuli are neutral. However, verbal reports of participants indicated that they tended to imagine rotations of their own body to put themselves into the position of the figure. Furthermore, it has been shown that there is no advantage if the figures in a mental body rotation task show pictures of oneself instead of another person [32,38]. This implies that participants adopt a strategy to mentally rotate their own body to be congruent with a depicted figure, regardless of whether they implicitly associate themselves with the figure or not. Such first person imagery is most likely enriched by embodied kinesthetic action imagery where proprioceptive information is more relevant than in visual action imagery [12].

When comparing the absolute values of both studies (see Figures 2 and 4), it becomes apparent that the median LISAS in Study 1 (from 1.25 to 2.5) are at least tendentially higher than in Study 2 (from 1 to 1.5). Most likely, this resulted from the additional familiarization blocks in Study 2 (4 × 64 trials) compared to just eight trials in Study 1. Additional analyses in Study 2 (see Supplementary Materials) indicated significant improvements over the familiarization blocks. Hence, repeated familiarization with the task leads to lower LISAS.

Limitations and Perspectives

The *angle of the raised limb* may affect both speed and accuracy of the judgment in mental body rotations. Unfortunately, this effect was not consistent between abstract and realistic images, nor between arms and legs. Future studies may wish to investigate whether the angle between the raised limb and the body has an impact on mental body rotation scores.

It remains unclear whether the mental body rotation task involves internal visual imagery, external visual imagery, or kinesthetic imagery. None of these dimensions of the VMIQ-2 [7] correlated with the mental body rotation scores. Still, it remains likely that at least one of these dimensions are related to action imagery ability that is used during mental body rotations [4]. Future studies may focus on participants' *strength of representation* during mental body rotations. For instance, after mental body rotations, participants could be asked to report how strongly they focused on these modalities (i.e., visual or kinesthetic) and perspectives (i.e., internal or external) using a provided scale [62].

Our findings have implications for the assessment and measurement of *action imagery ability*. For a more sensitive measure of action imagery ability, we suggest using both leg and arm stimuli to increase the overall difficulty of the task. Furthermore, we suggest using realistic stimuli rather than abstract stimuli, which may prevent misunderstandings of the perspective in some participants (see participant exclusions in both studies). To render out confounding factors in the measure of action imagery ability (such as other cognitive abilities like working memory), we recommend taking the front view head-up condition as a baseline measure that does not require any movement (only perspective taking, but no rotation of one's own body) to calculate difference measures.

The present study was not designed to test the effects in different populations. Hence, to provide greater generalizability of the findings, future studies may investigate selected subpopulations. For instance, action imagery ability may differ between gender [63] or movement expertise [63]. Furthermore, patients with pain (or amputations) specific to either legs or arms could be investigated.

From an applied point of view, improved measurements of action imagery ability may be helpful in various fields such as physiotherapy [64], neurorehabilitation [65,66], sports [67], speech and language therapy [68], and music [69]. In action imagery practice (or mental practice) which designates the repetitive use of action imagery to improve motor performance, it has been proposed that high action imagery abilities boost the practice effects [70,71]. Using an objective measure of action imagery ability, individualized interventions may compensate for potential imagery deficits [72].

5. Conclusions

The results support the assumption that action imagery is involved in solving the mental body rotation task with left-right judgements [17,31]. An increase in LISA scores was caused by vertical and horizontal rotations that required the imagination of a rotational action. Although it appears likely that action imagery is involved in mental rotations, the object of rotation during action imagery remains unresolved. One does not need to necessarily imagine rotating oneself to be congruent with a depicted figure. In the imagination, the depicted figure could also be rotated. The latter would be a mental object transformation rather than action imagery. However, mental object transformations should not be influenced by the abstractness of the stimuli. Therefore, the increase in LISA scores in abstract stimuli compared to realistic stimuli strongly supports the assumption of action imagery processes in mental body rotations. Such imagery processes may be similar to processes engaged for action execution [1,17], thereby providing further support for embodied theories of cognition [2].

Supplementary Materials: The following supporting information can be downloaded at: https://www.mdpi.com/article/10.3390/brainsci12111500/s1, RTs (Figure S1) and error rates (Figure S2) in Study 1, correlations between RTs and error rates (Figures S3 and S4), learning effects in Study 2 (Figure S5), and explorative analyses of age and gender (Table S1, Figures S6–S8).

Author Contributions: Conceptualization, S.F.D.; methodology, S.F.D.; software, S.F.D. and E.J.M.; validation, S.F.D.; formal analysis, S.F.D.; investigation, S.F.D. and E.J.M.; resources, P.M.P.; data curation, S.F.D.; writing—original draft preparation, S.F.D.; writing—review and editing, S.F.D., E.J.M. and P.M.P.; visualization, S.F.D. All authors have read and agreed to the published version of the manuscript.

Funding: This research received no external funding.

Institutional Review Board Statement: The study was conducted in accordance with the Declaration of Helsinki and approved by the Institutional Ethics Committee of the UMIT TIROL—private university of health sciences and health technology (protocol code 3021 at the 5 January 2022).

Informed Consent Statement: Informed consent was obtained from all subjects involved in the study.

Data Availability Statement: The authors confirm that the data supporting the finding of this study are available within the article and its Supplementary Materials. Data exclusions, manipulations, and all measures in the study are reported in the manuscript. Stimulus material and data are available in the following link: https://osf.io/ymf8w.

Acknowledgments: Many thanks go to the students involved in data collection: Anna-Laura Blasbichler, Sharleen Büchel, Jara Campbell, Christoph Heckelsmüller, Christina Ortner, Julia Reinstaller, Vanessa Rupprechter and Elisabeth Wolf.

Conflicts of Interest: The authors declare that they have no conflict of interest.

References

1. Jeannerod, M. Neural Simulation of Action: A Unifying Mechanism for Motor Cognition. *Neuroimage* **2001**, *14*, S103–S109. [CrossRef] [PubMed]
2. Iachini, T. Mental Imagery and Embodied Cognition: A Multimodal Approach. *J. Ment. Imag.* **2011**, *35*, 1–66.
3. Kosslyn, S.M.; Brunn, J.; Cave, K.R.; Wallach, R.W. Individual Differences in Mental Imagery Ability: A Computational Analysis. *Cognition* **1984**, *18*, 195–243. [CrossRef]
4. Dahm, S.F. On the Assessment of Motor Imagery Ability: A Research Commentary. *Imagin. Cogn. Pers.* **2020**, *39*, 397–408. [CrossRef]
5. Dahm, S.F.; Rieger, M. Is Imagery Better than Reality? Performance in Imagined Dart Throwing. *Hum. Mov. Sci.* **2019**, *66*, 38–52. [CrossRef]
6. Kilteni, K.; Andersson, B.J.; Houborg, C.; Ehrsson, H.H. Motor Imagery Involves Predicting the Sensory Consequences of the Imagined Movement. *Nat. Commun.* **2018**, *9*, 1617. [CrossRef]
7. Roberts, R.; Callow, N.; Hardy, L.; Markland, D.; Bringer, J. Movement Imagery Ability: Development and Assessment of a Revised Version of the Vividness of Movement Imagery Questionnaire. *J. Sport Exerc. Psychol.* **2008**, *30*, 200–221. [CrossRef]
8. Grush, R. The Emulation Theory of Representation: Motor Control, Imagery, and Perception. *Behav. Brain Sci.* **2004**, *27*, 377–396. [CrossRef]
9. Lorey, B.; Naumann, T.; Pilgramm, S.; Petermann, C.; Bischoff, M.; Zentgraf, K.; Stark, R.; Vaitl, D.; Munzert, J. How Equivalent Are the Action Execution, Imagery, and Observation of Intransitive Movements? Revisiting the Concept of Somatotopy during Action Simulation. *Brain Cogn.* **2013**, *81*, 139–150. [CrossRef]
10. Barsalou, L.W. Perceptual Symbol Systems. *Behav. Brain Sci.* **1999**, *22*, 577–660. [CrossRef]
11. Barsalou, L.W. Grounded Cognition: Past, Present, and Future. *Top. Cogn. Sci.* **2010**, *2*, 716–724. [CrossRef] [PubMed]
12. Lorey, B.; Bischoff, M.; Pilgramm, S.; Stark, R.; Munzert, J.; Zentgraf, K. The Embodied Nature of Motor Imagery: The Influence of Posture and Perspective. *Exp. Brain Res.* **2009**, *194*, 233–243. [CrossRef] [PubMed]
13. Gallese, V. Embodied Simulation: From Neurons to Phenomenal Experience. *Phenom. Cogn. Sci.* **2005**, *4*, 23–48. [CrossRef]
14. Wolpert, D.M.; Diedrichsen, J.; Flanagan, J.R. Principles of Sensorimotor Learning. *Nat. Rev. Neurosci.* **2011**, *12*, 739–751. [CrossRef]
15. Bart, V.K.E.; Koch, I.; Rieger, M. Inhibitory Mechanisms in Motor Imagery: Disentangling Different Forms of Inhibition Using Action Mode Switching. *Psychol. Res.* **2021**, *85*, 1418–1438. [CrossRef]
16. Rieger, M.; Dahm, S.F.; Koch, I. Inhibition in Motor Imagery: A Novel Action Mode Switching Paradigm. *Psychon. Bull. Rev.* **2017**, *24*, 459–466. [CrossRef]
17. Steggemann, Y.; Engbert, K.; Weigelt, M. Selective Effects of Motor Expertise in Mental Body Rotation Tasks: Comparing Object-Based and Perspective Transformations. *Brain Cogn.* **2011**, *76*, 97–105. [CrossRef]
18. Hyun, J.-S.; Luck, S.J. Visual Working Memory as the Substrate for Mental Rotation. *Psychon. Bull. Rev.* **2007**, *14*, 154–158. [CrossRef]
19. Seepanomwan, K.; Caligiore, D.; Cangelosi, A.; Baldassarre, G. Generalisation, Decision Making, and Embodiment Effects in Mental Rotation: A Neurorobotic Architecture Tested with a Humanoid Robot. *Neural Netw.* **2015**, *72*, 31–47. [CrossRef]
20. Xue, J.; Li, C.; Quan, C.; Lu, Y.; Yue, J.; Zhang, C. Uncovering the Cognitive Processes Underlying Mental Rotation: An Eye-Movement Study. *Sci. Rep.* **2017**, *7*, 10076. [CrossRef]
21. Parsons, L.M. Imagined Spatial Transformation of One's Body. *J. Exp. Psychol. Gen.* **1987**, *116*, 172–191. [CrossRef] [PubMed]
22. Barhoun, P.; Fuelscher, I.; Do, M.; He, J.L.; Bekkali, S.; Cerins, A.; Youssef, G.J.; Williams, J.; Enticott, P.G.; Hyde, C. Mental Rotation Performance in Young Adults with and without Developmental Coordination Disorder. *Hum. Mov. Sci.* **2021**, *77*, 102787. [CrossRef] [PubMed]
23. Voyer, D.; Jansen, P. Motor Expertise and Performance in Spatial Tasks: A Meta-Analysis. *Hum. Mov. Sci.* **2017**, *54*, 110–124. [CrossRef] [PubMed]

24. Zacks, J.M.; Mires, J.; Tversky, B.; Hazeltine, E. Mental Spatial Transformations of Objects and Perspective. *Spat. Cogn. Comput.* **2000**, *2*, 315–332. [CrossRef]
25. Böffel, C.; Müsseler, J. Perceived Ownership of Avatars Influences Visual Perspective Taking. *Front. Psychol.* **2018**, *9*, 743. [CrossRef]
26. Zacks, J.M.; Michelon, P. Transformations of Visuospatial Images. *Behav. Cogn. Neurosci. Rev.* **2005**, *4*, 96–118. [CrossRef]
27. Vingerhoets, G.; de Lange, F.P.; Vandemaele, P.; Deblaere, K.; Achten, E. Motor Imagery in Mental Rotation: An FMRI Study. *NeuroImage* **2002**, *17*, 1623–1633. [CrossRef]
28. Tomasino, B.; Rumiati, R.I.; Umiltà, C.A. Selective Deficit of Motor Imagery as Tapped by a Left-Right Decision of Visually Presented Hands. *Brain Cogn.* **2003**, *53*, 376–380. [CrossRef]
29. Feng, T.; Zhang, Z.; Ji, Z.; Jia, B.; Li, Y. Selective Effects of Sport Expertise on the Stages of Mental Rotation Tasks with Object-Based and Egocentric Transformations. *Adv. Cogn. Psychol.* **2017**, *13*, 248–256. [CrossRef]
30. Habacha, H.; Lejeune-Poutrain, L.; Molinaro, C. Realistic Stimuli Reveal Selective Effects of Motor Expertise during a Mental Body Rotation Task. *Am. J. Psychol.* **2017**, *130*, 47–62. [CrossRef]
31. Jost, L.; Jansen, P. Manual Training of Mental Rotation Performance: Visual Representation of Rotating Figures Is the Main Driver for Improvements. *Q. J. Exp. Psychol.* **2022**, *75*, 695–711. [CrossRef] [PubMed]
32. Kaltner, S.; Riecke, B.E.; Jansen, P. Embodied Mental Rotation: A Special Link between Egocentric Transformation and the Bodily Self. *Front. Psychol.* **2014**, *5*, 505. [CrossRef] [PubMed]
33. Schwabe, L.; Lenggenhager, B.; Blanke, O. The Timing of Temporoparietal and Frontal Activations during Mental Own Body Transformations from Different Visuospatial Perspectives. *Hum. Brain Mapp.* **2009**, *30*, 1801–1812. [CrossRef]
34. Barhoun, P.; Fuelscher, I.; Kothe, E.J.; He, J.L.; Youssef, G.J.; Enticott, P.G.; Williams, J.; Hyde, C. Motor Imagery in Children with DCD: A Systematic and Meta-Analytic Review of Hand-Rotation Task Performance. *Neurosci. Biobehav. Rev.* **2019**, *99*, 282–297. [CrossRef]
35. Vandierendonck, A. A Comparison of Methods to Combine Speed and Accuracy Measures of Performance: A Rejoinder on the Binning Procedure. *Behav. Res. Methods* **2017**, *49*, 653–673. [CrossRef]
36. Vandierendonck, A. On the Utility of Integrated Speed-Accuracy Measures When Speed-Accuracy Trade-off Is Present. *J. Cogn.* **2021**, *4*, 26. [CrossRef]
37. Vannuscorps, G.; Pillon, A.; Andres, M. Effect of Biomechanical Constraints in the Hand Laterality Judgment Task: Where Does It Come From? *Front. Hum. Neurosci.* **2012**, *6*, 299. [CrossRef] [PubMed]
38. van Elk, M.; Blanke, O. Imagined Own-Body Transformations during Passive Self-Motion. *Psychol. Res.* **2014**, *78*, 18–27. [CrossRef]
39. Peters, M. Neuropsychological Identification of Motor Problems: Can We Learn Something from the Feet and Legs That Hands and Arms Will Not Tell Us? *Neuropsychol. Rev.* **1990**, *1*, 165–183. [CrossRef]
40. Oldfield, R.C. The Assessment and Analysis of Handedness: The Edinburgh Inventory. *Neuropsychologia* **1971**, *9*, 97–113. [CrossRef]
41. *The Belmont Report: Ethical Principles and Guidelines for the Protection of Human Subjects of Research*; DHEW Publication; The Commission for Sale by the Supt. of Docs., U.S. Government Printing Office: Bethesda, MD, USA; Washington, DC, USA, 1978.
42. Muraki, E.J.; Dahm, S.F.; Pexman, P.M. Meaning in Hand: Investigating Shared Mechanisms of Motor Imagery and Sensorimotor Simulation in Language Processing. *PsyarXiv* **2022**, preprint.
43. Faul, F.; Erdfelder, E.; Lang, A.-G.; Buchner, A. G*Power 3: A Flexible Statistical Power Analysis Program for the Social, Behavioral, and Biomedical Sciences. *Behav. Res. Methods* **2007**, *39*, 175–191. [CrossRef] [PubMed]
44. Peirce, J.; Gray, J.R.; Simpson, S.; MacAskill, M.; Höchenberger, R.; Sogo, H.; Kastman, E.; Lindeløv, J.K. PsychoPy2: Experiments in Behavior Made Easy. *Behav. Res. Methods* **2019**, *51*, 195–203. [CrossRef] [PubMed]
45. Bridges, D.; Pitiot, A.; MacAskill, M.R.; Peirce, J.W. The Timing Mega-Study: Comparing a Range of Experiment Generators, Both Lab-Based and Online. *PeerJ* **2020**, *8*, e9414. [CrossRef]
46. Wickham, H.; Averick, M.; Bryan, J.; Chang, W.; McGowan, L.D.; François, R.; Grolemund, G.; Hayes, A.; Henry, L.; Hester, J.; et al. Welcome to the Tidyverse. *J. Open Source Softw.* **2019**, *4*, 1686. [CrossRef]
47. Kassambara, A. Rstatix: Pipe-Friendly Framework for Basic Statistical Tests. Rstatix, 2021. Available online: https://rpkgs.datanovia.com/rstatix/ (accessed on 5 January 2022).
48. RStudio Team. *RStudio: Integrated Development Environment for R*; RStudio Team: Boston, MA, USA, 2022.
49. Wickham, H. *Ggplot2: Elegant Graphics for Data Analysis*; Springer International Publishing: Berlin/Heidelberg, Germany, 2016; ISBN 9783319242750.
50. Schwarzer, R. Optimistische Kompetenzerwartung: Zur Erfassung Einer Personellen Bewältigungsressource. *Diagnostica* **1994**, *40*, 105–123.
51. Dahm, S.F.; Bart, V.K.E.; Pithan, J.M.; Rieger, M. Deutsche Übersetzung und Validierung des VMIQ-2 zur Erfassung der Lebhaftigkeit von Handlungsvorstellungen. *Z. Für Sportpsychol.* **2019**, *26*, 151–158. [CrossRef]
52. Dahm, S.F. Validation of a Computer-Based Version of the Vividness of Movement Imagery Questionnaire. *PTAD* **2022**, 1–13. [CrossRef]
53. Serdar, C.C.; Cihan, M.; Yücel, D.; Serdar, M.A. Sample Size, Power and Effect Size Revisited: Simplified and Practical Approaches in Pre-Clinical, Clinical and Laboratory Studies. *Biochem. Med.* **2021**, *31*, 010502. [CrossRef]

54. Blender Online Community. *Blender—A 3D Modelling and Rendering Package*; Stichting Blender Foundation: Amsterdam, The Netherlands, 2018.
55. Mathôt, S.; March, J. Conducting Linguistic Experiments Online with OpenSesame and OSWeb. *Lang. Learn* **2022**, 1–32. [CrossRef]
56. Lange, K.; Kühn, S.; Filevich, E. "Just Another Tool for Online Studies" (JATOS): An Easy Solution for Setup and Management of Web Servers Supporting Online Studies. *PLoS ONE* **2015**, *10*, e0130834. [CrossRef]
57. Burton, L.J.; Fogarty, G.J. The Factor Structure of Visual Imagery and Spatial Abilities. *Intelligence* **2003**, *31*, 289–318. [CrossRef]
58. Shepard, R.N.; Cooper, L.A. *Mental Images and Their Transformations*; The MIT Press: Cambridge, MA, USA, 1986; p. 364, ISBN 9780262192002.
59. Marks, D.F. Visual Imagery Differences in the Recall of Pictures. *Br. J. Psychol.* **1973**, *64*, 17–24. [CrossRef] [PubMed]
60. Williams, S.E.; Guillot, A.; Di Rienzo, F.; Cumming, J. Comparing Self-Report and Mental Chronometry Measures of Motor Imagery Ability. *Eur. J. Sport Sci.* **2015**, *15*, 703–711. [CrossRef] [PubMed]
61. Zell, E.; Strickhouser, J.E.; Sedikides, C.; Alicke, M.D. The Better-than-Average Effect in Comparative Self-Evaluation: A Comprehensive Review and Meta-Analysis. *Psychol. Bull.* **2020**, *146*, 118–149. [CrossRef]
62. Dahm, S.F.; Rieger, M. Cognitive Constraints on Motor Imagery. *Psychol. Res.* **2016**, *80*, 235–247. [CrossRef]
63. Pietsch, S.; Jansen, P. Different Mental Rotation Performance in Students of Music, Sport and Education. *Learn. Individ. Differ.* **2012**, *22*, 159–163. [CrossRef]
64. McInnes, K.; Friesen, C.; Boe, S. Specific Brain Lesions Impair Explicit Motor Imagery Ability: A Systematic Review of the Evidence. *Arch. Phys. Med. Rehabil.* **2016**, *97*, 478–489. [CrossRef]
65. Boe, S.; Gionfriddo, A.; Kraeutner, S.; Tremblay, A.; Little, G.; Bardouille, T. Laterality of Brain Activity during Motor Imagery Is Modulated by the Provision of Source Level Neurofeedback. *NeuroImage* **2014**, *101*, 159–167. [CrossRef]
66. Guillot, A.; Debarnot, U.; Louis, M.; Hoyek, N.; Collet, C. *Motor Imagery and Motor Performance: Evidence from the Sport Science Literature*; Oxford University Press: Oxford, UK, 2010; ISBN 9780191701412.
67. Simonsmeier, B.A.; Androniea, M.; Buecker, S.; Frank, C. The Effects of Imagery Interventions in Sports: A Meta-Analysis. *Int. J. Sport Exerc. Psychol.* **2021**, *14*, 186–207. [CrossRef]
68. Durand, E.; Masson-Trottier, M.; Sontheimer, A.; Ansaldo, A.I. Increased Links between Language and Motor Areas: A Proof-of-Concept Study on Resting-State Functional Connectivity Following Personalized Observation, Execution and Mental Imagery Therapy in Chronic Aphasia. *Brain Cogn.* **2021**, *148*, 105659. [CrossRef] [PubMed]
69. Bernardi, N.F.; De Buglio, M.; Trimarchi, P.D.; Chielli, A.; Bricolo, E. Mental Practice Promotes Motor Anticipation: Evidence from Skilled Music Performance. *Front. Hum. Neurosci.* **2013**, *7*, 451. [CrossRef] [PubMed]
70. Isaac, A.R.; Marks, D.F.; Russell, D.G. An Instrument for Assessing Imagery of Movement: The Vividness of Movement Imagery Questionnaire (VMIQ). *J. Ment. Imag.* **1986**, *10*, 23–30.
71. Milton, J.; Small, S.L.; Solodkin, A. Imaging Motor Imagery: Methodological Issues Related to Expertise. *Methods* **2008**, *45*, 336–341. [CrossRef]
72. Adams, I.L.J.; Lust, J.M.; Wilson, P.H.; Steenbergen, B. Development of Motor Imagery and Anticipatory Action Planning in Children with Developmental Coordination Disorder—A Longitudinal Approach. *Hum. Mov. Sci.* **2017**, *55*, 296–306. [CrossRef]

brain sciences

MDPI

Article

Profiles of Motor-Cognitive Interference in Parkinson's Disease—The Trail-Walking-Test to Discriminate between Motor Phenotypes

Thomas J. Klotzbier [1,*], Nadja Schott [1] and Quincy J. Almeida [2]

[1] Institute of Sport and Movement Science, University of Stuttgart, 70569 Stuttgart, Germany
[2] Movement Disorders Research & Rehabilitation Centre, Department of Kinesiology, Wilfrid Laurier University, Waterloo, ON N2L 3C5, Canada
* Correspondence: thomas.klotzbier@inspo.uni-stuttgart.de

Abstract: Background and Aims. Most research on Parkinson's disease (PD) focuses on describing symptoms and movement characteristics. Studies rarely focus on the early detection of PD and the search for suitable markers of a prodromal stage. Early detection is important, so treatments that may potentially change the course of the disease can be attempted early on. While gait disturbances are less pronounced in the early stages of the disease, the prevalence, and severity increase with disease progression. Therefore, postural instability and gait difficulties could be identified as sensitive biomarkers. The aim was to evaluate the discriminatory power of the Trail-Walking Test (TWT; Schott, 2015) as a potential diagnostic instrument to improve the predictive power of the clinical evaluation concerning the severity of the disease and record the different aspects of walking. Methods. A total of 20 older healthy (M = 72.4 years, SD = 5.53) adults and 43 older adults with PD and the motor phenotypes postural instability/gait difficulty (PIGD; M = 69.7 years, SD = 8.68) and tremor dominant (TD; M = 68.2 years, SD = 8.94) participated in the study. The participants performed a motor-cognitive dual task (DT) of increasing cognitive difficulty in which they had to walk a given path (condition 1), walk to numbers in ascending order (condition 2), and walk to numbers and letters alternately and in ascending order (condition 3). Results. With an increase in the cognitive load, the time to complete the tasks (seconds) became longer in all groups, $F(1.23, 73.5) = 121$, $p < 0.001$, $\eta^2_p = 0.670$. PIGD showed the longest times in all conditions of the TWT, $F(2, 60) = 8.15$, $p < 0.001$, $\eta^2_p = 0.214$. Mutual interferences in the cognitive and motor domain can be observed. However, clear group-specific patterns cannot be identified. A differentiation between the motor phenotypes of PD is especially feasible with the purely motor condition (TWT-M; AUC = 0.685, $p = 0.44$). Conclusions. PD patients with PIGD must be identified by valid, well-evaluated clinical tests that allow for a precise assessment of the disease's individual fall risk, the severity of the disease, and the prognosis of progression. The TWT covers various aspects of mobility, examines the relationship between cognitive functions and walking, and enables differentiation of the motor phenotypes of PD.

Keywords: dual-task; Trai-Walking Test; gait disorder; diagnosis; motor-cognitive interference; Parkinson's disease

check for updates

Citation: Klotzbier, T.J.; Schott, N.; Almeida, Q.J. Profiles of Motor-Cognitive Interference in Parkinson's Disease—The Trail-Walking-Test to Discriminate between Motor Phenotypes. *Brain Sci.* **2022**, *12*, 1217. https://doi.org/10.3390/brainsci12091217

Academic Editor: Daniele Corbo

Received: 20 July 2022
Accepted: 6 September 2022
Published: 9 September 2022

1. Introduction

In addition to the motor symptoms, various aspects of cognitive impairment in Parkinson's disease (PD) patients can have a negative impact on the ability to balance in static and dynamic situations [1,2]. The extent of the cognitive impairment is heterogeneous in those affected and worsens in the course of the disease parallel to the motor symptoms [3]. The prevalence of comorbid dementia is estimated at 26–44% [4,5], with main deficits being found in the executive/attentional, memory, and visuospatial domains [4], which can magnify their gait problems. In their multicenter study, Hely and colleagues [6] reported that at least 83% of survivors had dementia after 20 years. In particular, walking with additional motor or cognitive tasks to be performed in parallel seems difficult for those affected [7–9]. This is particularly significant because walking in the real world—usually under dual task (DT) conditions—requires attention to various changing environmental features to avoid tripping and slipping and to recover quickly from unavoidable postural disturbances. Therefore, it is not surprising that deficits in attention and executive functioning (EF) are independently associated with risk for postural instability, impairments in activities of daily living (ADL), and future falls [10–12]. Individuals with PD who have cognitive decline appear to be more susceptible to gait impairments due to their inability to use cognitive resources required to plan and control movements, especially when the automaticity of well-learned movements (gait) is compromised and where increased conscious control is required [13].

Although the diversity of DTs found in the literature makes a comparison between the studies difficult, the gait of individuals with PD is more influenced during the performance of more complex secondary tasks [14]. A meta-analysis by Ruffegeau and colleagues [15] demonstrated sufficient evidence to conclude that DT conditions involving EF skills significantly hinder walking in people with PD, despite variation between study paradigms. Moreover, DT paradigms with additional cognitive tasks can be helpful to parse apart the tremor dominant (TD) patients that will progress slower than those that are assigned to the phenotype postural instability/gait difficulty (PIGD) and will progress faster [16]. However, there is some controversy in this regard. While some studies report that the TD phenotype has a better prognosis and a lower rate of disease progression compared to the PIGD phenotype, others claim that there is no difference in long-term outcomes [17–20].

While most studies with the DT paradigm use relatively simple straight walking as a locomotion task—and also the guidelines of the Canadian Consortium on Neurodegeneration in Aging (CCNA) only address recommendations for straight walking [21]—complex locomotion tasks in which the walking speed is adjusted and the walking direction change seem to be particularly demanding and sensitive for PD patients to produce dual-task costs (DTC). While simple information processing processes can solve straightforward walking, cognitive flexibility and the ability to change tasks explain the speed of cornering [22] and walking with direction changes [23]. During complex walking situations (walking with direction changes), the increased cognitive and sensory processing required to plan gait modifications may strongly impact the walking performance [24]. Difficulties in turning around the body axis are one of the most common complaints among people with PD, may cause extreme gait slowness and loss of balance and may result from an overloaded or inefficient cognitive system in PD when planning complex gait adjustments. For this reason, and since walking performance is most affected by internal disturbances [25,26], a mobile version of the Trail-Making Test (TMT) is used in this study (Trail-Walking Test; [27]) as a motor-cognitive DT that demands EF with varying degrees of difficulty.

Research usually focuses on the early detection of PD and the search for suitable markers of a prodromal stage [28]. So far, there is no gold standard for the operationalization of gait disorders in PD. The walking test (Item 29) integrated into the motor part of the Unified Parkinson's Disease Rating Scale (UPDRS; Movement Disorder Society Task Force on Rating Scales for Parkinson's Disease, [29]) is frequently used in clinical settings. However, the PIGD score does not include the classification of freezing of gait, does not capture the performance of tandem gait, and lacks details about the range of postural deficits [30]. In order to improve the predictive power of the clinical examination concerning the risk of falling or to differentiate between the PD motor phenotypes, it appears necessary to record not only reactive and supportive aspects of balance control but also anticipatory, arbitrary, and cognitive aspects of locomotion.

The aim was to compare single task (ST) and DT conditions concerning a possible detection of PIGD and whether it is possible to differentiate the groups based on the TWT, what might be important for the timing and duration of therapy initiation and may help to assist with the prognosis and the tailoring of treatment. Based on the difficulties in the mentioned motor and cognitive domains in PD patients, we hypothesized that overall, individuals with PIGD perform more poorly than the control group and the TD group in all conditions of the TWT. We also assume that individuals with PIGD exhibit proportionally greater DTC under more complex, attention-demanding motor-cognitive DTs (TWT conditions) relative to TD patients or healthy older adults (see also [31] in people with mild cognitive impairment, MCI).

2. Materials and Methods

2.1. Participants

Table 1 shows the inclusion and exclusion criteria of the control group, PD patients with TD, and PD patients with PIGD. A total of 20 healthy older adults and 43 older adults with PD voluntarily participated in the study. Based on the Unified Parkinson Disease Rating Scale—Part 3; UPDRS; [32]—two motor phenotypes of PD patients were distinguished: the tremor dominant (TD) and the postural instability (PIGD) motor phenotype (classification according to Jankovic et al. [33]; see Table 2). Participants were invited in writing or orally (by telephone) to the Sun Life Financial Research and Rehabilitation Centre for Movement Disorders (MDRC) at Wilfrid Laurier University in Waterloo, Canada. The subjects with PD were asked to postpone their medication intake by 12 h before visiting the clinic to participate in the study without medication ("off-state") to minimize the confounding effect of dopaminergic medication on cognitive and motor performance, especially gait speed [34,35].

Table 1. Inclusion and exclusion criteria for the control group, PD patients with TD, and PD patients with PIGD.

	Control	TD	PIGD
Inclusion criteria	- Either gender - Age 50 to 80 years of age - Ability to walk for 10 min continuously unassisted - Able to understand English instructions - Normal or corrected vision	- Either gender - Age 50 to 80 years of age - Diagnosed with idiopathic PD by a Neurologist - Ratio of mean tremor score/mean PIGD score (Jankovic-based classification) is 1.5 or more (with the use of UPDRS-II) - Ability to walk for 10 min continuously unassisted - Able to understand English instructions - Normal or corrected vision	- Either gender - Age 50 to 80 years of age - Diagnosed with idiopathic PD by a Neurologist - Ratio of mean tremor score/mean PIGD score (Jankovic-based classification) was less than or equal to 1.0 (with the use of UPDRS-II) - Ability to walk for 10 min continuously unassisted - Able to understand English instructions - Normal or corrected vision
Exclusion criteria	- A neurological disease other than PD - Had brain surgery in the past including implanted deep-brain-stimulation - Have significant co-morbidities likely to affect gait, e.g., history of stroke, - Peripheral neuropathy - Visual impairments that cannot be corrected - Clinically diagnosed with dementia (as stated in the patient's information chart from the patient database at the Sun Life Financial Movement Disorders Research and Rehabilitation) - Are not able to comply with the protocol.	- Tremor score <4 or PIGD score >3 (with the use of UPDRS-II) - A neurological disease other than PD - Had brain surgery in the past including implanted deep-brain-stimulation - Have significant co-morbidities likely to affect gait, e.g., history of stroke, - Peripheral neuropathy - Visual impairments that cannot be corrected - Are unable to walk in the OFF state - Clinically diagnosed with dementia (as stated in the patient's information chart from the patient database at the Sun Life Financial Movement Disorders Research and Rehabilitation) - Are not able to comply with the protocol.	- Tremor score >3 or PIGD score <4 (with the use of UPDRS-II) - A neurological disease other than PD - Had brain surgery in the past including implanted deep-brain-stimulation - Have significant co-morbidities likely to affect gait, e.g., history of stroke, - Peripheral neuropathy - Visual impairments that cannot be corrected - Are unable to walk in the OFF state - Clinically diagnosed with dementia (as stated in the patient's information chart from the patient database at the Sun Life Financial Movement Disorders Research and Rehabilitation) - Are not able to comply with the protocol.

Table 2. Items of the UPDRS to classify the motor subtypes Tremor Dominant (TD) and Postural Instability (PIGD) in PD patients.

<table>
<tr><th colspan="4">UPDRS</th></tr>
<tr><th>N Items</th><th>Tremor-Dominant (TD)</th><th>N Items</th><th>Postural Instability and Gait Difficulty (PIGD)</th></tr>
<tr><td colspan="4">Part 2—Activities of daily living (ADL)</td></tr>
<tr><td>1</td><td>2.16 Tremor</td><td>3</td><td>2.13 Falling (independent of rigidity)
2.14 Freezing during walking
2.15 Walking</td></tr>
<tr><td colspan="4">Part 3—Motor examination</td></tr>
<tr><td>7</td><td>3.20 Rest Tremor, F
3.20 Rest Tremor, RH
3.20 Rest Tremor, LH
3.20 Rest Tremor, RF
3.20 Rest Tremor, LF
3.21 Action or posture tremor of the hands, L
3.21 Action or posture tremor of the hands, R</td><td>2</td><td>3.29 Gait
3.30 Postural Stability *</td></tr>
</table>

Note. F = face; RH = right hand; LH = left hand; RF = right foot; LF = left foot; R = right; L = left; * Reaction to sudden rearward displacement by pulling on the patient's shoulders; standing straight with eyes open and feet slightly apart (the patient is prepared). The ratio of the mean TD scores (8 items) to the mean PIGD scores (5 items) was used to classify the motor subtypes: TD (ratio ≤ 1.5), PIGD (ratio ≥ 1).

2.2. *Instruments*

2.2.1. Sociodemographic Information, Cognitive Performance, and Fall-Associated Self-Efficacy

Demographic information, medical history, physical activity, and the number of falls in the last year were collected using questionnaires. In addition, the height and weight of the participants were measured, and the body mass index (BMI, kg/m^2) was calculated.

Since cognitive status influences strategies for allocating attentional resources [36,37] and Johansson et al. [38] have shown in a recently published study that PD patients with MCI use a posture-first strategy and had larger DTCs on gait than PD non-MCI patients, it is important to consider the cognitive status. Although it is recommended to use a comprehensive cognitive assessment battery rather than individual global cognitive measures to assess the cognitive state, we used the well-established Montreal Cognitive Assessment (MoCA; score range: 0–30; [39]) to test general cognitive performance. This instrument appears to be sensitive to slight cognitive loss (mild cognitive impairment; MCI) in cognitively intact older adults [40,41].

The paper-pencil-based Trail-Making Test [42] comprises 25 circles to be connected (∅13 mm), which are numbers (Part A; visuomotor skills, visual processing speed) and numbers or letters (Part B; working memory, cognitive flexibility, executive functions, and visuo-spatial skills). The aim is to connect the numbers in ascending order from 1 to 25 (Part A) and the numbers in ascending order from 1 to 13 alternately with the letters in alphabetical order from A to L (Part B) in the shortest possible time without error. In addition, we introduced a motor speed condition with the task of following a predefined path connecting 25 circles [43] with (a) the idea of calculating the purely cognitive performance of the task of connecting numbers or numbers and letters without the influence of motor performance (moving the stylus, which can be difficult, especially for people with TD) and (b) to calculate the cognitive DTC.

The Activities Specific Balance Confidence (ABC) Scale [44] assessed fall-associated self-efficacy. On a scale of 0–100%, the participants should estimate their confidence to carry out 16 activities without becoming unbalanced. High percentages stand for a high fall-associated self-efficacy.

The Timed Up and Go (TUG; [45]) test is one of the most common tests used to examine balance, gait speed, and functional ability related to the performance of basic activities of daily living (ADL) in older populations. The TUG measures the time (seconds) it takes a participant to stand up from a chair, walk 3 m at a comfortable speed, walk around a cone, walk back, and sit down on the chair. With a "cut-off" value of 14 s or more, the TUG is considered a good predictor for identifying healthy individuals at risk of falling [46,47].

The Geriatric Depression Scale (GDS) according to Sheikh and Yesavage [48] comprises in short form 15 questions. The GDS allows for early detection of possible depression in aging patients. Scores between 0–5 are normal, scores between 6–10 indicate mild to moderate depression, and scores between 11–15 indicate severe depression.

2.2.2. Rating Scale for Parkinson's Disease

The UPDRS [32] is divided into the areas of (1) cognitive functions, behavior, and mood, (2) activities of daily living (ADL), (3) motor examination, and (4) complications of treatment. The motor dimension of UPDRS was used to determine PD cardinal symptoms of tremor, rigor, bradykinesia, and postural instability. Based on the evaluation and the classification algorithm according to Jankovic et al. [33], this scale allows for a classification of the mentioned PD motor phenotypes: tremor dominant type (TD = mean value of points for tremor/mean value of points for PIGD $\leq$1.5) and motor phenotype with postural instability and gait disorder (PIGD = mean value of points for tremor/mean value of points for PIGD $\geq$1.0). The following items were used to evaluate the two motor phenotypes:

2.2.3. Trail-Walking Test

The TWT [27] was used to assess motor-cognitive interference under change in direction walk conditions. In this motor-cognitive DT, 15 hats with banderoles are placed randomly in a 16 m^2 area (4 $\times$ 4 m). A 30 cm (diameter) circle was drawn around each hat. The TWT has three different conditions. In the first condition (TWT-M, ST), participants were instructed to follow a line connecting 15 circles (purely motor task). In condition 2 (TWT-A, DT), participants were asked to navigate to numbered targets in sequential and ascending order from 1 to 8. In a third and more cognitive demanding DT (TWT-B), participants were instructed to step on targets with an ascending, alternating number(1-8)-letter(A-G) sequence (1-A-2-B-3...-8) (see Supplementary Materials: Figure S1). Participants were asked to perform the task as accurately but as quickly as possible in all conditions. No priority was given to one domain or the other. Time per trial was taken with a stopwatch to the nearest 0.01 s, and motor errors (e.g., knocking over a hat or not stepping inside the circle) and shifting and sequencing errors (e.g., navigating to the wrong number/letter; [48]) were recorded. Sequence and shift errors were corrected promptly by the test administrator asking the participant to return to the last correct circle. Errors are reported and accounted for in the required times since the correction of errors takes extra time [43]. Each condition was performed three times.

2.3. Procedure

The participants were informed about the objectives and contents of the planned study and the test procedure, test duration, and possible risks of data collection. Before the data collection was carried out, a written declaration of consent was obtained. The methods used in these studies are in accordance with the ethical principles of the Helsinki Declaration [49], national legislation and relevant international norms and standards. The implementation of the procedures was randomized to avoid possible sequence effects. All tests were performed in a quiet environment to avoid distractions and to exclude possible interfering variables in the experimental situation. The majority of the participants could be tested within the planned 90 min. In order to keep the effect of fatigue to a minimum, a rest time of 1–3 min between the conditions and trials was made. None of the participants experienced complications during data collection. Previously trained test administrators carried out the data collection. The research project received ethical approval (REB # 4791

Project, "Motor-cognitive interference in dual tasks: allocation of resources in Parkinson's Disease patients" REB Clearance Issued: 19 February 2016). Participants were recruited in March 2016. Data collection also took place in March 2016.

2.4. Data Analysis

All statistics were performed with SPSS v.27 (SPSS, Chicago, IL, USA). We first explored dependent variables to examine missing values, normality of distributions (Kolmogorov–Smirnov tests), and presence of outliers (defined by the Explore command of SPSS v.27).

For the sample description, between group differences for continuous variables (e.g., age, height, weight, BMI, physical activity) were calculated using ANOVAs; partial eta^2 was calculated as an effect strength measure (Conventions of Cohen, [50]: 0.01 small effect; 0.06 medium effect; 0.14 strong effect). In addition, categorical demographic variables were tested with a Chi^2 test (e.g., sex).

To test the effect of different cognitive conditions and difficulty levels, a 3 (group: Control, TD, PIGD) × 3 (condition: TWT-M, TWT-A, TWT-B) ANOCVA with repeated measurements and duration of disease as covariate was performed for the times in the TWT. The between-subject factor is group, and the within-subject factor is condition (TWT-M, TWT-A, TWT-B). Group differences within a condition (e.g., TWT-M) were calculated with ANOVA. For the calculation of the dual-task costs (DTC), a 3 (group: Control, TD, PIGD) × 3 (condition: TWT-M, TWT-A, TWT-B) × 2 (interferences: motor vs. cognitive) ANOVA with repeated measurements was performed for the TWT. With significant results, post-hoc tests (Bonferroni correction) were used to check which factor levels significantly differ. An alpha value of 0.05 was used for all statistical tests (also for post-hoc analyses; [51]). In addition to the significance value ($p < 0.05$, * significant; $p < 0.01$; ** highly significant; $p < 0.001$, *** highly significant), the effect sizes for all ANOVAs were indicated using the partial eta^2.

The times in the conditions of the TWT were measured using a stopwatch and expressed as 0.01 s. For the times in the TWT conditions (TWT-M; TWT-A, and TWT-B), the mean values (X) of the three runs were used:

$$\overline{X} = \frac{1}{n}\sum_{i=1}^{1} Xi = \frac{X1 + X2 + X3}{n} \quad (1)$$

To calculate the DTCs, the performance under DT condition is related to the performance under ST condition. Since higher values indicate worse performance (times in the TWT), a negative sign was inserted. Negative DTCs indicate a deterioration compared to the ST condition [52]. Therefore, the motor and cognitive DTCs for the TWT are calculated as follows:

$$\text{DTC} = \frac{\text{Performance in DT} - \text{Performance in ST}}{\text{Performance in ST}} \times 100 \quad (2)$$

TWT

$$\text{motorDTC for TWTA } (\%) = \frac{-(\text{TWTA} - \text{TWTM})}{\text{TWTM}} \times 100 \quad (3)$$

$$\text{motorDTC for TWTB } (\%) = \frac{-(\text{TWTB} - \text{TWT})}{\text{TWTM}} \times 100 \quad (4)$$

$$\text{cognitiveDTC for A } (\%) = \frac{-(\text{TWTA} - (\text{TMTA} - \text{TMTM})}{(\text{TMTA} - \text{TMTM})} \times 100 \quad (5)$$

$$\text{cognitiveDTC for B } (\%) = \frac{-(\text{TWTB} - (\text{TMTB} - \text{TMTM})}{(\text{TMTB} - \text{TMTM})} \times 100 \quad (6)$$

The Trail-Making Test [42] was used to evaluate cognitive ST performance. Due to the different lengths of the conditions in the TMT (TMT-M: 185.4 cm; TMT-A: 185.4 cm and TMT-B: 243.8 cm) [53], the velocities in all conditions are first calculated as follows:

$$\text{Velocity TMT condition } \left(\frac{\text{cm}}{\text{s}}\right) = \frac{\text{Length of the condition}}{\text{Time for TMT condition}} \quad (7)$$

The velocity was normalized to the length of 200 cm (required time for 200 cm):

$$\text{Time for TMT condition (s)} = \frac{200}{\text{Velocity TMT condition}} \quad (8)$$

In addition, a "two-way" intra-class correlation coefficient (ICC) was calculated to quantify the consistency within the three trials of each TWT condition and the groups [54]. The test-retest reliability was assessed using the Standard Error of the Measurement [SEM = (SD $\times$ $\sqrt{(1 - \text{ICC})}$)] and the Minimum Detectable Change with a confidence interval of 95% [MDC95 = (1.96 $\times$ SEM $\times$ $\sqrt{(2)}$)] [55]. In order to be able to compare both measures, these were additionally expressed as percentages (SEM% and MDC95%; [56]).

ROC (Receiver Operating Characteristic) analyses were performed to determine the diagnostic quality of the TWT, where sensitivity, specificity, and area under the curve (AUC) were considered (for the interpretation of the values, see [57]). The Youden index was used to determine which threshold was best suited to differentiate the groups [58].

3. Results

3.1. Characteristics of the Study Population

Sixteen PD patients with dominant tremor (TD, M = 68.2 years, SD = 8.94), 27 PD patients with postural instability and gait difficulty (PIGD, M = 69.7 years, SD = 8.68) and 20 healthy older adults (control, M = 72.4 years, SD = 5.53) participated in this study. In the MoCA, no differences between the groups can be observed. Compared to TD and the control group, an increased frequency of falls can be observed in persons with PIGD. Accordingly, fall-associated self-efficacy is significantly lower in PIGD. The proportion of mild to moderate depression is also higher in PIGD than in both other groups (see Table 3).

Table 3. Characteristics of PD patients differentiated into motor phenotypes TD and PIGD, including mean values (standard deviation) and test values of UPDRS-III.

	Control	**TD Ratio ≤ 1.5**	**PIGD Ratio ≥ 1.0**	**Stat. Analyses**
	(n = 20)	(n = 16)	(n = 27)	
sex	7 men, 13 women	11 men, 5 women	23 men, 4 women	CHI2(2) = 14.2 **
age (years)	72.4 (5.53)	68.2 (8.94)	69.7 (8.68)	$F(2, 60) = 1.32$, $\eta^2_p = 0.042$
BMI (kg/m^2) Under-, Normal-, Obesity (n)	27.9 (4.73); 0, 6, 6, 5	24.2 (4.40) ^{t}; 2, 6, 6, 1	27.4 (4.07); 0, 8, 12, 5	$F(2, 57) = 3.79$ *, $\eta^2_p = 0.117$
UPDRS-III (Score; max = 108)	-	22.53 (7.47)	23.7 (7.97)	$t(41) = 0.646$, $d = 0.147$
Duration of the disease (years)	-	6.19 (4.92)	4.93 (4.37)	$t(41) = 0.388$, $d = 0.275$
Activities Specific Balance Confidence (ABC)Scale, %	95.3 (3.77)	89.7 (8.24)	79.3 (19.9) ^{t}	$F(2, 59) = 8.05$ **, $\eta^2_p = 0.214$
Fall experience last year (n persons; n in %, n falls)	3 persons (15%); 3 falls	4 persons (26.7%); 6 falls	9 persons (33%); 31 falls	$F(2, 14) = 1.68$, $\eta^2_p = 0.222$
Timed Up-and-Go test (TUG), seconds	8.55 (1.29)	9.19 (2.28)	11.1 (2.99) ^{t}	$F(2, 61) = 7.29$ *, $\eta^2_p = 0.198$
Montreal Cognitive Assessment (MoCA), score	27.9 (1.48)	26.9 (3.23)	27.6 (1.95)	$F(2, 60) = 0.854$, $\eta^2_p = 0.028$
	2 participants with a score below 26	4 participants with a score below 26	6 participants with a score below 26	$CHI^2(2) = 1.61$
Education (years)	10.5 (0.76)	13.7 (4.17)	13.5 (3.40) ^{t}	$F(2, 61) = 6.62$ **, $\eta^2_p = 0.189$
Geriatric Depression Scale (GDS), n	20 normal 0 mild to moderate 0 severe	15 normal 1 mild to moderate 0 severe	22 normal 5 mild to moderate 0 severe	$CHI^2(2) = 3.21$

Note. ** $p < 0.01$; * $p < 0.05$; ^{t} significant difference to control group ($p < 0.05$).

3.2. Reliability of Measurement Repetition in the TWT

Table 4 shows the relative and absolute reliability measures (ICC, SEM, MDC95). The inter-trial reliability is medium to excellent for all conditions and groups, with ICC values between 0.87 and 0.98. The reliability of the trials is between 0.87 and 0.98. In total, the SEM is between 0.22–3.20 s. The SEM% is low in all conditions and groups (0.51–4.05%). In 100% of the observations, a SEM% $\leq$ 10% can be found. The SEM varies between 0.26–2.68 s for the control group, 0.22–3.20 s for TD, and 0.43–2.18 s for PIGD. In total, the MDC95 is between 0.62–88.8 s for the absolute times in the TWT. The MDC95% fluctuated between 1.41–11.5% for the whole sample and is thus below $\leq$30%.

Table 4. Intra-class correlation (ICC) and absolute inter-trial reliability (SEM) across the three TWT conditions.

	Control			TD			PIGD		
	ICC (95% CI)	SEM/ SEM (%)	MDC95/ MDC95%	ICC (95% CI)	SEM/ SEM (%)	MDC95/ MDC95%	ICC (95% CI)	SEM/ SEM (%)	MDC95/ MDC95%
TWT-M	0.974 (0.95–0.99)	0.26/ 0.68	0.76/ 1.89	0.987 (0.97–0.99)	0.22/ 0.51	0.62/ 1.41	0.959 (0.92–0.98)	0.43/ 0.74	1.18/ 2.05
TWT-A	0.894 (0.78–0.96)	1.15/ 2.31	3.19/ 6.39	0.959 (0.90–0.98)	0.72/ 1.33	1.99/ 3.68	0.939 (0.88–0.97)	1.03/ 1.63	2.86/ 4.54
TWT-B	0.870 (0.72–0.94)	2.68/ 4.05	7.43/ 11.21	0.886 (0.74–0.96)	3.20/ 4.14	8.88/ 11.48	0.918 (0.84–0.96)	2.18/ 2.59	6.02/ 7.17

Note. In order to calculate the reliability measures, 3 trials (measurement time points) were included; CI = confidence interval; SEM = standard error of measurement; MDC = minimal detectable change.

3.3. Times as Performance Measure in the TWT

The times in TWT-M and TWT-A are normally distributed in all groups ($p < 0.05$). The times in the TWT-B tend to be normally distributed ($p = 0.069$). Age ($r = 215$, $p = 0.043$) correlates significantly with the times in TWT-B. Sex tends to have a significant influence on performance in TWT-B ($p = 0.093$), with higher times observed for women (women: M = 79.4, SE = 3.74; men: M = 68.7, SE = 5.02). Education does not correlate with the performance in the conditions of the TWT ($r = -0.064$–0.035). Duration of disease, however, correlates with performance on TWT-A ($r = 0.221$, $p = 0.04$) and TWT-B ($r = 0.249$, $p = 0.03$), but not with performance on TWT-M ($r = 0.128$, $p = 0.16$). Across groups, we observe 2 outliers. One participant in the PIGD group with a value of M = 136.3 in the TWTB and one participant in the TWT-M with a value of M = 127.7, also in the PIGD group.

A 3 (condition: TWT-M, TWT-A, TWT-B) $\times$ 3 (group: Control, TD, PIGD) ANCOVA with repeated measurement of times for the TWT and duration of disease as covariate shows significant main effect for condition, $F(1.24, 82.3) = 42.3$, $p < 0.001$, $\eta^2_p = 0.426$, and group, $F(2, 657) = 5.55$, $p < 0.05$, $\eta^2_p = 0.138$. The post-hoc analysis shows that times are significantly higher in TWT-B (M = 74.1, SE = 3.01) than in TWT-A (M = 55.9, SE = 1.50; $p < 0.001$, 95%CI = 13.1–23.3) or in the purely motor condition (TWT-M: M = 45.1, SE = 1.88; $p < 0.001$; 95%CI = 22.6–35.5) for all subjects. PIGD (M = 65.9, SE = 2.58) differ from TD (M = 56.2, SE = 3.74; $p = 0.108$, 95%CI = −1.44–20.9) and significantly differ from the control group (M = 52.9, SE = 3.81; $p < 0.05$, 95%CI = 0.735–25.1). TD patients are not significantly different from the control group ($p = 0.630$, 95%CI = −16.3–5.24). A significant interaction of condition x group does not exist, $F(2.88, 82.3) = 0.254$, $p = 0.851$, $\eta^2_p = 0.009$). This shows that all groups walk slower with increasing cognitive load and therefore need longer (see Figure 1). A difference in the times in the TWT can thus be observed, in particular between PIGD and the control group. The covariate duration of disease has no significant effect, $F(1, 57) = 0.582$, $p = 0.449$, $\eta^2_p = 0.010$.

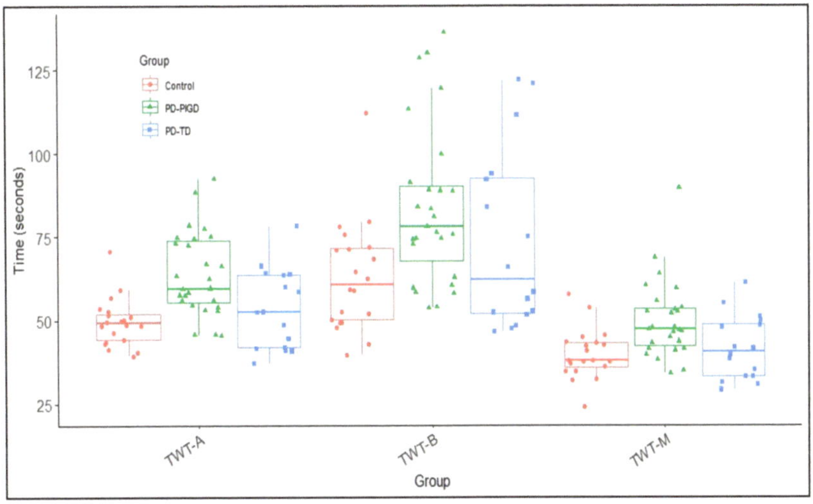

Figure 1. Mean and standard deviation of groups (PIGD, TD & control) and TWT conditions (TWT-A, TWT-B & TWT-M) based on times.

3.4. Motor-Related Cognitive Costs and Cognitive-Related Motor Costs in the TWT

Table 5 summarizes the mean values and standard deviations of the calculated DTC for the TWT.

Table 5. Mean values and standard deviation of DTC in the TWT divided into the PD phenotypes and healthy controls.

	PIGD (n = 27)	TD (n = 16)	Control (n = 20)	Statistical Analysis
Motor DTC TWT-A	−29.9 (18.4)	−29.7 (18.7)	−26.6 (17.6)	$F(2, 60) = 0.220$, $p = 0.803$, $\eta^2_p = 0.007$
Motor DTC TWT-B	−65.4 (31.9)	−75.4 (43.1)	−58.1 (26.8)	$F(2, 33) = 1.16$, $p = 0.320$, $\eta^2_p = 0.037$
Cognitive DTC TWT-A	−431 (543)	−430 (417)	−317 (284)	$F(2, 33) = 0.445$, $p = 0.643$, $\eta^2_p = 0.015$
Cognitive DTC TWT-B	−61.2 (62.2)	−118 (126)	−103 (90.8)	$F(2, 33) = 2.33$, $p = 0.106$, $\eta^2_p = 0.072$

Note. DTC = dual-task costs; the empirical mean values and standard deviations are shown; PIGD: Parkinson-Postural Instability and Gait Difficulty; TD: Parkinson-Tremor Dominant; Control: older adults without Parkinson diagnosis.

Regarding the proportional DTC a 3 (group: Control, TD, PIGD) × 2 (condition: TWT-A, TWT-B) × 2 (interference: cognitive, motor domain) ANOVA with repeated measurement for the times in TWT was calculated. The results show significant major effects for condition, $F(1, 60) = 19.5$, $p < 0.001$, $\eta^2_p = 0.245$, and interference, $F(1, 60) = 44.6$, $p < 0.001$, $\eta^2_p = 0.426$. A significant interaction effect can be observed for condition x interference, $F(1, 60) = 32.9$, $p < 0.001$, $\eta^2_p = 0.354$. Post-hoc analysis shows that with low cognitive load, (TWT-A: M = −211, SE = 28.7) the performance losses are greater than they are with high cognitive load (TWT-B: M = −80.3, SE = 5.52; $p < 0.001$; 95%CI = −189.9–71.5).

Figure 2a shows the distribution of cognitive and motor interference in TWT-A in individuals with PD (PIGD & TD) and healthy controls. Most participants show mutual interferences with performance losses, especially in the cognitive task. Interferences in the motor task are low across all groups. The level of motor and cognitive interference and the range is comparable in all groups. Some participants show minor interference in the motor task but improvements in the cognitive task performance. However, group-specific patterns cannot be identified. In the condition with a high additional cognitive load (TWT-B, Figure 2b), the cognitive interferences are lower than in the TWT-A. Mutual interference can also be observed in TWT-B across groups. A few participants show low or positive interferences in the cognitive but deterioration in the motor task performance (cognitive-motor interference or cognitive task prioritization). However, clear group-specific patterns also cannot be identified.

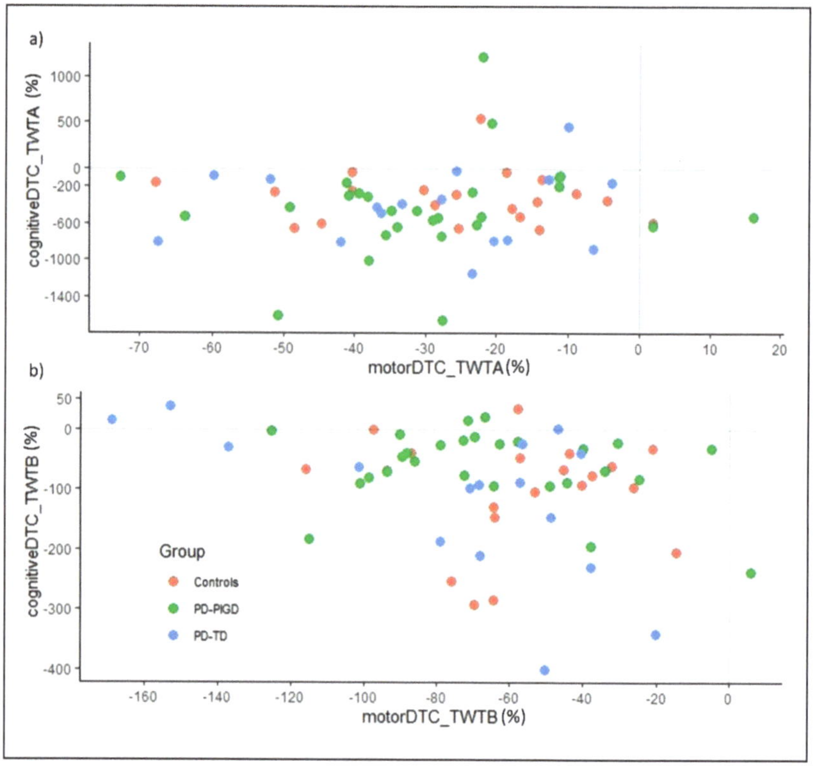

Figure 2. (**a**) Pattern of motor-cognitive DTC in TWT-A based on times in PIGD, TD, and control. (**b**) Pattern of motor-cognitive DTC in TWT-B based on times in PIGD, TD, and control. Motor and cognitive DTCs are calculated using the following formula: DTC (%) = ((performance DT − performance in ST)/performance in ST) $\times$100).

Based on the calculated velocities, the TWT conditions allow for an appropriate differentiation between the motor phenotype PIGD and the control group (AUC > 0.8; see Table 5; grey marked cells). The TWT-A allows for a good differentiation (AUC = 0.831; sensitivity = 0.852; specificity = 0.800). However, the differentiation between phenotypes, PIGD and TD, is not satisfactory by any TWT conditions (AUC < 0.7). Only the TWT-M condition shows a significant result here as evidence of the accuracy of the test procedure (see Table 6; value in bold). The TWT-A tends to be significant (see Table 6; value in bold). Additionally, a distinction between TD and the control group is not sufficiently precise with any of the TWT conditions (AUC < 0.7).

Table 6. Statistics and receiver operating characteristic curve thresholds for the TWT (velocities in the TWT; motor DTC) to differentiate between PIGD, TD, and the control group.

Condition	Groups	n	Youden Index	Sensitivity	Specificity	Threshold	AUC	p
TWT-M	PIGD vs. TD	27/16	0.326	0.889	0.438	1.05	0.685	**0.044**
	PIGD vs. Control	27/20	0.530	0.630	0.900	0.891	0.791	**<0.001**
	TD vs. Control	16/20	0.288	0.436	0.850	0.914	0.553	0.588
TWT-A	PIGD vs TD	27/16	0.352	0.852	0.500	0.778	0.662	0.079
	PIGD vs. Control	27/20	0.652	0.852	0.800	0.776	0.831	**<0.001**
	TD vs. Control	16/20	0.400	0.500	0.900	0.711	0.638	0.161
TWT-B	PIGD vs. TD	27/16	0.303	0.741	0.563	0.593	0.623	0.183
	PIGD vs. Control	27/20	0.541	0.741	0.800	0.567	0.783	<0.001
	TD vs. Control	16/20	0.325	0.375	0.950	0.503	0.613	0.252
Motor DTC TWT-A	PIGD vs. TD	27/16	0.234	0.296	0.938	−16.52	0.588	0.340
	PIGD vs. Control	27/20	0.356	0.556	0.800	−37.81	0.659	0.064
	TD vs. Control	16/20	0.300	0.500	0.800	−38.13	0.597	0.324
Motor DTC TWT-B	PIGD vs. TD	27/16	−0.093	0.407	0.500	−241.1	0.479	0.821
	PIGD vs. Control	27/20	−0.089	0.111	0.800	−133.7	0.513	0.880
	TD vs. Control	16/20	0.188	0.938	0.250	−378.7	0.541	0.679
Cognitive DTC TWT-A	PIGD vs. TD	27/16	0.264	0.889	0.625	−823.6	0.528	0.763
	PIGD vs. Control	27/20	−0.219	0.481	0.300	−493.3	0.431	0.426
	TD vs. Control	16/20	−0.375	0.625	0.000	−786.5	0.413	0.373
Cognitive DTC TWT-B	PIGD vs. TD	27/16	0.215	0.778	0.438	−35.53	0.567	0.466
	PIGD vs. Control	27/20	0.344	0.444	0.900	2.65	0.615	0.182
	TD vs. Control	16/20	0.225	0.375	0.850	−3.94	0.544	0.656

Note. n = number of cases; p = significance value; PD = Parsinson Disease; PIGD = Postural Instability/Gait Difficulty; TD = Tremor Dominant; DTC = dual-task costs; TWT = Trail-Walking Test; AUC (AUROC) = Area Under the Receiver Operating Characteristic Curve; For continuous variables, limit values were determined from the optimal combination of sensitivity and specificity using the Youden index; the relevant data mentioned in the text are highlighted in the table by the grey cells and significant results are highlighted in bold.

On the other hand, the motor and cognitive DTC do not allow the groups to be differentiated. Sensitivity and specificity are insufficient to distinguish the groups from each other. As a result, the areas under the curve of the ROC analyses as a measure of accuracy are too small.

4. Discussion

This study aimed to evaluate the TWT as a potential method for quantifying postural instability and gait disturbances in PD patients and distinguishing between PD motor phenotypes. As expected, all participants in the study were slower under DT conditions [59]. The effect was greater in PIGD patients than in the control and TD groups. The greater the cognitive load, the greater the influence on walking performance. However, the difference between the groups became smaller with increasing cognitive load. The largest differences between the groups were found in the TWT-M (purely motor condition).

The TWT performance differs both overall and within the three groups as expected. Times increase with increasing cognitive load and is in line with the studies by Spildooren et al. [60], Wild et al. [61], and Kelly et al. [62]. They demonstrated increased difficulties and balance problems with locomotion tasks under DT conditions in PD (cf. the meta-analysis by Raffegeau et al. [15]). In particular, walking speed is significantly influenced in these studies. A significant difference between the two PD phenotypes can only be observed in the TWT-M and the task with low cognitive load (TWT-A). A distinction between PIGD and the control group becomes significant in all conditions. However, a

95% CI between 0.735—25.1 s shows that the confidence interval is very large. The values are very heterogeneous. The increased time by 25.1 s would be clinically relevant, but an increased processing time by 0.735 s is clinically less relevant. The difference between TD and the control group does not become significant in any condition. This can be explained by the fact that the TWT primarily claims aspects of mobility [27]. In the condition of high cognitive load, the control group also shows problems with the automatic execution of walking and increased walking times, so the group differences become smaller. Based on the calculated AUC values, a good discriminatory power is demonstrated to distinguish PIGD from individuals without PD (control group), with only TWT-A showing sufficient sensitivity (85.2%) and specificity (80%). Only moderate to poor AUC values can be found to differentiate between the two PD groups. These results suggest that one of the underlying mechanisms for gait dysfunction is cognition and slowed walking in complex situations may result from an overloaded or inefficient cognitive system in PIGD.

Regarding the motor and cognitive DTC, differences between the conditions of the TWT (TWT-A & TWT-B) can be observed. Higher motor DTCs (−66.3%) can be observed in condition TWT-B compared to −28.7% for TWT-A ($p < 0.001$). The magnitude of the motor DTC is larger than the studies summarized in the overview article by Kelly et al. [9]. In the studies by Kelly and colleagues, a range between −1% to −59% motor DTC is reported. This indicates that the TWT is significantly more demanding and requires more cognitive resources than walking straight ahead [63,64] or walking with a 180-degree turn [60,65]. In contrast, higher cognitive DTC can be observed in the TWT-A condition, with −392% compared to −94.1% for TWT-B ($p < 0.001$). In comparison to the few studies that also calculate DTC for the cognitive task (Galletly & Brauer, [66], with +31% and +72%, points to an improvement and prioritization of the cognitive task; O'Shea, Morris & Iansek, [67], with −5%; Yogev et al. [14], with −42%), in the present study cognitive DTCs can also be found, which are many times larger. With −392% (TWT-A) and −28.7% (TWT-B) cognitive DTC, high cognitive performance decrements can be observed, especially under low cognitive load. Moreover, there is a difference between the conditions of TWT in motor and cognitive DTC. All groups in the condition with high cognitive load (TWT-B) show greater motor-related cognitive DTCs compared to cognitive-related motor DTCs ($p < 0.001$). In the condition with low cognitive load (TWT-A), larger motor-related cognitive DTCs can be observed across all groups ($p < 0.001$). One possible explanation for the large cognitive DTC in TWT-A is that the relatively simple counting in ascending order is possible despite resource allocation toward the motor task, and the task can still be accomplished. In the TWT-B, on the other hand, the cognitive task (numbers and letters running alternately and in ascending order) requires more cognitive resources, which means that the limited attention resources [68] must be shifted in the direction of the cognitive task in order to complete TWT-B. Thus, a strategic allocation of attention resources is necessary to complete the TWT as successfully as possible. Contrary to theoretical expectations, no group-specific patterns of this allocation can be observed in this study. PD patients and the control group appear to have similar motor-cognitive interference patterns in complex locomotor tasks. Both PD groups and the control group show a risky allocation of resources towards cognitive tasks ("posture second" strategy) in DTs with high cognitive load (TWT-B) [36]. In the DT with lower cognitive load (TWT-A), on the other hand, an allocation of resources towards motor tasks with high cognitive DTC can be observed ("posture first" strategy) (see [61]).

There are some limitations in this study that need to be mentioned. Our sample size is very small. It is difficult to draw meaningful conclusions from analyses based on groups of 16, 27, and 20 participants. The study is cross-sectional. With longitudinal studies, changes over time can be mapped, and the prognosis can be improved. An explanation of the insufficient differentiation between the groups is that the classification into the mentioned motor phenotypes by the classification algorithm according to Jankovic et al. [33] only reflects the relationship between the cardinal symptoms (tremor and postural instability). For example, TD with strong tremors also showed significant constraints in balance control [69]. If the UPDRS is used to classify motor phenotypes, the scale (especially the UPDRS III;

motor analysis) is fundamental. While some neurologists and researchers advocate the scale, others consider the scale to be a less representative snapshot of the current physical condition of PD patients. The test is based on subjective assessment by a neurologist and is highly dependent on the examiner's expertise. In addition, the DTC was calculated based on the required times. Other measures are probably needed to show differences between the two PD motor phenotypes [70]. Gait parameters and their changes under DTs could allow for a more differentiated conclusion of the motor differences between these phenotypes [28,71] and improve the prognosis in the progression of PIGD.

5. Conclusions

Currently, there is no gold standard for assessing postural instability and gait disorders that encompasses all aspects of cognitive and physical characteristics in PD [71]. DTs, compared to STs, are often more sensitive in assessing these gait disorders [31,72]. Problems that remain undetected in the single-task condition can emerge with the use of DT paradigms.

PD therapy is primarily based on early detection and treatment of symptoms. The aim is to maintain the independence of those affected by the disease as long as possible to preserve the quality of life. Thus, patients with gait disorders must be identified by valid, well-evaluated clinical tests that precisely assess the individual fall risk and severity of the disease. Unfortunately, there is currently no gold standard for assessing postural instability and gait disorders that address all aspects of PD's cognitive and physical characteristics [71]. The TWT covers various aspects of mobility and examines the relationship between cognitive functions and walking [27]. Especially the pure motor condition shows high ICC values and a SEM% below 1. Based on the results and concerning the sensitivity and specificity of the procedure, a differentiation into PD motor subtypes can be made as expected, especially with the purely motor condition of the TWT (TWT-M) and based on times. In future studies, it would be interesting to examine whether walking with directional changes (TWT) and an additional motor task (e.g., box-checking task; analogous to the studies by Heinzel, Maechtel, Hasmann, Hobert, Heger, Berg & Maetzler, [73]) generates more apparent differences in the DTCs between PD and a control group.

Supplementary Materials: The following supporting information can be downloaded at: https://www.mdpi.com/article/10.3390/brainsci12091217/s1, Figure S1: The conditions of the Trail-Walking-Test.

Author Contributions: Conceptualization, N.S. and Q.J.A.; Data curation, N.S.; Formal analysis, T.J.K. and N.S.; Investigation, T.J.K., N.S. and Q.J.A.; Methodology, T.J.K., N.S. and Q.J.A.; Supervision, Q.J.A.; Visualization, T.J.K. and N.S.; Writing—original draft, T.J.K. and N.S.; Writing—review & editing, Q.J.A. All authors have read and agreed to the published version of the manuscript.

Funding: This research received no external funding. The APC was funded by the University of Stuttgart through the Institutional Open Access Program (IOAP).

Institutional Review Board Statement: All participants were informed of the nature and aim of the study and signed a consent form. All procedures followed the Declaration of Helsinki with ethical standards, legal requirements, and international norms. We have ethics approval from the Wilfried Laurier University, Waterloo, Canada, for using this protocol (REB # 4791 Project, "Motor-cognitive interference in dual tasks: allocation of resources in Parkinson's Disease patients" REB Clearance Issued: 19 February 2016).

Informed Consent Statement: Informed consent was obtained from all subjects involved in the study.

Data Availability Statement: All relevant data are within the study, and raw data are available on request.

Acknowledgments: The authors thank the volunteers who participated in the study.

Conflicts of Interest: The authors have no financial or personal relationships with any other person or organization that could improperly influence or otherwise influence their work in this study. The authors declare no conflict of interest.

References

1. Aarsland, D.; Bronnick, K.; Williams-Gray, C.; Weintraub, D.; Marder, K.; Kulisevsky, J.; Burn, D.; Barone, P.; Pagonabarraga, J.; Allcock, L.; et al. Mild cognitive impairment in Parkinson disease: A multicenter pooled analysis. *Neurology* **2010**, *75*, 1062–1069. [CrossRef] [PubMed]
2. Watson, G.S.; Leverenz, J.B. Profile of cognitive impairment in Parkinson's disease. *Brain Pathol.* **2010**, *20*, 640–645. [CrossRef] [PubMed]
3. Kehagia, A.A.; Barker, R.A.; Robbins, T.W. Neuropsychological and clinical heterogeneity of cognitive impairment and dementia in patients with Parkinson's disease. *Lancet Neurol.* **2010**, *9*, 1200–1213. [CrossRef]
4. Aarsland, D.; Andersen, K.; Larsen, J.P.; Lolk, A. Prevalence and characteristics of dementia in Parkinson disease: An 8-year prospective study. *Arch. Neurol.* **2003**, *60*, 387–392. [CrossRef]
5. Hobson, P.; Meara, J. The detection of dementia and cognitive impairment in a community population of elderly people with Parkinson's disease by use of the CAMCOG neuropsychological test. *Age Ageing* **1999**, *28*, 39–43. [CrossRef] [PubMed]
6. Hely, M.A.; Reid, W.G.; Adena, M.A.; Halliday, G.M.; Morris, J.G. The Sydney multicenter study of Parkinson's disease: The inevitability of dementia at 20 years. *Mov. Disord.* **2008**, *23*, 837–844. [CrossRef]
7. Amboni, M.; Barone, P.; Ms, L.I.; Lista, I.; Tranfaglia, R.; Fasano, A.; Picillo, M.; Vitale, C.; Santangelo, G.; Agosti, V.; et al. Gait patterns in parkinsonian patients with or without mild cognitive impairment. *Mov. Disord.* **2012**, *27*, 1536–1543. [CrossRef]
8. Marchese, R.; Bove, M.; Abbruzzese, G. Effect of cognitive and motor tasks on postural stability in Parkinson's disease: A posturographic study. *Mov. Disord.* **2003**, *18*, 652–658. [CrossRef]
9. Kelly, V.E.; Eusterbrock, A.J.; Shumway-Cook, A. A review of dual-task walking deficits in people with Parkinson's disease: Motor and cognitive contributions, mechanisms, and clinical implications. *Parkinson's Dis.* **2012**, *2012*, 918719. [CrossRef]
10. Pelicioni, P.H.; Menant, J.C.; Henderson, E.J.; Latt, M.D.; Brodie, M.A.; Lord, S.R. Mild and marked executive dysfunction and falls in people with Parkinson's disease. *Braz. J. Phys. Ther.* **2021**, *25*, 437–443. [CrossRef]
11. Mak, M.K.; Wong, A.; Pang, M.Y. Impaired executive function can predict recurrent falls in Parkinson's disease. *Arch. Phys. Med. Rehabil.* **2014**, *95*, 2390–2395. [CrossRef] [PubMed]
12. Foster, E.R.; Hershey, T. Everyday Executive Function is Associated with Activity Participation in Parkinson Disease without Dementia. *OTJR Occup. Particip. Health* **2011**, *31*, S16–S22. [CrossRef] [PubMed]
13. Poldrack, R.A.; Sabb, F.W.; Foerde, K.; Tom, S.M.; Asarnow, R.F.; Bookheimer, S.Y.; Knowlton, B.J. The neural correlates of motor skill automaticity. *J. Neurosci.* **2005**, *25*, 5356–5364. [CrossRef] [PubMed]
14. Yogev, G.; Giladi, N.; Peretz, C.; Springer, S.; Simon, E.S.; Hausdorff, J.M. Dual tasking, gait rhythmicity, and Parkinson's disease: Which aspects of gait are attention-demanding? *Eur. J. Neurosci.* **2005**, *22*, 1248–1256. [CrossRef]
15. Raffegeau, T.E.; Krehbiel, L.M.; Kang, N.; Thijs, F.J.; Altmann, L.J.; Cauraugh, J.H.; Hass, C.J. A meta-analysis: Parkinson's disease and dual-task walking. *Parkinsonism Relat. Disord.* **2019**, *62*, 28–35. [CrossRef]
16. Aleksovski, D.; Miljkovic, D.; Bravi, D.; Antonini, A. Disease progression in Parkinson subtypes: The PPMI dataset. *Neurol. Sci.* **2018**, *39*, 1971–1976. [CrossRef]
17. Eggers, C.; Pedrosa, D.; Kahraman, D.; Maier, F.; Lewis, C.J.; Fink, G.R.; Schmidt, M.; Timmermann, L. Parkinson Subtypes Progress Differently in Clinical Course and Imaging Pattern. *PLoS ONE* **2012**, *7*, e46813. [CrossRef]
18. Ba, F.; Obaid, M.; Wieler, M.; Camicioli, R.; Martin, W.W. Parkinson Disease: The Relationship between Non-motor Symptoms and Motor Phenotype. *Can. J. Neurol. Sci./J. Can. Des Sci. Neurol.* **2016**, *43*, 261–267. [CrossRef]
19. Huang, X.; Ng, S.E.; Chia, N.Y.; Setiawan, F.; Tay, K.Y.; Au, W.L.; Tan, E.K.; Tan, L.S. Non-motor symptoms in early Parkinson's disease with different motor subtypes and their associations with quality of life. *Eur. J. Neurol.* **2019**, *26*, 400–406. [CrossRef]
20. Ren, J.; Hua, P.; Pan, C.; Li, Y.; Zhang, L.; Zhang, W.; Xu, P.; Zhang, M.; Liu, W. Non-Motor Symptoms of the Postural Instability and Gait Difficulty Subtype in De Novo Parkinson's Disease Patients: A Cross-Sectional Study in a Single Center. *Neuropsychiatr. Dis. Treat.* **2020**, *16*, 2605–2612. [CrossRef]
21. Cullen, S.; Montero-Odasso, M.; Bherer, L.; Almeida, Q.; Fraser, S.; Hunter, S.; Li, K.; Liu-Ambrose, T.; McGibbon, C.A.; McIlroy, W.; et al. Guidelines for Gait Assessments in the Canadian Consortium on Neurodegeneration in Aging (CCNA). *Can. Geriatr. J.* **2018**, *21*, 157–165. [CrossRef] [PubMed]
22. Lowry, K.A.; Brach, J.S.; Nebes, R.D.; Studenski, S.A.; VanSwearingen, J.M. Contributions of Cognitive Function to Straight- and Curved-Path Walking in Older Adults. *Arch. Phys. Med. Rehabil.* **2012**, *93*, 802–807. [CrossRef] [PubMed]
23. Mazaheri, M.; Roerdink, M.; Bood, R.J.; Duysens, J.; Beek, P.J.; Peper, C.E. Attentional costs of visually guided walking: Effects of age, executive function and stepping-task demands. *Gait Posture* **2014**, *40*, 182–186. [CrossRef] [PubMed]
24. Schott, N.; Klotzbier, T. The motor–cognitive connection: Indicator of future developmental success in children and adolescents? In *Physical Activity and Educational Achievement*; Routledge: London, UK, 2018; pp. 133–151.
25. Beurskens, R.; Bock, O. Age-related decline of peripheral visual processing: The role of eye movements. *Exp. Brain Res.* **2012**, *217*, 117–124. [CrossRef] [PubMed]
26. Al-Yahya, E.; Dawes, H.; Smith, L.; Dennis, A.; Howells, K.; Cockburn, J. Cognitive motor interference while walking: A systematic review and meta-analysis. *Neurosci. Biobehav. Rev.* **2011**, *35*, 715–728. [CrossRef]
27. Schott, N. Trail Walking Test zur Erfassung der motorisch-kognitiven Interferenz bei älteren Erwachsenen. *Z. Für Gerontol. Geriatr.* **2015**, *48*, 722–733. [CrossRef]

28. Belghali, M.; Chastan, N.; Cignetti, F.; Davenne, D.; Decker, L.M. Loss of gait control assessed by cognitive-motor dual-tasks: Pros and cons in detecting people at risk of developing Alzheimer's and Parkinson's diseases. *GeroScience* **2017**, *39*, 305–329. [CrossRef]
29. Movement Disorder Society Task Force on Rating Scales for Parkinson's Disease. The unified Parkinson's disease rating scale (UPDRS): Status and recommendations. *Mov. Disord.* **2003**, *18*, 738–750. [CrossRef]
30. Bloem, B.R.; Marinus, J.; Almeida, Q.; Dibble, L.; Nieuwboer, A.; Post, B.; Schrag, A.; Movement Disorders Society Rating Scales Committee. Measurement instruments to assess posture, gait, and balance in Parkinson's disease: Critique and recommendations. *Mov. Disord.* **2016**, *31*, 1342–1355. [CrossRef]
31. Klotzbier, T.J.; Schott, N. Cognitive-Motor Interference during Walking in Older Adults with Probable Mild Cognitive Impairment. *Front. Aging Neurosci.* **2017**, *9*, 350. [CrossRef] [PubMed]
32. Goetz, C.G.; Fahn, S.; Martinez-Martin, P.; Poewe, W.; Sampaio, C.; Stebbins, G.T.; Stern, M.B.; Tilley, B.C.; Dodel, R.; Dubois, B.; et al. Movement Disorder Society-sponsored revision of the Unified Parkinson's Disease Rating Scale (MDS-UPDRS): Process, format, and clinimetric testing plan. *Mov. Disord.* **2007**, *22*, 41–47. [CrossRef]
33. Jankovic, J.; McDermott, M.; Carter, J.; Gauthier, S.; Goetz, C.; Golbe, L.; Huber, S.; Koller, W.; Olanow, C.; Shoulson, I.; et al. Variable expression of Parkinson's disease A base-line analysis of the DAT ATOP cohort. *Neurol.* **1990**, *40*, 1529. [CrossRef]
34. Blin, O.; Ferrandez, A.M.; Pailhous, J.; Serratrice, G. Dopa-sensitive and dopa-resistant gait parameters in Parkinson's disease. *J. Neurol. Sci.* **1991**, *103*, 51–54. [CrossRef]
35. Morris, M.E.; Huxham, F.; McGinley, J.; Dodd, K.; Iansek, R. The biomechanics and motor control of gait in Parkinson disease. *Clin. Biomech.* **2001**, *16*, 459–470. [CrossRef]
36. Bloem, B.R.; Grimbergen, Y.A.; van Dijk, J.G.; Munneke, M. The "posture second" strategy: A review of wrong priorities in Parkinson's disease. *J. Neurol. Sci.* **2006**, *248*, 196–204. [CrossRef]
37. Yogev-Seligmann, G.; Hausdorff, J.M.; Giladi, N. Do we always prioritize balance when walking? Towards an integrated model of task prioritization. *Mov. Disord.* **2012**, *27*, 765–770. [CrossRef]
38. Johansson, H.; Ekman, U.; Rennie, L.; Peterson, D.S.; Leavy, B.; Franzén, E. Dual-Task Effects During a Motor-Cognitive Task in Parkinson's Disease: Patterns of Prioritization and the Influence of Cognitive Status. *Neurorehabilit. Neural Repair* **2021**, *35*, 356–366. [CrossRef] [PubMed]
39. Nasreddine, Z.S.; Phillips, N.A.; Bédirian, V.; Charbonneau, S.; Whitehead, V.; Collin, I.; Cummings, J.L.; Chertkow, H. The Montreal Cognitive Assessment, MoCA: A Brief Screening Tool for Mild Cognitive Impairment. *J. Am. Geriatr. Soc.* **2005**, *53*, 695–699. [CrossRef]
40. Duffin, J.T.; Collins, D.R.; Coughlan, T.; O'Neill, D.; Roche, R.A.P.; Commins, S. Subtle memory and attentional deficits revealed in an Irish stroke patient sample using domain-specific cognitive tasks. *J. Clin. Exp. Neuropsychol.* **2012**, *34*, 864–875. [CrossRef]
41. Kenny, R.A.; Coen, R.F.; Frewen, J.; Donoghue, O.A.; Cronin, H.; Savva, G.M. Normative Values of Cognitive and Physical Function in Older Adults: Findings from The Irish Longitudinal Study on Ageing. *J. Am. Geriatr. Soc.* **2013**, *61*, S279–S290. [CrossRef]
42. Reitan, R. Validity of TMT as an indication of organic brain damage. *Percept. Mot. Ski.* **1958**, *8*, 271–276. [CrossRef]
43. Schott, N.; El-Rajab, I.; Klotzbier, T. Cognitive-motor interference during fine and gross motor tasks in children with Developmental Coordination Disorder (DCD). *Res. Dev. Disabil.* **2016**, *57*, 136–148. [CrossRef]
44. Powell, L.E.; Myers, A.M. The activities-specific balance confidence (ABC) scale. *J. Gerontol. Ser. A Biol. Sci. Med. Sci.* **1995**, *50*, M28–M34. [CrossRef] [PubMed]
45. Podsiadlo, D.; Richardson, S. The timed "Up & Go": A test of basic functional mobility for frail elderly persons. *J. Am. Geriatr. Soc.* **1991**, *39*, 142–148.
46. Shumway-Cook, A.; Brauer, S.; Woollacott, M. Predicting the Probability for Falls in Community-Dwelling Older Adults Using the Timed Up & Go Test. *Phys. Ther.* **2000**, *80*, 896–903. [CrossRef] [PubMed]
47. Allison, L.K.; Painter, J.A.; Emory, A.; Whitehurst, P.; Raby, A. Participation Restriction, Not Fear of Falling, Predicts Actual Balance and Mobility Abilities in Rural Community-Dwelling Older Adults. *J. Geriatr. Phys. Ther.* **2013**, *36*, 13–23. [CrossRef] [PubMed]
48. Klusman, L.E.; Cripe, L.I.; Dodrill, C.B. Analysis of Errors on the Trail Making Test. *Percept. Mot. Ski.* **1989**, *68* (Suppl. S3), 1199–1204. [CrossRef]
49. World Medical Association. World Medical Association Declaration of Helsinki: Ethical principles for medical research involving human subjects. *JAMA* **2013**, *310*, 2191–2194. [CrossRef]
50. Cohen, J. *Statistical Power Analysis for the Behavioral Sciences*; Lawrence Earlbaum Associates: Hillsdale, NJ, USA, 1988.
51. Tabachnick, B.G.; Fidell, L.S.; Ullman, J.B. *Using Multivariate Statistics*; Pearson: Boston, MA, USA, 2007; Volume 5, pp. 481–498.
52. Eplummer, P.; A Eskes, G. Measuring treatment effects on dual-task performance: A framework for research and clinical practice. *Front. Hum. Neurosci.* **2015**, *9*, 225. [CrossRef]
53. Gaudino, E.A.; Geisler, M.W.; Squires, N.K. Construct validity in the trail making test: What makes part B harder? *J. Clin. Exp. Neuropsychol.* **1995**, *17*, 529–535. [CrossRef]
54. Fleiss, J.L. Reliability of Measurement. In *The Design and Analysis of Clinical Experiments*; John Wiley & Sons, Inc.: Hoboken, NJ, USA, 1999; pp. 1–32. [CrossRef]

55. Weir, J.P. Quantifying test-retest reliability using the intraclass correlation coefficient and the sem. *J. Strength Cond. Res.* **2005**, *19*, 231–240. [CrossRef] [PubMed]
56. Atkinson, G.; Nevill, A.M. Statistical Methods for Assessing Measurement Error (Reliability) in Variables Relevant to Sports Medicine. *Sports Med.* **1998**, *26*, 217–238. [CrossRef] [PubMed]
57. Swets, J.A. Measuring the accuracy of diagnostic systems. *Science* **1988**, *240*, 1285–1293. [CrossRef] [PubMed]
58. Hilden, J.; Glasziou, P. Regret graphs, diagnostic uncertainty and Youden's Index. *Stat. Med.* **1996**, *15*, 969–986. [CrossRef]
59. Lord, S.; Rochester, L.; Hetherington, V.; Allcock, L.M.; Burn, D. Executive dysfunction and attention contribute to gait interference in 'off' state Parkinson's Disease. *Gait Posture* **2010**, *31*, 169–174. [CrossRef]
60. Spildooren, J.; Vercruysse, S.; Desloovere, K.; Vandenberghe, W.; Kerckhofs, E.; Nieuwboer, A. Freezing of gait in Parkinson's disease: The impact of dual-tasking and turning. *Mov. Disord.* **2010**, *25*, 2563–2570. [CrossRef]
61. Wild, L.B.; De Lima, D.B.; Balardin, J.B.; Rizzi, L.; Giacobbo, B.L.; Oliveira, H.B.; Argimon, I.; Peyré-Tartaruga, L.A.; Rieder, C.R.; Bromberg, E. Characterization of cognitive and motor performance during dual-tasking in healthy older adults and patients with Parkinson's disease. *J. Neurol.* **2013**, *260*, 580–589. [CrossRef]
62. Kelly, V.E.; Shumway-Cook, A. The ability of people with Parkinson's disease to modify dual-task performance in response to instructions during simple and complex walking tasks. *Exp. Brain Res.* **2014**, *232*, 263–271. [CrossRef] [PubMed]
63. LaPointe, L.L.; Stierwalt, J.A.; Maitland, C.G. Talking while walking: Cognitive loading and injurious falls in Parkinson's disease. *Int. J. Speech-Lang. Pathol.* **2010**, *12*, 455–459. [CrossRef]
64. Yogev, G.; Plotnik, M.; Peretz, C.; Giladi, N.; Hausdorff, J.M. Gait asymmetry in patients with Parkinson's disease and elderly fallers: When does the bilateral coordination of gait require attention? *Exp. Brain Res.* **2007**, *177*, 336–346. [CrossRef]
65. Lord, S.; Baker, K.; Nieuwboer, A.; Burn, D.; Rochester, L. Gait variability in Parkinson's disease: An indicator of non-dopaminergic contributors to gait dysfunction? *J. Neurol.* **2011**, *258*, 566–572. [CrossRef] [PubMed]
66. Galletly, R.; Brauer, S.G. Does the type of concurrent task affect preferred and cued gait in people with Parkinson's disease? *Aust. J. Physiother.* **2005**, *51*, 175–180. [CrossRef]
67. O'Shea, S.; Morris, M.; Iansek, R. Dual Task Interference during Gait in People with Parkinson Disease: Effects of Motor Versus Cognitive Secondary Tasks. *Phys. Ther.* **2002**, *82*, 888–897. [CrossRef]
68. Tombu, M.; Jolicœur, P. A central capacity sharing model of dual-task performance. *J. Exp. Psychol. Hum. Percept. Perform.* **2003**, *29*, 3–18. [CrossRef] [PubMed]
69. Maetzler, W.; Hausdorff, J.M. Motor signs in the prodromal phase of Parkinson's disease. *Mov. Disord.* **2012**, *27*, 627–633. [CrossRef]
70. Mirelman, A.; Ben Or Frank, M.; Melamed, M.; Granovsky, L.; Nieuwboer, A.; Rochester, L.; Del Din, S.; Avanzino, L.; Pelosin, E.; Bloem, B.R.; et al. Detecting Sensitive Mobility Features for Parkinson's Disease Stages Via Machine Learning. *Mov. Disord.* **2021**, *36*, 2144–2155. [CrossRef] [PubMed]
71. Bridenbaugh, S.A.; Kressig, R.W. Motor cognitive dual tasking. *Z. Für Gerontol. Geriatr.* **2015**, *48*, 15–21. [CrossRef]
72. Vance, R.C.; Healy, D.G.; Galvin, R.; French, H.P. Dual Tasking with the Timed "Up & Go" Test Improves Detection of Risk of Falls in People with Parkinson Disease. *Phys. Ther.* **2015**, *95*, 95–102. [CrossRef]
73. Heinzel, S.; Maechtel, M.; Hasmann, S.E.; Hobert, M.A.; Heger, T.; Berg, D.; Maetzler, W. Motor dual-tasking deficits predict falls in Parkinson's disease: A prospective study. *Parkinsonism Relat. Disord.* **2016**, *26*, 73–77. [CrossRef]

brain sciences

MDPI

Article

The Remapping of Peripersonal Space in a Real but Not in a Virtual Environment

Francesca Ferroni [1,*], Vittorio Gallese [1], Agata Marta Soccini [2], Nunzio Langiulli [1], Francesca Rastelli [1], Donato Ferri [3], Francesco Bianchi [3] and Martina Ardizzi [1]

1 Unit of Neuroscience, Department of Medicine and Surgery, University of Parma, 43126 Parma, Italy
2 Computer Science Department, University of Torino, 10124 Turin, Italy
3 EY Advisory, 00187 Rome, Italy
* Correspondence: francesca.ferroni1@unipr.it; Tel.: +39-0521-903873; Fax: +39-0521-903879

Abstract: One of the most surprising features of our brain is the fact that it is extremely plastic. Among the various plastic processes supported by our brain, there is the neural representation of the space surrounding our body, the peripersonal space (PPS). The effects of real-world tool use on the PPS are well known in cognitive neuroscience, but little is still known whether similar mechanisms also govern virtual tool use. To this purpose, the present study investigated the plasticity of the PPS before and after a real (Experiment 1) or virtual motor training with a tool (Experiment 2). The results show the expansion of the PPS only following real-world tool use but not virtual use, highlighting how the two types of training potentially rely on different processes. This study enriches the current state of the art on the plasticity of PPS in real and virtual environments. We discuss our data with respect to the relevance for the development of effective immersive environment for trainings, learning and rehabilitation.

Keywords: action; multisensory integration; plasticity; space; tool use; virtual reality

check for updates

Citation: Ferroni, F.; Gallese, V.; Soccini, A.M.; Langiulli, N.; Rastelli, F.; Ferri, D.; Bianchi, F.; Ardizzi, M. The Remapping of Peripersonal Space in a Real but Not in a Virtual Environment. *Brain Sci.* **2022**, *12*, 1125. https://doi.org/10.3390/brainsci12091125

Academic Editor: Daniele Corbo

Received: 28 July 2022
Accepted: 21 August 2022
Published: 24 August 2022

1. Introduction

Through evolution, the brain has functionally developed a modular representation of space creating an imaginary sector immediately surrounding our body, known as peripersonal space (PPS) [1–3]. This region of space is a multisensory space where visual or auditory inputs related to the environment are integrated with tactile and proprioceptive information concerning specific body parts [1,4–6]. This spatial representation remaps plastically after different experiences, e.g., it expands after tool use [7–11], it contracts after immobilization [12], it blurs after sensory deprivation [13], or it is modulated by the type of social context [14,15]. Several studies have described the changes in the representation of PPS after interacting with physical tools [16–18]. However, we often interact also with remote tools to control computers, such as a keyboard, computer mouse or joysticks. As an example of the variety of applications, more and more surgeons interact daily with robotic devices that allow them to perform complex surgical procedures in a more accurate and efficient manner without being in direct contact with the patient [19–21]. In these contexts, our relationships with objects/people in the space surrounding our body–PPS–are affected. Indeed, although still few in number, some studies have started to investigate how PPS can be influenced by remote interactions. It has been in fact demonstrated that PPS is shaped by remote interaction mediated by the computer mouse [22] or surgical robots [23,24]. Specifically, Sengül et al. [24] have demonstrated that the active use of a robotic tool changes the integration of multisensory information, assessed by means of the cross-modal congruency task in PPS, comparable with the remapping of PPS during real-world tool use. Moreover, Bassolino and colleagues [22] showed that the space where a pointing tool (i.e., a computer mouse) was actually held, close to hand, was extended to the space where it operated (i.e., the computer screen), even though these spaces were not physically connected. Alongside increasingly cutting-edge technologies that integrate real-objects and remote virtual interactions, immersive virtual technologies are finding widespread applications in our hybrid (i.e., real–virtual) world. Today, the virtual reality environment is a valid technology applied to several activities, including neuropsychological rehabilitation [25]; psychopathological exposure therapy [26]; and military, educational and surgical training [27,28]. In these virtual immersive environments, people have the ability to move around in the space with the potential to manipulate or interact with objects much as they could in the real world. Despite all these technical advances in the field of virtual immersive technologies and a recent growing interest of neuroscience in virtual environments, e.g., [15,29–33], little attention has been paid to the neural and behavioral mechanisms underlying the interactions with these devices and the consequent remapping of the physical milieu surrounding our body. Thus, taking into account the potentially relevant implications on our current world and the physical space in which we act every day, it is relevant to investigate what happens during a virtual interaction with a tool in one's own PPS. To fill this gap, we have investigated the plasticity of PPS after two different motor training in a real (Experiment 1) and a virtual immersive environment (Experiment 2). Based on previous observations [30,34], in both tasks, we assess the extension of PPS in an immersive virtual reality environment. With respect to previous studies in this field [35,36] that adopted not widely used tasks as accurate proxies for measuring PPS, we decided to use a multisensory task to measure PPS in order to have more generalizable results with the current literature in this field. Moreover, we adopted a within-subjects experimental design to compare the performances of participants during real (i.e, real-world tool use) and virtual (virtual-world tool use) trainings. Taking into account evidence from several previous studies on PPS remapping after tool use [7,8,11,22], we expected (i) to confirm the validity of the visuo-tactile PPS task [29,30] in detecting PPS boundaries; (ii) to find a PPS expansion after the real-world tool use, accordingly with the current wide literature e.g., [7,8,37]; (iii) to detect a similar expansion, or slightly smaller, after the virtual motor training than that observed during the real training.

2. Materials and Methods

A total of 22 participants took part in this study (9 males, M = 24.57 years, SE = 5.24). Participants performed both Experiment 1 and Experiment 2 (within subjects design), administered in a balanced manner among participants. Participants' handedness was assessed by the Edinburgh handedness inventory [38] (M = 0.83, SE = 0.18). All participants had normal touch and normal or corrected-to-normal vision. The study was approved by the Local Ethical Committee (AVEN) and was carried out in accordance with the Declaration of Helsinki (1964 and subsequent amendments).

2.1. Procedure

The experimental procedure, identical for Experiments 1 and 2, consisted of three sessions all carried out on the same day. First, participants performed the visuo-tactile Peripersonal Space task (Session 1) in order to measure the individual PPS at baseline. After this session, they took part in Session 2 (i.e., training phase, different for Experiment 1 and 2; see below). Lastly, participants were submitted again to the PPS task (Session 3) in order to measure PPS boundaries after the Session 2.

2.1.1. Session 1 and 3

The location of participants' PPS boundary was measured with the adapted version of the visuo-tactile task widely adopted to measure PPS extent [30,39–44] implemented by our group using Unity3D 2020 for Meta Quest 2. Participants were seated behind a table and were asked to wear the head-mounted display while holding the controllers in both hands. They were instructed to ignore the presentation of the approaching or receding visual stimuli, travelling at the velocity of 75 cm/s, and respond as fast as possible to the vibrotactile stimulation, administered on the right controller, by pressing the button on the left controller. The visual stimulus was a tridimensional virtual red ball, 6.5 cm in diameter, looming toward the right hand of the participant. The ball travelled in virtual space from far to near or vice versa in the case of receding stimuli. Tactile stimuli of 10 ms of duration were delivered at 5 different temporal delays from the onset of the looming and receding visual stimuli (after 2165, 1732, 1299, 866, and 433 ms), resulting in 5 different distances from the body (D1–D5, ranging from 37.12 to 167.03 cm from the participant, in 32.5 cm intervals), following the procedure adopted by Masson and colleagues [29]. Trials were equally divided into two blocks for a total of 220 trials, lasting about 8 min each. Each trial was repeated if participants failed to respond to the tactile target. For a detailed description of the procedure, please refer to Supplementary Materials (Section 1).

2.1.2. Session 2

Session 2 differed between the two experiments. Specifically, in Experiment 1, during Session 2 participants were instructed to move 50 small colored objects (green and red), placed on two marked areas of the table, in the far space (85 cm from participants' chest) [8]. Participants sat along the short side of the table and were requested to use a tool to grab and move one object at a time across the two areas. All objects were moved from one marked area to the other and then repositioned on the initial area for a total of 100 movements (Figure 1). In Experiment 2, participants performed the virtual version of the motor training adopted in our previous study [8], described above. Participants were instructed to move 50 small colored virtual objects (green and red), placed on two marked areas of the virtual table in far space, using a virtual 75 cm-long garbage clamp held pressing a button of the right controller, as in Experiment 1. Participants had to wear a pair of white surgical gloves in order to promote a sense of embodiment with the virtual white hands, as we did not use an avatar with humanoid appearance and with phenotypic characteristics of the participants' real hand. In both experiments, Session 2 lasted around 10 min (Figure 1).

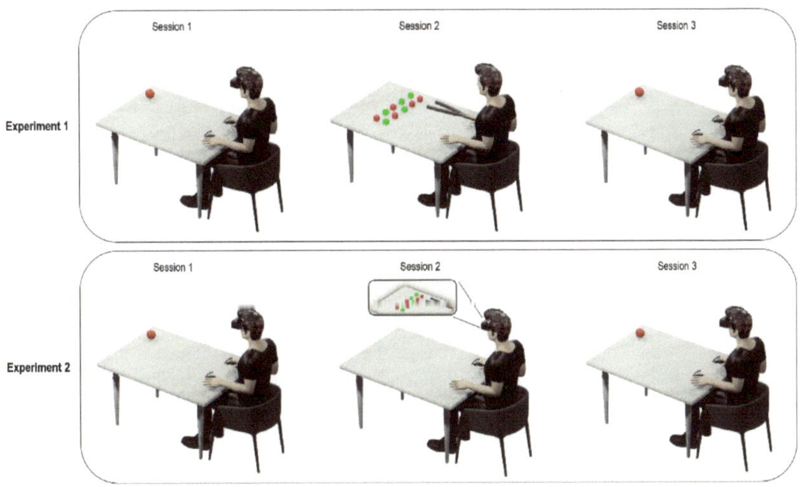

Figure 1. Graphical representation of the experimental procedure. For both experiments, Session 1 and Session 3 show the experimental setting of the visuo-tactile Peripersonal Space (PPS) task. Session 2 shows the qualitative representation of the training phase for both Experiments 1 and 2.

3. Data analysis and Results

3.1. Multisensory Tactile RTs

We first performed an analysis of variance (ANOVA) to check the different modulations of the looming compared with receding stimuli on tactile reaction times (RTs) independent of condition (Session 1, Session 3) or experiment (Experiment 1, Experiment 2) at different distances (D1, D2, D3, D4, D5). Specifically, data were entered in a repeated-measures ANOVA with two within-subject factors, Visual stimuli (Looming, Receding) and Distance (D1, D2, D3, D4, D5). For RT measurement, please consult the Supplementary Materials (Section 2). The ANOVA showed a significant effect of visual stimuli ($F_{(1,80)} = 14.80$, $p < 0.001$, $\eta^2_p = 0.42$). Indeed, it has been repeatedly shown that the present task is especially sensitive to approaching as compared with receding stimuli [7,14,39,42,45]; therefore, we here focused on results concerning the Looming visual stimuli only (Receding visual stimuli data are reported in Supplementary Table S1), as in previous studies on PPS, e.g., [42]. The function describing the relationship between tactile RTs and the perceived position of the visual stimuli in space showed that tactile RTs progressively sped up as the perceived visual stimuli's distance from the body decreased, as expected. Specifically, the ANOVA conducted on Looming stimuli only revealed a significant main effect of Distance ($F_{(1,4)} = 11.10$, $p < 0.001$, $\eta^2_p = 0.36$). Newman–Keuls post hoc tests showed that RTs at D1 (when the visual stimuli were perceived far from the body; M = 560.53 ms, SE = 20.40) and D2 (M = 535.31 ms, SE = 12.91) were significantly slower compared with RTs at D3 (when the visual stimuli were perceived close to the body; M = 515.53 ms, SE = 12.23), D4 (M = 499.28 ms, SE = 12.15) and D5 (M = 498.48, SE = 12.70; all $p_s < 0.03$). This was considered a preliminary step in order to proceed to considering only Looming stimuli as experimental variables.

3.2. Unisensory Tactile RTs

Additionally, an ANOVA across sensory modalities was carried out to confirm that multisensory looming trials (regardless of distance) were faster than unisensory tactile trials, and thus, as expected, visual presentations facilitated tactile responses e.g., [46]. Specifically, we performed a repeated-measures ANOVA on Looming and Unisensory RTs to tactile targets measured across sensory modalities to confirm the multisensory facilitation effect independently from Distance (D1, D2, D3, D4, D5), Condition (Session 1, Session 3) or Experiment (Experiment 1, Experiment 2). Thus, RTs were entered in a repeated-measures ANOVA with sensory modality (multisensory, unisensory) as the within-subjects factor. The ANOVA showed a significant main effect of sensory modality ($F_{(1,20)} = 55.77$, $p < 0.001$, $\eta^2_p = 0.74$), showing that multisensory RTs (multisensory: M = 521.83, SE = 12.58) were faster than the unisensory tactile ones (unisensory: M = 604.11, SE = 16.64), demonstrating a clear multisensory facilitation effect in line with previous studies e.g., [9,46].

3.3. Peripersonal Space Estimation

Lastly, to estimate the individual boundary of PPS, PSE (point of subjective equality) of the psychometric function describing visuo-tactile RTs as a function of visuo-tactile distance was measured via the Spearman–Karber (SK) method [47,48] in line with recent studies on PPS [9,29,46]. For more specific details about the implemented procedure, please refer to the Supplementary Materials (Section 2). PSE values estimated using Looming RTs in Session 1 (PSE-pre) and in Session 3 (PSE-post) were entered into ANOVA with Condition (PSE-pre, PSE-post) and Experiment (Experiment 1, Experiment 2) as within-subjects factors. Results showed a significant interaction Experiment by Condition ($F_{(1,21)} = 9.37$, $p = 0.005$, $\eta^2_p = 0.31$). Newman–Keuls post-hoc carried out on the significant interaction Experiment by Condition revealed that PSE-pre values (M = 1277.50 ms, SE = 42.31) were significantly higher than the PSE-post ones (M = 1167.63 ms, SE = 33.60, $p = 0.03$) only for the Real Training (Figure 2), thus revealing a peripersonal space expansion only after the real-world tool use (Experiment 1). No differences were found between PSE-pre and PSE-post values after the virtual-world tool use (Experiment 2) or in any other comparison (all $p_s > 0.08$).

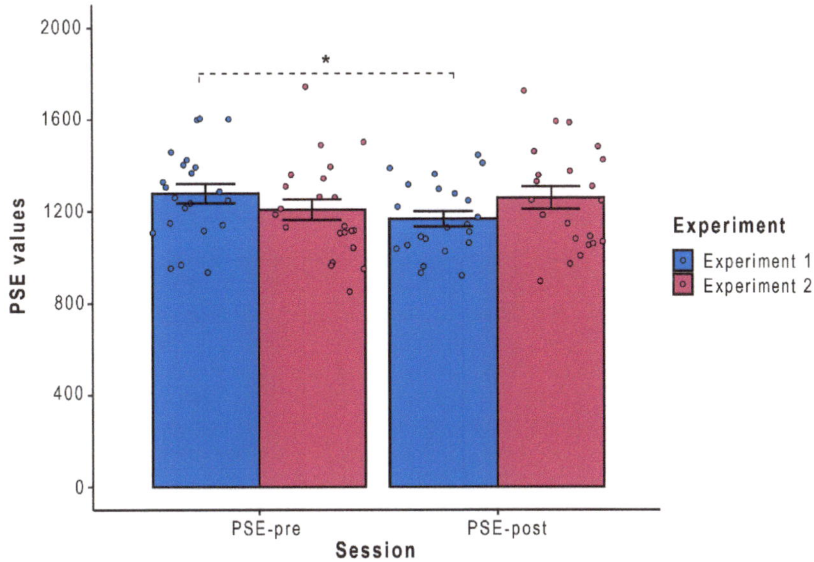

Figure 2. Point of subjective equality (PSE) values measured in Session 1 and Session 3, for both Experiments. Error bars depicted SE; * = $p < 0.05$.

3.4. Slopes Estimation

The slope's values (hereafter DL, difference limen, estimated via the SK method) measured in Session 1 (DL-pre) and in Session 3 (DL-post) were entered into ANOVA with Condition (DL-pre, DL-post) and Experiment (Experiment 1, Experiment 2) as within-subjects factors. No significant results were found (all $p_s > 0.09$).

For all the analyses, whenever appropriate, significant differences were explored performing Newman–Keuls post-hoc comparison. Partial eta-squared (η^2_p) was calculated as effect size measure.

4. Discussion

The present study investigated the plasticity of PPS after a real (Experiment 1) and a virtual motor training session (Experiment 2) executed with a tool in the extrapersonal space. To reach this goal, participants performed a visuo-tactile interaction task, e.g., [15,29] to identify the distance at which an approaching visual stimulus speeded up tactile processing as a proxy for the boundary of PPS, both before and after the two different training sessions with a tool. Our results confirm the validity of the visuo-tactile PPS task [30,34]. Indeed, independently from Experiments 1 and 2, we showed that a virtual approaching visual stimulus sped up participants' reaction time to match a tactile stimulation on their body at Session 1. This visuo-tactile interaction effect depended on the distance between the virtual stimulus and the body of the participant, as a significant facilitation emerged specifically when the virtual object was closer than a certain distance, which can be measured as a proxy of the location of the boundary of the individual's PPS, e.g., [15,30]. Importantly, we found a PPS expansion only after the real-world tool use (Experiment 1), as shown by the lower PSE-post values than the PSE-pre ones only in the case of Experiment 1. Differently, and contrary to our expectations, an effect of virtual tool use in the remapping of PPS was not found, as no significant change was pointed out between PSE-pre and -post values in the case of Experiment 2.

Today, we live in an unprecedented condition in which the space around our bodies can abstract from physical contingency and might have a virtual immanence. The data of the present work show that under the same minimal conditions (the appearance and type of interactions in the two experiments were in fact the same), the virtual environment is less effective than the real one in stimulating the plasticity of PPS. Multiple factors, which must be taken into account when developing immersive virtual environments, could explain this difference.

Firstly, participants were not familiar with the use of the virtual tool. Indeed, no preliminary familiarization phase with the virtual tool was carried out. Despite it is well known that the boundaries of PPS enlarge even after an interaction with a real unfamiliar object [49], it is possible that the use of a virtual unfamiliar tool requires an extra-familiarization phase to elicit a PPS expansion. Knowing that, the higher the level of practice with a real tool, the longer the extension of PPS [11,50], it is possible that a virtual training relies on different temporal dynamics than those of the real training, requiring a longer time to elicit the expansion.

Secondly, we did not use an avatar with humanoid appearance and with phenotypic characteristics of the participants' body. Participants simply saw virtual white hands moving in synchrony with their own hands while wearing white surgical gloves. In agreement with studies showing that action self-attribution in virtual environment resulted from an interaction between bottom-up and top-down processes e.g., [51], the here implemented dynamic visual congruency was not enough to induce the integration of the virtual tool into the participants' body representation. Considering also previous studies highlighting the impact of the bodily self on the encoding of virtual environment e.g., [52–54], to develop effective virtual immersive environment the inclusion of an avatar body in a first-person perspective should be considered.

Lastly, in the present setup, a tactile vibration was delivered by the controller when participants grasped the virtual object with the virtual pliers. However, other proprioceptive and sensory feedbacks resulting from the movement of the real objects (e.g., weight changes at the end of the clamp) were missing in the case of virtual objects. Often surgeons and robotics researchers report this lack of sensory feedback, thus highlighting a significant limitation in robot-assisted minimally invasive surgery [55]. To date, the issue of haptic and proprioceptive feedbacks in robotic surgery systems is still a major technological limitation that does not allow surgeons to feel the interaction between the tool and the anatomy, operating with obscured sensory feedback [56]. Providing sensory feedback in virtual training can produce a strong change in PPS representation and a relevant remapping of visuo-tactile integration, not just for virtual tool use training but also in action observation task [36].

At first glance, it is impossible to define which of these hypothesized factors might be the most plausible one. To our knowledge, this is the first study in which the expansion of the PPS is tested in a virtual immersive environment, by adopting a multisensory interaction task as a proxy for the PPS. As a benchmark, we can certainly confirm that the minimum conditions proposed here (and currently available to the common virtual reality devices) are not sufficient to induce PPS plasticity. It is of interest to emphasize how these conclusions can lead to improvements in both neuroscientific research and the application of virtual reality [57]. Indeed, on the one hand, the use of virtual settings in neuroscientific research can shed light on the processes required to induce plasticity in the PPS, which to date have been little investigated [42]. By individually modifying the factors evidenced here (i.e., familiarity, body presence and sensory feedback) in future studies, it will be possible to determine what are the necessary and sufficient conditions to induce a PPS expansion in virtual immersive environments. There is no doubt that virtual environment is the only context in which it is possible to manipulate each of the aforementioned factors individually. On the other hand, virtual technologies are finding widespread applications in our hybrid-world and they are progressively more popular in various fields (e.g., information services, video-games, people services), especially now that spatial computing and the metaverse are promising to be the next standard paradigm of the internet and therefore become pervasive in everyday life. Understanding what are the mandatory features that our brain requires to allow a virtual environment to induce plastic change can greatly enhance the performativity of the designed environments. Indeed, although much is already known about the neural basis underlying the representation of PPS e.g., [58–61], the neural mechanisms underlying the PPS plasticity elicited in real and virtual environments still remain to be well understood. The resulting increased concreteness of virtual environments may turn out to be of pivotal importance for the design of future learning, training, and rehabilitative protocols [62,63].

5. Conclusions

The present study provides the first evidence of the expansion of the PPS only following a real world tool use but not a virtual one, highlighting how the two types of training potentially rely on different processes. The unity of our bodily experiences with respect to an ever-changing and evolving world is a fundamental condition of human interaction with the external environment, even when the environment is virtual. Indeed, the way in which the environment offers more and more variation affects the impact of the space (around us) on humans and vice versa. Thus, in order to maintain this mutual interaction, it is necessary to investigate the virtual environment from a neuroscientific perspective so that it is truly a space for (inter)-action.

Supplementary Materials: The following supporting information can be downloaded at: https://www.mdpi.com/article/10.3390/brainsci12091125/s1, Table S1: Data of Receding RTs at five different distances [64,65].

Author Contributions: Conceptualization, F.F., V.G. and M.A.; Methodology, F.F. and M.A.; Software, A.M.S.; Formal analysis, F.F. and N.L.; Investigation, F.F. and F.R.; Resources, M.A.; Data curation, F.F., V.G., D.F., F.B. and M.A.; Writing—original draft preparation, F.F. and M.A.; Supervision, M.A.; Project administration, M.A.; Funding acquisition, M.A. All authors have read and agreed to the published version of the manuscript.

Funding: This research has financially been supported by the Programme "FIL-Quota Incentivante" of University of Parma and co-sponsored by Fondazione Cariparma to M.A and by Ernst and Young Business School grant to M.A.

Institutional Review Board Statement: The study was conducted in accordance with the Declaration of Helsinki, and approved by the Local Ethical Committee (AVEN), protocol code 35488 (892/2020/DISP/UNIPR), on18/09/2020.

Informed Consent Statement: Informed consent was obtained from all subjects involved in the study.

Data Availability Statement: The data presented in this study are available on request from the corresponding author.

Acknowledgments: We thank Luca Sergey Rinaldi who kindly assisted with the collection of the data, and Andrea Forino for his contribution in the design of the experimental set-up.

Conflicts of Interest: The authors have no financial or ethical conflict of interest to declare.

References

1. Rizzolatti, G.; Fadiga, L.; Fogassi, L.; Gallese, V. The space around us. *Science* **1997**, *277*, 190–191. [CrossRef] [PubMed]
2. Làdavas, E. Functional and dynamic properties of visual peripersonal space. *Trends Cogn. Sci.* **2002**, *6*, 17–22. [CrossRef]
3. Graziano, M.S.; Cooke, D.F. Parieto-frontal interactions, personal space, and defensive behavior. *Neuropsychologia* **2006**, *44*, 845–859. [CrossRef] [PubMed]
4. Fogassi, L.; Gallese, V.; Di Pellegrino, G.; Fadiga, L.; Gentilucci, M.; Luppino, G.; Matelli, M.; Pedotti, A.; Rizzolatti, G. Space coding by premotor cortex. *Exp. Brain Res.* **1992**, *89*, 686–690. [CrossRef]
5. Fogassi, L.; Gallese, V.; Fadiga, L.; Luppino, G.; Matelli, M.; Rizzolatti, G. Coding of peripersonal space in inferior premotor cortex (area F4). *J. Neurophysiol.* **1996**, *76*, 141–157. [CrossRef]
6. Gross, C.G.; Graziano, M.S.A. Review: Multiple Representations of Space in the Brain. *Neuroscientist* **1995**, *1*, 43–50. [CrossRef]
7. Canzoneri, E.; Ubaldi, S.; Rastelli, V.; Finisguerra, A.; Bassolino, M.; Serino, A. Tool-use reshapes the boundaries of body and peripersonal space representations. *Exp. Brain Res.* **2013**, *228*, 25–42. [CrossRef]
8. Ferroni, F.; Ardizzi, M.; Ferri, F.; Tesanovic, A.; Langiulli, N.; Tonna, M.; Marchesi, C.; Gallese, V. Schizotypy and individual differences in peripersonal space plasticity. *Neuropsychologia* **2020**, *147*, 107579. [CrossRef]

9. Ferroni, F.; Ardizzi, M.; Magnani, F.; Langiulli, N.; Rastelli, F.; Lucarini, V.; Giustozzi, F.; Volpe, R.; Marchesi, C.; Tonna, M.; et al. Tool-use Extends Peripersonal Space Boundaries in Schizophrenic Patients. *Schizophr. Bull.* **2022**, sbac067. [CrossRef]
10. Iriki, A.; Tanaka, M.; Iwamura, Y. Coding of modified body schema during tool use by macaque postcentral neurones. *NeuroReport* **1996**, *7*, 2325–2330.
11. Serino, A.; Bassolino, M.; Farnè, A.; Làdavas, E. Extended Multisensory Space in Blind Cane Users. *Psychol. Sci.* **2007**, *18*, 642–648. [CrossRef] [PubMed]
12. Bassolino, M.; Finisguerra, A.; Canzoneri, E.; Serino, A.; Pozzo, T. Dissociating effect of upper limb non-use and overuse on space and body representations. *Neuropsychologia* **2015**, *70*, 385–392. [CrossRef] [PubMed]
13. Noel, J.-P.; Park, H.-D.; Pasqualini, I.; Lissek, H.; Wallace, M.; Blanke, O.; Serino, A. Audio-visual sensory deprivation degrades visuo-tactile peri-personal space. *Conscious. Cogn.* **2018**, *61*, 61–75. [CrossRef]
14. Teneggi, C.; Canzoneri, E.; Di Pellegrino, G.; Serino, A. Social modulation of peripersonal space boundaries. *Curr. Biol.* **2013**, *23*, 406–411. [CrossRef]
15. Pellencin, E.; Paladino, M.P.; Herbelin, B.; Serino, A. Social perception of others shapes one's own multisensory peripersonal space. *Cortex* **2018**, *104*, 163–179. [CrossRef] [PubMed]
16. Holmes, N.P.; Sanabria, D.; Calvert, G.A.; Spence, C. Tool-use: Capturing multisensory spatial attention or extending multisensory peripersonal space? *Cortex* **2007**, *43*, 469–489. [CrossRef]
17. Maravita, A.; Iriki, A. Tools for the body (schema). *Trends Cogn. Sci.* **2004**, *8*, 79–86. [CrossRef]
18. Maravita, A.; Spence, C.; Kennett, S.; Driver, J. Tool-use changes multimodal spatial interactions between vision and touch in normal humans. *Cognition* **2002**, *83*, B25–B34. [CrossRef]
19. Horgan, S.; Vanuno, D. Robots in laparoscopic surgery. *J. Laparoendosc. Adv. Surg. Tech.* **2001**, *11*, 415–419. [CrossRef] [PubMed]
20. Ayav, A.; Bresler, L.; Brunaud, L.; Boissel, P. Early results of one-year robotic surgery using the Da Vinci system to perform advanced laparoscopic procedures. *J. Gastrointest. Surg.* **2004**, *8*, 720–726. [CrossRef]
21. Maeso, S.; Reza, M.; Mayol, J.; Blasco, J.A.; Guerra, M.; Andradas, E.; Plana, M.N. Efficacy of the da vinci surgical system in abdominal surgery compared with that of laparoscopy: A systematic review and meta-analysis. *Ann. Surg.* **2010**, *252*, 254–262. [CrossRef] [PubMed]
22. Bassolino, M.; Serino, A.; Ubaldi, S.; Làdavas, E. Everyday use of the computer mouse extends peripersonal space representation. *Neuropsychologia* **2010**, *48*, 803–811. [CrossRef] [PubMed]
23. Rognini, G.; Sengül, A.; Aspell, J.E.; Salomon, R.; Bleuler, H.; Blanke, O. Visuo-tactile integration and body ownership during self-generated action. *Eur. J. Neurosci.* **2013**, *37*, 1120–1129. [CrossRef]
24. Sengül, A.; Van Elk, M.; Rognini, G.; Aspell, J.E.; Bleuler, H.; Blanke, O. Extending the body to virtual tools using a robotic surgical interface: Evidence from the crossmodal congruency task. *PLoS ONE* **2012**, *7*, e49473. [CrossRef] [PubMed]
25. Rizzo, A.A.; Schultheis, M.; Kerns, K.A.; Mateer, C. Analysis of assets for virtual reality applications in neuropsychology. *Neuropsychol. Rehabil.* **2004**, *14*, 207–239. [CrossRef]
26. Powers, M.B.; Emmelkamp, P.M.G. Virtual reality exposure therapy for anxiety disorders: A meta-analysis. *J. Anxiety Disord.* **2008**, *22*, 561–569. [CrossRef] [PubMed]
27. Fox, J.; Arena, D.; Bailenson, J.N. Virtual Reality: A Survival Guide for the Social Scientist. *J. Media Psychol.* **2009**, *21*, 95–113. [CrossRef]
28. Wang, P.; Becker, A.; Jones, I.; Glover, A.; Benford, S.; Greenhalgh, C.; Vloeberghs, M. A virtual reality surgery simulation of cutting and retraction in neurosurgery with force-feedback. *Comput. Methods Programs Biomed.* **2006**, *84*, 11–18. [CrossRef]
29. Masson, C.; van der Westhuizen, D.; Noel, J.-P.; Prevost, A.; van Honk, J.; Fotopoulou, A.; Solms, M.; Serino, A. Testosterone administration in women increases the size of their peripersonal space. *Exp. Brain Res.* **2021**, *239*, 1639–1649. [CrossRef]
30. Serino, A.; Noel, J.-P.; Mange, R.; Canzoneri, E.; Pellencin, E.; Ruiz, J.B.; Bernasconi, F.; Blanke, O.; Herbelin, B. Peripersonal space: An index of multisensory body-environment interactions in real, virtual, and mixed realities. *Front. ICT* **2018**, *4*, 31. [CrossRef]
31. Iachini, T.; Coello, Y.; Frassinetti, F.; Ruggiero, G. Body Space in Social Interactions: A Comparison of Reaching and Comfort Distance in Immersive Virtual Reality. *PLoS ONE* **2014**, *9*, e111511. [CrossRef] [PubMed]
32. Iachini, T.; Coello, Y.; Frassinetti, F.; Senese, V.P.; Galante, F.; Ruggiero, G. Peripersonal and interpersonal space in virtual and real environments: Effects of gender and age. *J. Environ. Psychol.* **2016**, *45*, 154–164. [CrossRef]
33. Noel, J.P.; Pfeiffer, C.; Blanke, O.; Serino, A. Peripersonal space as the space of the bodily self. *Cognition* **2015**, *144*, 49–57. [CrossRef] [PubMed]
34. Buck, L.E.; Park, S.; Bodenheimer, B. Determining Peripersonal Space Boundaries and Their Plasticity in Relation to Object and Agent Characteristics in an Immersive Virtual Environment. In Proceedings of the 2020 IEEE Conference on Virtual Reality and 3D User Interfaces (VR), Atlanta, GA, USA, 22–26 March 2020; pp. 332–342. [CrossRef]
35. Sengül, A.; Rognini, G.; van Elk, M.; Aspell, J.E.; Bleuler, H.; Blanke, O. Force feedback facilitates multisensory integration during robotic tool use. *Exp. Brain Res.* **2013**, *227*, 497–507. [CrossRef]
36. Sengül, A.; van Elk, M.; Blanke, O.; Bleuler, H. Congruent Visuo-Tactile Feedback Facilitates the Extension of Peripersonal Space. *Lect. Notes Comput. Sci.* **2018**, *10894*, 673–684. [CrossRef]
37. Galigani, M.; Castellani, N.; Donno, B.; Franza, M.; Zuber, C.; Allet, L.; Garbarini, F.; Bassolino, M. Effect of tool-use observation on metric body representation and peripersonal space. *Neuropsychologia* **2020**, *148*, 107622. [CrossRef]
38. Oldfield, R.C. The assessment and analysis of handedness: The Edinburgh inventory. *Neuropsychologia* **1971**, *9*, 97–113. [CrossRef]

39. Canzoneri, E.; Magosso, E.; Serino, A. Dynamic sounds capture the boundaries of peripersonal space representation in humans. *PLoS ONE* **2012**, *7*, e44306. [CrossRef]
40. Noel, J.-P.; Failla, M.D.; Quinde-Zlibut, J.M.; Williams, Z.J.; Gerdes, M.; Tracy, J.M.; Zoltowski, A.R.; Foss-Feig, J.H.; Nichols, H.; Armstrong, K.; et al. Visual-Tactile Spatial Multisensory Interaction in Adults with Autism and Schizophrenia. *Front. Psychiatry* **2020**, *11*, 578401. [CrossRef]
41. Serino, A.; Noel, J.-P.; Galli, G.; Canzoneri, E.; Marmaroli, P.; Lissek, H.; Blanke, O. Body part-centered and full body-centered peripersonal space representations. *Sci. Rep.* **2015**, *5*, 18603. [CrossRef]
42. Serino, A.; Canzoneri, E.; Marzolla, M.; di Pellegrino, G.; Magosso, E. Extending peripersonal space representation without tool-use: Evidence from a combined behavioral-computational approach. *Front. Behav. Neurosci.* **2015**, *9*, 4. [CrossRef] [PubMed]
43. Noel, J.; Marta, Ł.; Wallace, M.; Serino, A. Multisensory simultaneity judgment and proximity to the body. *J. Vis.* **2018**, *16*, 21. [CrossRef] [PubMed]
44. Lee, H.-S.; Hong, S.-J.J.; Baxter, T.; Scott, J.; Shenoy, S.; Buck, L.; Bodenheimer, B.; Park, S. Altered Peripersonal Space and the Bodily Self in Schizophrenia: A Virtual Reality Study. *Schizophr. Bull.* **2021**, *47*, 927–937. [CrossRef] [PubMed]
45. Canzoneri, E.; Marzolla, M.; Amoresano, A.; Verni, G.; Serino, A. Amputation and prosthesis implantation shape body and peripersonal space representations. *Sci. Rep.* **2013**, *3*, 2844. [CrossRef]
46. Noel, J.-P.; Paredes, R.; Terrebonne, E.; Feldman, J.I.; Woynaroski, T.; Cascio, C.J.; Seriès, P.; Wallace, M.T. Inflexible Updating of the Self-Other Divide During a Social Context in Autism: Psychophysical, Electrophysiological, and Neural Network Modeling Evidence. *Biol. Psychiatry Cogn. Neurosci. Neuroimaging* **2021**, *7*, 756–764. [CrossRef]
47. Bausenhart, K.M.; Di Luca, M.; Ulrich, R. Assessing Duration Discrimination: Psychophysical Methods and Psychometric Function Analysis. In *Timing and Time Perception: Procedures, Measures, & Applications*; Brill: Leiden, The Netherlands, 2018; pp. 52–78.
48. Miller, J.; Ulrich, R. On the analysis of psychometric functions: The Spearman-Kärber method. *Percept. Psychophys.* **2001**, *63*, 1399–1420. [CrossRef]
49. Farnè, A.; Iriki, A.; Làdavas, E. Shaping multisensory action-space with tools: Evidence from patients with cross-modal extinction. *Neuropsychologia* **2005**, *43*, 238–248. [CrossRef]
50. Biggio, M.; Bisio, A.; Avanzino, L.; Ruggeri, P.; Bove, M. This racket is not mine: The influence of the tool-use on peripersonal space. *Neuropsychologia* **2017**, *103*, 54–58. [CrossRef]
51. Tsakiris, M.; Haggard, P. The rubber hand illusion revisited: Visuotactile integration and self-attribution. *J. Exp. Psychol. Hum. Percept. Perform.* **2005**, *31*, 80. [CrossRef]
52. Brechet, L.; Mange, R.; Herbelin, B.; Theillaud, Q.; Gauthier, B.; Serino, A.; Blanke, O. First-person view of one's body in immersive virtual reality: Influence on episodic memory. *PLoS ONE* **2019**, *14*, e0197763. [CrossRef] [PubMed]
53. Bréchet, L.; Hausmann, S.B.; Mange, R.; Herbelin, B.; Blanke, O.; Serino, A. Subjective feeling of re-experiencing past events using immersive virtual reality prevents a loss of episodic memory. *Brain Behav.* **2020**, *10*, e01571. [CrossRef] [PubMed]
54. Gauthier, B.; Bréchet, L.; Lance, F.; Mange, R.; Herbelin, B.; Faivre, N.; Bolton, T.A.; Van De Ville, D.; Blanke, O. First-person body view modulates the neural substrates of episodic memory and autonoetic consciousness: A functional connectivity study. *NeuroImage* **2020**, *223*, 117370. [CrossRef]
55. Okamura, A.M. Haptic feedback in robot-assisted minimally invasive surgery. *Curr. Opin. Urol.* **2009**, *19*, 102–107. [CrossRef]
56. Simaan, N.; Yasin, R.M.; Wang, L. Medical Technologies and Challenges of Robot-Assisted Minimally Invasive Intervention and Diagnostics. *Annu. Rev. Control Robot. Auton. Syst.* **2018**, *1*, 465–490. [CrossRef]
57. Soccini, A.M.; Ferroni, F.; Ardizzi, M. From Virtual Reality to Neuroscience and Back: A Use Case on Peripersonal Hand Space Plasticity. In Proceedings of the 2020 IEEE International Conference on Artificial Intelligence and Virtual Reality (AIVR), Utrecht, The Netherlands, 14–18 December 2020, pp. 394–396. [CrossRef]
58. Ferri, F.; Costantini, M.; Huang, Z.; Perrucci, M.G.; Ferretti, A.; Romani, G.L.; Northoff, G. Intertrial variability in the premotor cortex accounts for individual differences in peripersonal space. *J. Neurosci.* **2015**, *35*, 16328–16339. [CrossRef] [PubMed]
59. Noel, J.-P.; Bertoni, T.; Terrebonne, E.; Pellencin, E.; Herbelin, B.; Cascio, C.; Blanke, O.; Magosso, E.; Wallace, M.T.; Serino, A. Rapid recalibration of peri-personal space: Psychophysical, electrophysiological, and neural network modeling evidence. *Cereb. Cortex* **2020**, *30*, 5088–5106. [CrossRef] [PubMed]
60. Bernasconi, F.; Noel, J.-P.; Park, H.D.; Faivre, N.; Seeck, M.; Spinelli, L.; Schaller, K.; Blanke, O.; Serino, A. Audio-tactile and peripersonal space processing around the trunk in human parietal and temporal cortex: An intracranial EEG study. *Cereb. Cortex* **2018**, *28*, 3385–3397. [CrossRef]
61. Ronga, I.; Galigani, M.; Bruno, V.; Castellani, N.; Sebastiano, A.R.; Valentini, E.; Fossataro, C.; Neppi-Modona, M.; Garbarini, F. Seeming confines: Electrophysiological evidence of peripersonal space remapping following tool-use in humans. *Cortex* **2021**, *144*, 133–150. [CrossRef]
62. Soccini, A.M.; Cena, F. The Ethics of Rehabilitation in Virtual Reality: The Role of Self-Avatars and Deep Learning. In Proceedings of the 2021 IEEE International Conference on Artificial Intelligence and Virtual Reality (AIVR), Taichung, Taiwan, 15–17 November 2021; pp. 324–328. [CrossRef]
63. Soccini, A.M.; Clocchiatti, A.; Inamura, T. Effects of frequent changes in extended self- avatar movement on adaptation performance. *J Robot. Mechatron.* **2022**, *34*, 756–766. [CrossRef]

64. Hobeika, L.; Taffou, M.; Carpentier, T.; Warusfel, O.; Viaud-Delmon, I. Capturing the dynamics of peripersonal space by integrating expectancy effects and sound propagation properties. *J Neurosci Methods*. **2020**, *332*, 108534. [CrossRef]
65. Kandula, M.; Van der Stoep, N.; Hofman, D.; Dijkerman, H. On the contribution of overt tactile expectations to visuo-tactile interactions within the peripersonal space. *Exp brain Res.* **2017**, *235*, 2511–2522. [CrossRef] [PubMed]

brain sciences

MDPI

Article

Prediction of Cognitive Degeneration in Parkinson's Disease Patients Using a Machine Learning Method

Pei-Hao Chen [1,2], Ting-Yi Hou [3], Fang-Yu Cheng [2] and Jin-Siang Shaw [3,*]

1 Department of Neurology, MacKay Memorial Hospital, Taipei 104217, Taiwan
2 Institute of Long-Term Care, Mackay Medical College, New Taipei City 252, Taiwan
3 Institute of Mechatronic Engineering, National Taipei University of Technology, Taipei 106344, Taiwan
* Correspondence: jshaw@ntut.edu.tw

Abstract: This study developed a predictive model for cognitive degeneration in patients with Parkinson's disease (PD) using a machine learning method. The clinical data, plasma biomarkers, and neuropsychological test results of patients with PD were collected and utilized as model predictors. Machine learning methods comprising support vector machines (SVMs) and principal component analysis (PCA) were applied to obtain a cognitive classification model. Using 32 comprehensive predictive parameters, the PCA-SVM classifier reached 92.3% accuracy and 0.929 area under the receiver operating characteristic curve (AUC). Furthermore, the accuracy could be increased to 100% and the AUC to 1.0 in a PCA-SVM model using only 13 carefully chosen features.

Keywords: Parkinson's disease; machine learning; neuropsychological test; biomarker

check for updates

Citation: Chen, P.-H.; Hou, T.-Y.; Cheng, F.-Y.; Shaw, J.-S. Prediction of Cognitive Degeneration in Parkinson's Disease Patients Using a Machine Learning Method. *Brain Sci.* **2022**, *12*, 1048. https://doi.org/10.3390/brainsci12081048

Academic Editors: Daniele Corbo and Roberta Ferrucci

Received: 9 July 2022
Accepted: 6 August 2022
Published: 7 August 2022

1. Introduction

Parkinson's disease (PD) is a neurodegenerative disease. Clinical motor dysfunctions, such as resting tremors, rigidity, bradykinesia, postural instability, and inability to initiate motion, are commonly seen in patients with PD. In addition to motor dysfunction, patients with PD also tend to have cognitive impairments, such as mild cognitive impairment (MCI) and dementia. MCI and dementia may also affect motor dysfunction in PD patients, and there is a complicated relationship between motor function and cognition in patients with PD [1].

According to previous research, there is a high probability that patients with PD develop cognitive impairment that may affect their quality of life; this impairment predominantly involves the cognitive domains of attention, executive function, and visuospatial skills [2–4]. Biomarkers obtained mainly from neuroimaging data were extensively discussed for finding predictors of cognitive dysfunction in Parkinson's disease in a literature survey [5,6]. Indeed, it is crucial to identify the factors influencing cognitive decline that affect clinical prognosis and require early intervention [7].

Machine learning in artificial intelligence is popular in constructing a predictive model. In a study with 45 subjects, four machine learning models were developed to assess the ability to discriminate between PD patients with cognitive integrity (PDCI), mild cognitive impairment (PDMCI), and dementia (PDD). In an SVM model for classifying PDD and PDCI, the most relevant variables related to PD dementia were white matter, lateral ventricle, and hippocampus volume, and the prediction accuracy could reach 96.67% [8]. In another study with a cohort of 75 PD patients, a set of five biomarkers (cerebrospinal fluid (CSF) total tau levels, CSF phosphorylated tau levels, CSF Aβ42 levels, APOE genotype, and SPARE-AD imaging score) was adopted as the predictor of a logistic regression classifier, and 80% accuracy was achieved in discriminating PD patients with normal cognition from PD patients with dementia [9].

In this preliminary study, a cross-sectional investigation of clinical variables, neuropsychological test results, and plasma biomarkers [10–12] in patients with PD was conducted to identify features related to cognitive impairment. More specifically, machine learning was applied to obtain a predictive cognitive degeneration model and ascertain key predictors that help medical experts quickly identify a patient's cognitive condition and provide treatment.

2. Methods

2.1. Participants

This cross-sectional study recruited patients with PD from October 2019 to November 2019 and from July 2020 to November 2020. The patients were recruited at the Neurology Department of the MacKay Memorial Hospital (Taiwan).

The study was performed following the Declaration of Helsinki and was approved by the Institutional Review Board of Mackay Memorial Hospital in Taiwan (IRB Number: 18MMHIS152). Informed consent was obtained from all participants. A consecutive series of patients with PD were recruited in the Neurology outpatient clinics of a tertiary medical center in northern Taiwan from October 2019 to November 2020. All participants met the following criteria: (a) age > 30 years, (b) diagnosed with idiopathic PD according to the PD clinical diagnostic criteria of the Movement Disorder Association [13,14], and (c) no diagnosis of dementia (for those who have received more than six years of education, the Mini-Mental State Evaluation [MMSE] score must be >23 points; for those who have less than six years of education, the MMSE must be >13 points). Participants were excluded if (a) they had more than two incomplete tests or (b) uncontrolled medical conditions that cause severe physical and cognitive disabilities. A physician evaluated the presence of the exclusion criteria, and the process was shown in Figure 1.

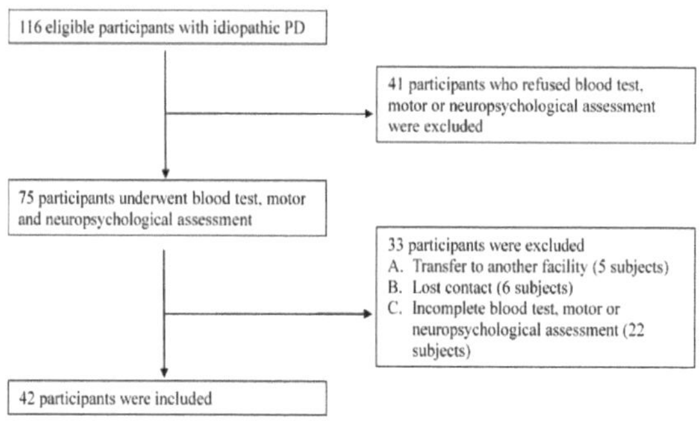

Figure 1. Exclusion flow chart.

2.2. Clinical Data

We collected clinical information from patients, including sex, age, course of the disease, education level, levodopa dose, Barthel index, Hoehn and Yahr stage, and Unified Parkinson's Disease Rating Scale (UPDRS) parts I–III subscale scores [15–19].

Trained nurses performed a comprehensive neuropsychological assessment of all patients. The assessment includes general cognition and specific cognitive domains involving the following examinations: (1) global cognition (MMSE and Clinical Dementia Rating-Sum of Boxes [CDR-SB]); (2) processing speed and working memory (Digits Recall Forward and Backward); (3) verbal learning and memory (California Verbal Language Test-II Short Form [CVLT-SF]); (4) semantic verbal fluency (animal naming); (5) language (Boston Naming Test); (6) attention and visuospatial processing (Trail Making Test A and B [TMT-A and TMT-B]); and (7) visuoperceptual and visuospatial processing (Benton Judgement of Line Orientation) [20–27].

2.3. Neurobiological Indicator

A blood sample of 10 mL was collected from each subject and centrifuged within one hour of collection. The plasma was separated and immediately frozen in test tubes at −80 °C. We then delivered frozen plasma on dry ice to MagQu Co., Ltd. (New Taipei City, Taiwan) and measured the levels of plasma α-syn, Aβ42, and t-tau using an immunomagnetic reduction assay.

2.4. Data Analysis

This study collected 29 clinical data and the three plasma biomarkers for each participant, as shown in Table 1. In addition, 42 patients with these complete data were included to build a classification model using support vector machine (SVM) and principal component analysis (PCA) in the Python Sklearn package. It was previously shown that the PCA–SVM method effectively classified PD–MCI from non-PD–MCI patients with high accuracy, provided good predictors were used [28].

Table 1. Twenty-nine clinical data and the three plasma biomarkers.

Hoehn–Yahr Stage	UPDRS I	UPDRS II	UPDRS III
LED (mg/day)	Gender	Age of visits	Age of onset
Disease duration	Education (years)	Barthel Index	MMSE
IADL	JLO	PSQI	EQ-5D index
EQ-5D VAS	GDS−15	GAD−7	TMT-A
TMT-B	Verbal fluency	Digits Forwards	Digits Backwards
CVLT-SF total recall	CVLT-SF Immediate	CVLT-SF delay	CVLT-SF recognition
BNT	α-syn (pg/mL)	Aβ42 (pg/mL)	t-tau (pg/mL)

Abbreviation: Aβ42, amyloid-β 42; BNT, Boston Naming Test; CVLT-SF, California Verbal Learning Test-Short Form; EQ-5D, EuroQol-5 dimensions; GAD-7, Generalized anxiety disorder scale 7-item; GDS-15, Geriatric depression scale 15-item; IADL, Instrumental activities of daily living; JLO, Judgment of Line Orientation; LED, Levodopa equivalent dose; MMSE, Mini-Mental State Examination; PSQI, Pittsburgh sleep quality index; SD, Standard Deviation; TMT, Trail Making Test; UPDRS, Unified Parkinson's Disease Rating Scale; VAS, visual analog scale; t-tau, total tau; α-syn, α-synuclein.

2.5. Data Normalization

Before using the SVM prediction model, it is necessary to preprocess the collected data to obtain a better data structure for training and avoid differences in the data distribution area, which affects the convergence speed and accuracy of the prediction model. Normalization is a standard preprocessing technique [29]. Min-max normalization is used in the data preprocessing. The data are scaled to between 0 and 1 through normalization without changing the distribution of the data [30] using the following transformation:

$$x_{norm} = \frac{x - x_{min}}{x_{max} - x_{min}} \quad (1)$$

where x_{max} is the maximum value, x_{min} is the minimum value, and x_{norm} is the normalized value between 0 and 1 for the dataset, x.

2.6. SVM

SVMs are a type of supervised learning method. It is to find a hyperplane between two-class categories. The SVMs try to find the decision boundary in the training data set to maximize the margin between the two classes to reduce the generalization error of the classifier. The maximum boundary hyperplane can be determined through various kernels to build a linear or nonlinear classification [31–33]. This study uses different kernel functions, including linear, RBF (radial basis function), and Poly (polynomial) functions to compare which model is better for predicting cognitive impairment.

2.7. PCA

PCA is an unsupervised learning method for feature extraction. Using the first few principal components (PCs) of the covariance matrix, normalized high-dimensional data can be projected into a lower-dimension space using orthogonal transformation while preserving the essential features [34–36]. More specifically, the dimensionality of the original dataset $X \in R^{n \times p}$ (i.e., n samples and p features) can be reduced to $X' \in R^{n \times s}$ by PCA with $s < p$. That is, X' with less dimension presents the data more concisely while retaining most of the key features (the cumulative energy of the first s eigenvalues of the covariance matrix is above a certain threshold, for example, 90%, of the total energy). The new features are then provided to the SVM with a lower dimension for predictive classification; hence, the training model can accelerate the calculation and improve the accuracy.

2.8. Area under the Receiver Operating Curve

The receiver operating characteristic curve (ROC curve) is drawn as a plot with the false positive rate (FPR) as the X-axis, and the true positive rate (TPR) as the Y-axis that illustrates the diagnostic ability of a classifier as its discrimination threshold is varied. The area under the ROC curve (AUC) measures the power of a classifier to distinguish between classes and is used as a summary of the ROC curve [37]. The higher the AUC, the better the model's performance at distinguishing between the positive and negative classes. When the AUC is equal to 0.5, the classifier cannot differentiate between positive and negative categories. Therefore, an AUC between 0.9 and 1 indicates that the predictive classifier has an excellent discriminatory ability.

3. Results

After one year of data collection, there were 116 patients with idiopathic PD. Of those, 41 patients refused blood and neuropsychological tests, five were transferred to another hospital, six lost contact, and 22 had incomplete data. Ultimately, only 42 patients had complete data. In this study, we used CDR-SB scores for the two classifications. The score interval for the patients without cognitive impairment (from normal to MCI) was ≤0.5, and the score interval for those with moderate and severe cognitive impairment was >0.5. After judging and categorizing, 16 patients were classified as not having cognitive impairment, and 26 patients had moderate to severe cognitive impairment. The demographic and collected data for these two groups were presented in Table 2.

It was worth mentioning why CDR-SB scores were used for the dichotomy of cognitive degeneration. A more quantitative representation of the CDR is provided by the sum of the severity ratings for the six cognitive and functional domains. CDR-SB provides a more quantitative measure of dementia severity than the global CDR. The CDR-SB frequently assesses Alzheimer's disease progression in clinical research [38,39] and has been used in patients with Parkinson's disease [40]. Owing to the increased range of values, the CDR-SB offers several advantages over the global score, including increased utility in tracking changes within and between stages of dementia severity. Unlike the other global cognitive testing (i.e., MMSE) in this study, CDR is not influenced by age, education, and gender.

First, all variables (p = 32) were included as feature inputs; 70% of the 42 patients were randomly selected as the training set and 30% as the verification set. Different kernel functions were used to train the SVM and PCA–SVM classification models. The validation accuracy under the full-parameter linear function reached 84.6%, and the AUC was 92.9%. After reducing the dimensionality of the original 32 features using PCA to six features, the accuracy increased to 92.3% for the same AUC rate. After PCA's dimensionality reduction, the overall forecast confidence improved, as shown in Table 3 and Figure 2.

Table 2. The demographic and data comparisons of the participants.

$N = 42$	Without Cognitive Impairment ($N = 16$)	Moderate and Severe Cognitive Impairment ($N = 26$)	p Value
Hoehn–Yahr stage	1.78 (0.73)	2.37 (0.61)	0.291
UPDRS I	2.38 (1.147)	4.15 (1.78)	0.078
UPDRS II	5.63 (2.391)	11.23 (5.88)	0.002
UPDRS III	12.63 (5.35)	20.65 (10.35)	0.013
LED (mg/day)	428.56 (229.13)	440.77 (241.8)	0.617
Gender	Male 8/50%	Male 10/38.46%	0.463
Age of visits	68.38 (8.57)	76.65 (7.27)	0.417
Age of onset	65.81 (8.72)	71.92 (8.19)	0.753
Disease duration	2.56 (2.39)	4.73 (3.52)	0.022
Education (years)	7.69 (3.22)	7.04 (4.96)	0.114
Barthel Index	156.25 (225)	88.27 (16.31)	0.019
MMSE	26.94 (2.24)	22.96 (3.96)	0.015
IADL	23.38 (1.26)	17.38 (6.76)	0.000
JLO	14.5 (4)	12.23 (4.86)	0.366
PSQI	5.38 (2.39)	7 (2.79)	0.71
EQ-5D index	0.77 (0.17)	0.75 (0.21)	0.78
EQ-5D VAS	68.88 (10.78)	66.54 (16.54)	0.335
GDS−15	2.5 (3.16)	3.54 (4.71)	0.067
GAD−7	1 (1.86)	2.08 (3.5)	0.068
TMT-A	27.19 (10.88)	36.62 (12.74)	0.494
TMT-B	72.06 (28.19)	87.96 (33.74)	0.15
Verbal fluency	11.56 (4.56)	9.27 (3.76)	0.426
Digits Forwards	7.38 (1.31)	6.12 (1.58)	0.21
Digits Backwards	5.19 (1.56)	3.58 (1.53)	0.897
CVLT-SF total recall	19.94 (5.89)	17.54 (4.42)	0.440
CVLT-SF immediate	6 (1.75)	4.96 (1.8)	0.784
CVLT-SF delay	4.69 (2.06)	3.81 (1.96)	0.696
CVLT-SF recognition	5.69 (2.44)	4.65 (2.45)	0.461
BNT	23.88 (2.99)	19.08 (6.46)	0.006
α-syn (pg/mL)	0.1 (0.05)	0.12 (0.05)	0.793
Aβ42 (pg/mL)	16.66 (0.45)	16.7 (0.59)	0.669
t-tau (pg/mL)	22.75 (2.63)	23.62 (3.63)	0.162

Table 3. Thirty-two parameter set to predict CDR-SB deterioration.

Classifier	Kernel	Feature Number	Accuracy	AUC
SVM	Linear	32	0.846	0.929
	RBF		0.769	0.857
	Poly		0.615	0.762
PCA-SVM	Linear	6	0.923	0.929
	RBF		0.769	0.857
	Poly		0.615	0.833

Second, from the above results, it suggested that a set of more concise predictors was possible. The six items (Hoehn–Yahr stage, IADL, Barthel Index, UPDRS I, II, and III) are related to essential motor and non-motor functions in PD patients. They are commonly used as clinical tools to assess PD patients. For the advanced neuropsychological tests, we selected four tests on executive functioning (TMT-B, Verbal fluency, Digits Forwards and Backwards) based on previous research [41–43] showing that executive dysfunction was joint in PD, especially early PD. The three biomarkers (α-syn, Aβ42, and t-tau), typically pathognomonic for the pathology of PD and AD, were also included which could predict executive dysfunction and cognitive decline in PD [12,44]. Therefore, a total of condensed 13 parameters were chosen as feature inputs as shown in Table 4. A randomly selected set of 70% of the 42 patients was used to train the prediction model, and the remaining 30% were used to verify the model performance. Different kernel functions were used to train the SVM and PCA–SVM classification models. The validation accuracy under the linear function in the SVM classification model reached 84.6%, and the AUC reached 100%. When reducing the dimensionality of the 13 features using PCA to three features, the accuracy under the linear function significantly improved to 100%. The AUC was maintained at 100%, as shown in Table 5 and Figure 3.

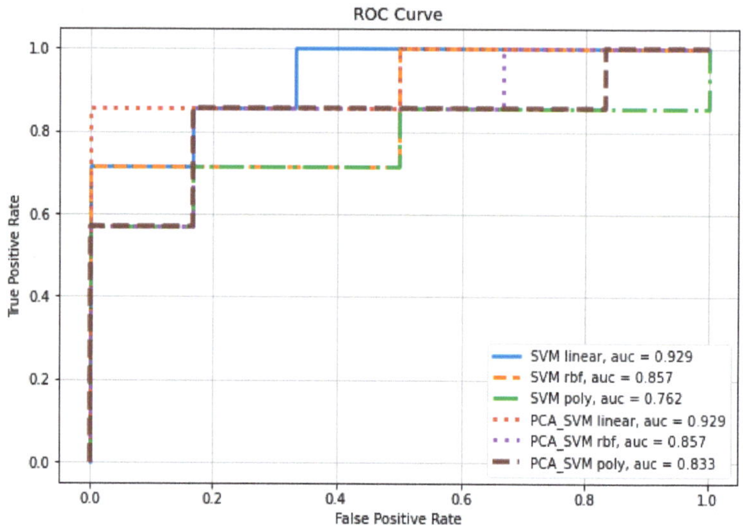

Figure 2. ROC curve and AUC results for each 32-parameter classifier of CDR-SB deterioration.

Table 4. Condensed thirteen parameters as the model predictors.

Hoehn–Yahr Stage	IADL	Barthel Index
UPDRS I	UPDRS II	UPDRS III
Verbal fluency	Digits Forwards	Digits Backwards
TMT-B	α-syn	Aβ42
t-tau		

Table 5. Thirteen selected parameters to predict CDR-SB deterioration.

Classifier	Kernel	Feature Number	Accuracy	AUC
SVM	Linear	13	0.846	1
	RBF		0.538	0.738
	Poly		0.846	0.976
PCA-SVM	Linear	3	1	1
	RBF		0.923	0.976
	Poly		0.692	0.905

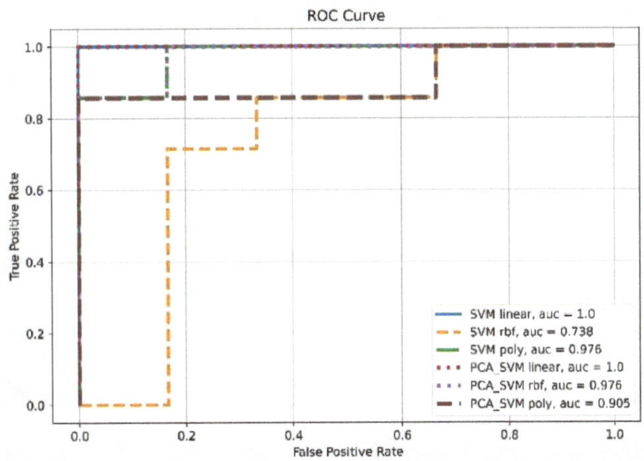

Figure 3. ROC curve and AUC result for each 13-parameter classifier of CDR-SB deterioration.

4. Discussion

In this study, machine learning was used to accurately classify the presence or absence of cognitive disorders in terms of CDR-SB scores in patients with idiopathic PD. In particular, we selected ten parameters related to clinical data and dynamic execution in neuropsychological tests and the three plasma biological indicators shown in Table 4 as the predictors that led to an accuracy rate and AUC for the PCA-SVM model as high as 100%. Therefore, dynamic execution and plasma biometrics are highly relevant for assessing the cognitive ability of PD patients. Compared to the two previously mentioned machine learning models for predicting cognitive degeneration [8,9], the developed PCA-SVM model produced the best prediction accuracy. In addition, literature on the use of the standard clinical assessment tools including neuropsychological tests for PD patients as cognitive predictors in a machine learning was limited and was even rarely seen using plasma biomarkers.

CDR is generally used to determine the severity of a patient's overall cognitive status, which is time-consuming and requires professional judgment. A patient's cognitive ability cannot be determined by questionnaires alone. However, only ten questionnaire items from clinical and neuropsychological tests and three plasma biological indicators were needed to train the predictive model through this training model. It is noted that questionnaires can be readily implemented after suitable personnel training and do not necessarily require professional medical persons; thus, it can reduce the time and burden on medical persons.

This study has some limitations. First, this was a cross-sectional study. Longitudinal studies are needed to identify the key indicators that can predict cognitive degeneration in a future time in PD patients, trace these predictors in the different disease stages, and clarify their roles in other cognition domains. Second, the sample size of this study was relatively small because of the need for neuropsychological evaluations and blood tests. Third, there was a lack of a control group of healthy subjects to compare the levels of these indicators. Finally, although all our participants fulfilled the diagnostic criteria of clinically established or probable PD, the possibility of overlapping clinical manifestation and misdiagnosis of progressive supranuclear palsy-parkinsonism predominant type (PSP-P) and postural instability and gait difficulty subtype of PD should be emphasized [45].

For future perspectives, increasing the sample size and conducting a longitudinal study are among the priorities. Specifically, increasing the sample size can further validate and support the developed model's performance in view of the small sample size in this study. Conducting longitudinal study can identify the key indicators and help develop a prediction model that can predict cognitive degeneration in a future time in PD patients, which is extremely important for medical experts to quickly identify a patient's cognitive condition and provide treatment in advance.

Author Contributions: Study conception and design: P.-H.C., F.-Y.C. and J.-S.S. Material preparation, data collection and analysis: P.-H.C., F.-Y.C. and T.-Y.H. Reviewing and supervision: P.-H.C. and J.-S.S. All authors have read and agreed to the published version of the manuscript.

Funding: This research was funded by the National Taipei University of Technology, Mackay Medical College, and MacKay Memorial Hospital (NTUT-MMH-109-07, MMH-MM-10811, MMH-TT-10905).

Institutional Review Board Statement: The study was conducted in accordance with the Declaration of Helsinki, and approved by the Institutional Review Board of Mackay Memorial Hospital in Taiwan (IRB Number: 18MMHIS152).

Informed Consent Statement: Informed consent was obtained from all subjects involved in the study.

Data Availability Statement: The data that support the findings of this study are available on request from the corresponding author. The data are not publicly available due to the privacy of participants.

Acknowledgments: The authors would like to thank all the participants and the research assistants, Shu-Ju Shiao, Wen-Chun Wu, and Ya-Yuan Yang for participating in the assessment.

Conflicts of Interest: The authors declare no conflict of interest.

References

1. Poewe, W. Non-motor symptoms in Parkinson's disease. *Eur. J. Neurol.* **2008**, *15*, 14–20. [CrossRef] [PubMed]
2. Dubois, B.; Pillon, B. Cognitive deficits in Parkinson's disease. *J. Neurol.* **1997**, *244*, 2–8. [CrossRef] [PubMed]
3. Caballol, N.; Marti, M.J.; Tolosa, E. Cognitive dysfunction and dementia in Parkinson disease. *Mov. Disord.* **2007**, *22* (Suppl. 17), S358–S366. [CrossRef] [PubMed]
4. Saredakis, D.; Collins-Praino, L.E.; Gutteridge, D.S.; Stephan, B.C.M.; Keage, H.A.D. Conversion to MCI and dementia in Parkinson's disease: A systematic review and meta-analysis. *Parkinsonism Relat. Disord.* **2019**, *65*, 20–31. [CrossRef] [PubMed]
5. Kalia, L.V. Biomarkers for cognitive dysfunction in Parkinson's disease. *Parkinsonism Relat. Disord.* **2018**, *46*, S19–S23. [CrossRef] [PubMed]
6. Vitale, A.; Villa, R.; Ugga, L.; Romeo, V.; Stanzione, A.; Cuocolo, R. Artificial intelligence applied to neuroimaging data in Parkinsonian syndromes: Actuality and expectations. *Math. Biosci. Eng.* **2021**, *18*, 1753–1773. [CrossRef]
7. Jones, J.D.; Kuhna, T.P.; Szymkowiczb, S.M. Reverters from PD-MCI to cognitively intact are at risk for future cognitive impairment: Analysis of the PPMI cohort. *Parkinsonism Relat. Disord.* **2018**, *47*, 3–7. [CrossRef]
8. Morales, D.A.; Vives-Gilabert, Y.; Gómez-Ansón, B.; Bengoetxea, E.; Larrañaga, P.; Bielza, C.; Pagonabarraga, J.; Kulisevsky, J.; Corcuera-Solano, I.; Delfino, M. Predicting dementia development in Parkinson's disease using Bayesian network classifiers. *Psychiatry Res.* **2013**, *213*, 92–98. [CrossRef]
9. Berlyand, Y.; Weintraub, D.; Xie, S.X.; Mellis, I.A.; Doshi, J.; Rick, J.; McBride, J.; Davatzikos, C.; Shaw, L.M.; Hurtig, H.; et al. An Alzheimer's disease-derived biomarker signature identifies Parkinson's disease patients with dementia. *PLoS ONE* **2016**, *11*, e0147319. [CrossRef]
10. Chang, C.W.; Yang, S.Y.; Yang, C.C.; Chang, C.W.; Wu, Y.R. Plasma and serum alpha-synuclein as a biomarker of diagnosis in patients with Parkinson's disease. *Front. Neurol.* **2019**, *10*, 1388. [CrossRef]
11. Chen, C.H.; Lee, B.C.; Lin, C.H. Integrated plasma and neuroimaging biomarkers associated with motor and cognition severity in Parkinson's disease. *J. Parkinsons Dis.* **2020**, *10*, 77–88. [CrossRef] [PubMed]
12. Lin, W.T.; Shaw, J.S.; Cheng, F.Y.; Chen, P.H. Plasma total tau predicts executive dysfunction in Parkinson's disease. *Acta Neurol. Scand.* **2022**, *145*, 30–37. [CrossRef] [PubMed]
13. Postuma, R.B.; Berg, D.; Stern, M.; Poewe, W.; Olanow, C.W.; Oertel, W.; Obeso, J.; Marek, K.; Litvan, I.; Lang, A.E.; et al. MDS clinical diagnostic criteria for Parkinson's disease. *Mov. Disord.* **2015**, *30*, 1591–1601. [CrossRef] [PubMed]
14. Hughes, A.J.; Daniel, S.E.; Kilford, L.; Lees, A.J. Accuracy of clinical diagnosis of idiopathic Parkinson's disease: A clinico-pathological study of 100 cases. *J. Neurol. Neurosurg. Psychiatry* **1992**, *55*, 181–184. [CrossRef]
15. Kahn, L.; Mathkour, M.; Lee, S.X.; Gouveia, E.E.; Hanna, J.A.; Garces, J.; Scullen, T.; McCormack, E.; Riffle, J.; Glynn, R.; et al. Long-term outcomes of deep brain stimulation in severe Parkinson's disease utilizing UPDRS III and modified Hoehn and Yahr as a severity scale. *Clin. Neurol. Neurosurg.* **2019**, *179*, 67–73. [CrossRef]

16. Siciliano, M.; Micco, R.D.; Trojano, L.; Stefano, M.D.; Baiano, C.; Passaniti, C.; Mase, A.D.; Russo, A.; Tedeschi, G.; Tessitore, A. Cognitive impairment is associated with Hoehn and Yahr stages in early, de novo Parkinson disease patients. *Parkinsonism Relat. Disord.* **2017**, *41*, 86–91. [CrossRef]
17. Winser, S.J.; Kannan, P.; Bello, U.M.; Whitney, S. Measures of balance and falls risk prediction in people with Parkinson's disease: A systematic review of psychometric properties. *Clin. Rehabil.* **2019**, *33*, 1949–1962. [CrossRef]
18. Sangarapillai, K.; Norman, B.M.; Almeida, Q.J. Boxing vs sensory exercise for Parkinson's disease: A double-blinded randomized controlled trial. *Neurorehabil. Neural. Repair.* **2021**, *35*, 769–777. [CrossRef]
19. Polo-Morales, A.; Alcocer-Salas, Á.; Rodríguez-Violante, M.; Pinto-Solís, D.; Solís-Vivanco, R.; Cervantes-Arriaga, A. Association between somatization and nonmotor symptoms severity in people with Parkinson disease. *J. Geriatr. Psychiatry Neurol.* **2021**, *34*, 60–65. [CrossRef]
20. Folstein, M.F.; Folstein, S.E.; McHugh, P.R. "Mini-mental state": A practical method for grading the cognitive state of patients for the clinician. *J. Psychiatr. Res.* **1975**, *12*, 189–198. [CrossRef]
21. Lynch, C.A.; Walsh, C.; Blanco, A.; Moran, M.; Coen, R.F.; Walsh, J.B.; Lawlor, B.A. The Clinical Dementia Rating Sum of Box Score in mild dementia. *Dement. Geriatr. Cogn. Disord.* **2006**, *21*, 40–43. [CrossRef] [PubMed]
22. Dennis, N.; Michael, P.A.P.; Jane, H. Learning nonwords: The Hebb repetition effect as a model of word learning. *Memory* **2018**, *26*, 852–857.
23. Woods, S.P.; Delis, D.C.; Scott, J.C.; Kramer, J.H.; Holdnack, J.A. The California Verbal Learning Test—Second edition: Test-retest reliability, practice effects, and reliable change indices for the standard and alternate forms. *Arch. Clin. Neuropsychol.* **2006**, *21*, 413–420. [CrossRef] [PubMed]
24. Nevado, A.; Río, D.D.; Martín-Aragoneses, M.T.; Prados, J.A.; López-Higes, R. Preserved semantic categorical organization in mild cognitive impairment: A network analysis of verbal fluency. *Neuropsychologia* **2021**, *157*, 107875-1–107875-8. [CrossRef] [PubMed]
25. Tallberg, I.M. The Boston Naming Test in Swedish: Normative data. *Brain Lang.* **2005**, *94*, 19–31. [CrossRef]
26. Nascimento Silva, R.D.; Afonso, S.V.; Felipe, L.R.; Oliveira, R.A.; Martins, L.J.P.; Souza, L.A.P.S.D. Dual-task intervention based on trail making test: Effects on Parkinson's disease. *J. Bodyw. Mov.* **2021**, *27*, 628–633. [CrossRef]
27. Qualls, C.E.; Bliwise, N.G.; Stringer, A.Y. Short forms of The Benton Judgment of Line Orientation Test: Development and psychometric properties. *Arch. Clin. Neuropsychol.* **2000**, *1*, 159–163.
28. Chen, P.H.; Lien, C.W.; Wu, W.C.; Lee, L.S.; Shaw, J.S. Gait-based machine learning for classifying patients with different types of mild cognitive impairment. *J. Med. Syst.* **2020**, *44*, 1–6. [CrossRef]
29. Jain, S.; Shukla, S.; Wadhvani, R. Dynamic selection of normalization techniques using data complexity measures. *Expert Syst. Appl.* **2018**, *106*, 252–262. [CrossRef]
30. Singh, D.; Singh, B. Investigating the impact of data normalization on classification performance. *Appl. Soft Comput.* **2020**, *97*, 105524-1–105524-23. [CrossRef]
31. Liu, Z.; Xu, H. Kernel parameter selection for support vector machine classification. *J. Algorithm Comput. Technol.* **2014**, *8*, 163–177. [CrossRef]
32. Gu, X.; Chen, C.; Yang, Y.; Miao, X.; Yao, B. Reliability prediction of further transit service based on support vector machine. *Meas Control.* **2020**, *54*, 845–855. [CrossRef]
33. Yu, Y.; Shao, M.; Jiang, L.; Ke, Y.; Wei, D.; Zhang, D.; Jiang, M.; Yang, Y. Quantitative analysis of multiple components based on support vector machine (SVM). *Optik* **2021**, *237*, 166759-1–166759-6. [CrossRef]
34. Ali, A.K.; Erçelebi, E. Automatic modulation classification using different neural network and PCA combinations. *Expert Syst. Appl.* **2021**, *178*, 114931-1–114931-17. [CrossRef]
35. Qaraei, M.; Abbaasi, S.; Ghiasi-Shirazi, K. Randomized nonlinear PCA networks. *Inf. Sci.* **2021**, *545*, 241–253. [CrossRef]
36. Zhang, J.; Zhou, D.; Chen, M. Monitoring multimode processes: A modified PCA algorithm with continual learning ability. *J. Process. Control.* **2021**, *103*, 76–86. [CrossRef]
37. Mandrekar, J.N. Receiver operating characteristic curve in diagnostic test assessment. *J. Thorac. Oncol.* **2010**, *5*, 1315–1316. [CrossRef]
38. Williams, M.M.; Storandt, M.; Roe, C.M.; Morris, J.C. Progression of Alzheimer's disease as measured by Clinical Dementing Rating Sum of Boxes scores. *Alzheimers Dement.* **2013**, *9*, S39–S44. [CrossRef]
39. Samtani, M.N.; Raghavan, N.; Novak, G.; Nandy, P.; Narayan, V.A. Disease progression model for Clinical Dementia Rating-Sum of Boxes in mild cognitive impairment and Alzheimer's subjects from the Alzheimer's Disease Neuroimaging Initiative. *Neuropsychiatr. Dis. Treat.* **2014**, *10*, 929–952. [CrossRef]
40. Wyman-Chick, K.A.; Scott, B.J. Development of clinical dementia rating scale cut-off scores for patients with parkinson's disease. *Mov. Disord. Clin. Pract.* **2015**, *2*, 243–248. [CrossRef]
41. Levy, G.; Jacobs, D.M.; Tang, M.X.; Côté, L.J.; Louis, E.D.; Alfaro, B.; Mejia, H.; Stern, Y.; Marder, K. Memory and executive function impairment predict dementia in Parkinson's disease. *Mov. Disord.* **2002**, *17*, 1221–1226. [CrossRef] [PubMed]
42. Kudlicka, A.; Clare, L.; Hindle, J.V. Executive functions in Parkinson's disease: Systematic review and meta-analysis. *Mov. Disord.* **2011**, *26*, 2305–2315. [CrossRef] [PubMed]
43. Lange, F.; Brückner, C.; Knebel, A.; Seer, C.; Kopp, B. Executive dysfunction in Parkinson's disease: A meta-analysis on the Wisconsin Card Sorting Test literature. *Neurosci. Biobehav. Rev.* **2018**, *93*, 38–56. [CrossRef] [PubMed]

44. Liu, C.; Cholerton, B.; Shi, M.; Ginghina, C.; Cain, K.C.; Auinger, P.; Parkinson Study Group DATATOP Investigators; Zhang, J. CSF tau and tau/Aβ42 predict cognitive decline in Parkinson's disease. *Parkinsonism Relat. Disord.* **2015**, *21*, 271–276. [CrossRef] [PubMed]
45. Alster, P.; Madetko, N.; Koziorowski, D.; Frieman, A. Progressive Supranuclear Palsy—Parkinsonism Predominant (PSP-P)—A clinical challenge at the boundaries of PSP and Parkinson's Disease (PD). *Front. Neurol.* **2020**, *11*, 180. [CrossRef] [PubMed]

brain sciences

MDPI

Article

Influence of Speech and Cognitive Load on Balance and Timed up and Go

Olivier Van Hove [1,*], Romain Pichon [2,3], Pauline Pallanca [4], Ana Maria Cebolla [5], Sarah Noel [5], Véronique Feipel [6], Gaël Deboeck [7] and Bruno Bonnechère [8,9]

1 Department of Pneumology, Erasme Hospital, 1070 Brussels, Belgium
2 Institut de Formation en Pédicurie-Podologie, Ergothérapie Kinésithérapie (IFPEK), 35000 Rennes, France; r.pichon@ifpek.org
3 M2S Laboratory—EA7470, University Rennes 2, 35000 Rennes, France
4 Department of Intensive Care, Erasme Hospital, 1070 Brussels, Belgium; pauline.pallanca@ulb.be
5 Laboratory of Neurophysiology and Movement Biomechanics, Université Libre de Bruxelles, 1070 Brussels, Belgium; ana.maria.cebolla.alvarez@ulb.be (A.M.C.); noux99@hotmail.com (S.N.)
6 Laboratory of Functional Anatomy, Université Libre de Bruxelles, 1070 Brussels, Belgium; veronique.feipel@ulb.be
7 Research Unit in Rehabilitation, Université Libre de Bruxelles, 1070 Brussels, Belgium; gael.deboeck@ulb.be
8 REVAL Rehabilitation Research Center, Faculty of Rehabilitation Sciences, University of Hasselt, 3590 Diepenbeek, Belgium; bruno.bonnechere@uhasselt.be
9 Technology-Supported and Data-Driven Rehabilitation, Data Sciences Institute, University of Hasselt, 3590 Diepenbeek, Belgium
* Correspondence: olivier.van.hove@erasme.ulb.ac.be

Abstract: The interaction between oral and/or mental cognitive tasks and postural control and mobility remains unclear. The aim of this study was to analyse the influence of speech production and cognitive load levels on static balance and timed up and go (TUG) during dual-task activities. Thirty healthy young subjects (25 ± 4 years old, 17 women) participated in this study. A control situation and two different cognitive arithmetic tasks were tested: counting backward in increments of 3 and 7 under oral (O) and mental (M) conditions during static balance and the TUG. We evaluated the dual-task cost (DTC) and the effect of speech production (SP) and the level of cognitive load (CL) on these variables. There was a significant increase in the centre of pressure oscillation velocity in static balance when the dual task was performed orally compared to the control situation The DTC was more pronounced for the O than for the M. The SP, but not the CL, had a significant effect on oscillation velocity. There was an increase in TUG associated with the cognitive load, but the mental or oral aspect did not seem to have an influence. Mobility is more affected by SP when the cognitive task is complex. This may be particularly important for the choice of the test and understanding postural control disorders.

Keywords: balance; dual task; cognitive loads; Wii balance board; timed up and go

check for updates

Citation: Van Hove, O.; Pichon, R.; Pallanca, P.; Cebolla, A.M.; Noel, S.; Feipel, V.; Deboeck, G.; Bonnechère, B. Influence of Speech and Cognitive Load on Balance and Timed up and Go. *Brain Sci.* **2022**, *12*, 1018. https://doi.org/10.3390/brainsci12081018

Academic Editors: Daniele Corbo and Yang Zhang

Received: 15 June 2022
Accepted: 29 July 2022
Published: 31 July 2022

1. Introduction

The interactions between postural control, motor control and cognitive load have been previously highlighted [1]. However, only a few studies have investigated the impact of the cognitive tasks separately, also referred to as cognitive cost, and the impact of speech production on gait and posture.

Postural control is a complex mechanism resulting from the integration of information from the vestibular, visual and proprioceptive systems [2]. Furthermore, there is a strong link between balance and respiratory functions. Breathing is known to influence balance: different studies have shown that increased tidal volume [3], increased inspiratory load [4] and inspiratory muscle fatigue [5] decrease postural control. On the other hand, cognitive load [6] or the absence of breathing (apnea) seems to improve the balance [7]. This is likely due to a cross effect between cognitive load and breathing, as it has been demonstrated that cognitive load decreases the tidal volume, which can lead to balance improvement [6]. An interesting condition that involves both breathing and cognition is speech. Speaking is characterized by shorter inspiration, longer expiration and an increased tidal volume and respiratory rate [8], leading to deteriorated postural control [9]. Interestingly, postural control and cognitive function are related [10] but the effects of cognition on postural control are still unclear. Indeed, some studies have shown that focusing on postural control could deteriorate balance [11]. Meanwhile, the addition of a cognitive load distraction improves it [12], but the postural sway increases with the difficulty of the task [1]. To summarize, a mild cognitive task improves balance, but postural control is negatively affected if the task is too challenging. The production of speech requires a cortical control of breathing [13]. Therefore, an oral cognitive task may be considered a dual-task activity; this may explain the increased postural sway during vocalization [9].

The motor control required during a mobility task is also influenced by cognitive load. In fact, a motor task may require a high-level process and interfere with cognition [14]. Thus, cognitive performance decreases when walking under a significant cognitive load (i.e., countback in increments of 7) [15]. During TUG, a decrease in mobility and cognitive performance is observed even in young subjects [16]. Controlling breathing for speech could also play a role in increasing cognitive-motor interference.

Speaking while performing a cognitive or motor task can also be considered a double task. Motor aspects of speech are affected by cognitive load. It increases articulatory coordination variability and movement [17]. There is thus a cognitive–motor interaction during speech production. There is also, for example, a complex relationship between hand movements and speech. This relation produces interference, facilitation, or null effects on hand motor tasks depending on conditions and their complexity [18]. For postural control, speaking increases sways [9,19,20] or interferes with the gait in both healthy subjects [21] and patients (e.g., stroke [22]).

Activities of daily living require the ability to perform several tasks simultaneously. It is therefore of the utmost importance to assess the subject's abilities to perform such complex tasks. In clinics, dual-task tests are the evaluation that mimics the best real-life conditions. These tests allow evaluating the cognitive-motor interference in several circumstances such as walking [23,24] or balance [18,19]. This cognitive-motor interference marks the limit in the ability of humans to manage several tasks simultaneously. This is the dual-tasking paradigm [25]. This suggests that the motor task, which requires attention (more or less important depending on the complexity of the tasks), and the cognitive task share, at least partially, the same brain systems. This implies a decrease in performance in one or both tasks [26]. The decrease in performance observed when adding a task is called the cost of dual tasking [27].

Responding verbally to a cognitive task while performing a task requiring postural control may resemble multitasking. Multiple interferences are possible, including cognitive load on speech, speech on postural control, cognitive load on postural control and postural control on cognitive tasks.

The mental and the oral task will have a different effect on postural control. We hypothesize that the oral task will have a negative effect on postural control, whether dynamic or static. While for the mental task the impact will be less or will improve the postural control. Nevertheless, a mental or oral task may have a different effect on our dynamic static test. This is what we will test in this study and the influence of the cognitive load will also be evaluated. These data could be important when choosing a test to evaluate postural control or establish a dual task training plan. However, also, they can allow a better understanding of balance disorders in the real life. In fact, in real life, we alternate static and dynamic phases as well as oral and mental tasks and light and heavy cognitive tasks. To our best knowledge, the influence of speech on balance and TUG during dual-task activities is still poorly understood. Therefore, the aim of this study was to analyse the influence of speech on static balance and TUG on healthy subjects during different cognitive tasks.

2. Materials and Methods

2.1. Participants

Thirty healthy participants participated in this study (23.7 ± 2.5 years; 64.9 ± 10.1 kg; 171.6 ± 8.2 cm; 22 ± 2.5 kg/m^2; 17 women). This study was approved by the Ethical Committee of Erasme Hospital (B4062021000062), and written informed consent was obtained from all subjects prior to their participation. Inclusion criteria were healthy subjects aged 18–40 years Exclusion criteria included neurological conditions, balance deficits or orthopaedic disorders in the last six months.

2.2. Protocol

2.2.1. Balance Assessment

Subjects were asked to stand on the middle of the Wii balance board (WBB) (45 × 26.5 cm) as quietly as possible with arms relaxed along the body and fix a target located on a wall two meters away. Participants were asked to not move from the WBB during the protocol to decrease the risk of bias inherent to body position while it has been shown that the position of the foot on the WBB did not influence the results [28]. The WBB is a valid tool to assess balance in different conditions [4,29]. A control situation and two different cognitive arithmetic tasks were tested. Each trial lasted for 60 s, and the order of the five trials was randomly determined. The WBB was connected to a laptop (Intel Core I5, Windows 7, 6 GB RAM) through a Bluetooth connection; data were retrieved using custom-written software based on the WiimoteLib software. The data collection frequency was set at 100 Hz.

2.2.2. Mobility Task

The mobility test consists of a TUG. It is a standardized test commonly used in clinics: the subject is seated on a chair, must stand up, walk 3 m, make a 180-degree turn and sit down again as quickly as possible [30]. The outcome is the time needed to achieve the task. A control situation and two different cognitive situations (see below) were tested.

2.2.3. Cognitive Task

Two different cognitive tasks were evaluated. In the first one, the subjects had to count backwards in increments of 3 (starting from a number randomly selected between 300 and 340, denoted 3), and in the second, the subjects had to count backwards in increments of 7 (still starting from a randomly selected number between 300 and 340, denoted 7).

Those two tasks were performed mentally (*M*) and orally (*O*). When performed orally, the ranges (difference between start and end numbers) were computed as well as the number of potential errors. To check that the participants performed the cognitive task under the mental condition, we asked the participants at the end of the mental tasks the final number they reached to assess the range and evaluate if the task was performed error-free.

For reference values, and to be as close as possible to the conditions of balance and mobility, we performed the countback 3 and 7 tests at rest in a seated position: 1 min for the balance test and 15 s for the TUG.

The order of the different conditions (oral or mental) and cognitive load (3 and 7) were randomly defined to avoid fatigue or familiarization. Subjects have 1 min of rest between the different trials and a 10 min wash-out period between the static and the dynamic evaluation. The complete flow of the study is presented in Figure 1.

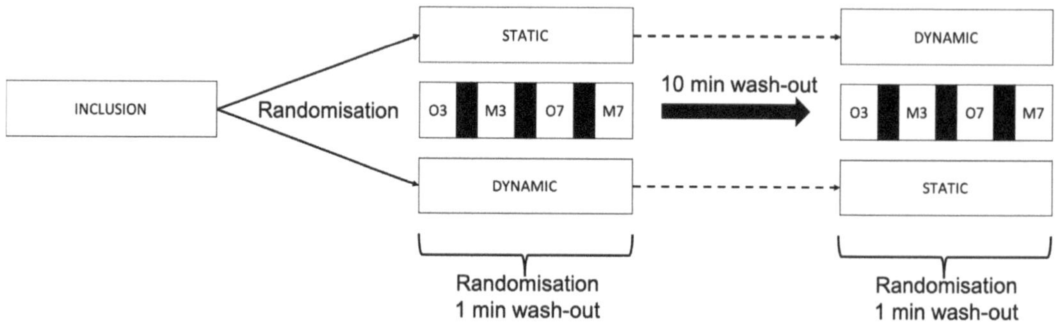

Figure 1. Study flow diagram.

2.3. Data Processing

For the balance assessment, several parameters were computed based on the centre of pressure (COP) displacement using a previously validated method [4]. CP anterior-posterior (CP AP) and mediolateral (CP ML) displacements were obtained from the four strain gauge loads located at the four corners of the WBB using Equations (1) and (2):

$$CPap = (FR + PR) - (FL + PL) \tag{1}$$

$$CPml = (FL + FR) - (PL + PR) \tag{2}$$

where PL, PR, FL and FR are the displacement values from the posterior left, posterior right, front left and front right WBB sensors, respectively. Previous works have shown that the time interval between samples of WBB was inconsistent, therefore, linear interpolation of the raw signals of WBB sensors was applied to obtain a regular sample rate. From those displacements, the nine studied parameters were computed, and descriptions of the computed variables and the equations are presented in Table 1. Data were analysed during the 5th and 55th seconds of each trial, as previous studies have shown that the signal is the most stable during this period.

Table 1. Descriptions of the variables used in this study and equations used to process the data.

Balance		
Name	**Description**	**Equation**
DOT	Total displacement of sway	$\sum_{i=1}^{N} \sqrt{CPap(i)^2 + CPml(i)^2}$
Area	The area of the 95% prediction ellipse (often referred to as the 95% confidence ellipse)	$\pi \times prod\left(2.4478 \times \sqrt{svd(eig(cov(CPap, CPml)))}\right)$
AP RoM	The distance between the maximum and minimum COP displacement in the antero-posterior direction	$\max(CPap) - \min(CPap)$
ML RoM	The distance between the maximum and minimum COP displacement in the medio lateral direction	$\max(CPml) - \min(CPml)$
AP SD	The dispersion of COP displacement from the mean position in the antero-posterior direction	$\frac{1}{N}\sqrt{\sum_{i=1}^{N} CPap(i)^2}$

Table 1. *Cont.*

ML SD	The dispersion of COP displacement from the mean position in the medio-lateral direction	$\frac{1}{N}\sqrt{\sum_{i=1}^{N} CPml(i)^2}$
AP velocity	The mean AP velocity of COP displacement	$\frac{f}{N}\sum_{i=1}^{N-1}\lvert CPap(i+1) - CPap(i)\rvert$
ML velocity	The mean ML velocity of COP displacement	$\frac{f}{N}\sum_{i=1}^{N-1}\lvert CPap(i+1) - CPap(i)\rvert$
TMV	The AP and ML displacements of the total COP sway divided by the total duration of the trial	$\frac{f}{N}\sum_{i=1}^{N-1}\sqrt{(CPap(i+1) - CPap(i))^2 + (CPml(i+1) - CPml(i))^2}$
	Motor and Cognitive Interaction	
Name	**Description**	**Equation**
CCR	Correct response rate	response rate per sec ond $\times$ percent of accuracy
DTC_{cogn}	Dual-task cost cognitive expressed in percent. A negative value indicates improvement, while a positive value indicates worse performance.	$\left(\text{single CCR score}^{-}\text{dual\_task CCR score}/\text{single CCR score}\right) \times 100$
DTC_{mob}	Dual-task cost mobility in percent. A negative value indicates improvement, while a positive value indicates worse performance.	$\left(\text{dual\_task mobility score}^{-}\text{single\_task mobility score}/\text{single\_task mobility score}\right) \times 100$
SP	The effect of speech production on postural control	$DTC_{Mob}\,(Mental) - DTC_{Mob}\,(Oral)$
CL_O	The effect of cognitive load level on postural control during oral tasks	$DTC_{Mob}\,(O3) - DTC_{Mob}\,(O7)$
CL_M	The effect of cognitive load level on postural control during mental tasks	$DTC_{Mob}\,(M3) - DTC_{Mob}\,(M7)$

To evaluate the motor and cognitive interaction, we assessed the dual-task cost (DTC). The different variables and formulas used to assess the influence of the dual task on cognitive and mobility or postural control performance are presented in Table 1. Two cognitive loads (3;7) and two conditions (O; M) were tested. CCR and DTC_{mob} were calculated for 3 and 7 at rest during the quiet standing and TUG. A negative value indicates improvement, while a positive value indicates worse performance.

We then evaluated the effect of speech production (SP) on postural control during quiet standing and TUG. Finally, we evaluated the effect of cognitive load (CL) on postural control during quiet standing and TUG for each condition (oral (CL_O) or mental (CL_M)) and for all balance parameters and TUG times.

2.4. Statistical Analysis

The normality of each parameter was checked using graphical methods (boxplots, histograms and Q–Q plots) and the homogeneity of variances using the Levene test. As the data were normally distributed, we used the parametric method. Two-way ANOVA was used to compare the effects of the conditions (i.e., control, oral and mental task), the cognitive loads (i.e., 3 and 7) and the interaction between these two factors. Bonferroni's corrections were adjusted for multiple comparisons in our post hoc analysis. Statistical analyses were performed at an overall significance level of 0.05. Statistics were analysed in RStudio (version 1.2.135) with R version 3.6.1.

To detect a difference of 15% in TMV between the different conditions (for static balance) with 80% power and a two-sided type I error of 5%, we calculated prior to the start of the study the need to include 29 subjects.

3. Results

We first discuss the results for the balance, then for the mobility.

First, we compared the results of the cognitive tasks and found no difference between the two conditions (mental or oral) for the simple (3-3 countback) and more complex task (7-7 countback). The results of the balance assessment for the different conditions are presented in Table 2.

Table 2. Mean (std) results for the studied parameters under the five different conditions. *p*-values are the results of the ANOVA.

Variables	Control	Oral 3	Mental 3	Oral 7	Mental 7	*p*-Values		
						Cond.	Cogn.	Inter.
DOT (mm)	1303 (438)	1109 (324)	1171 (464)	1195 (462)	1199 (521)	0.074	0.92	0.16
Area (mm²)	3488 (2956)	2775 (3120)	2476 (3552)	3472 (4700)	2825 (2963)	0.078	0.72	0.63
ML RoM (mm)	38 (19)	33 (30)	32 (21)	37 (28)	38 (20)	0.021	0.53	0.93
AP RoM (mm)	161 (91)	137 (96)	148 (74)	174 (121)	158 (83)	0.029	0.87	0.32
ML SD (mm)	5.6 (2.7)	5.3 (3.9)	5.4 (4.5)	5.4 (5.0)	6.4 (3.2)	0.056	0.54	0.99
AP SD (mm)	30.4 (17.9)	25.4 (12.8)	27.4 (19.9)	30.8 (21.9)	26.1 (14.7)	0.081	0.99	0.12
MVml (mm/s)	2.8 (0.5)	3.1 (0.6)	2.7 (0.5)	3.0 (0.6)	2.7 (0.4)	<0.001	0.34	0.38
MVap (mm/s)	6.1 (1.3)	7.8 (1.9)	6.0 (1.4)	7.9 (2.5)	6.0 (1.3)	<0.001	0.52	0.21
TMV (mm/s)	7.50 (1.53)	11.2 (3.70)	9.19 (6.35)	12.0 (5.95)	8.00 (2.90)	<0.001	0.82	0.19
TUG, s	4.82 (0.62)	5.77 (1.03)	5.66 (1.21)	6.25 (1.22)	5.80 (1.32)	<0.001	0.43	0.21

Cond. = conditions (control, oral and mental), Cogn. = cognition (3 or 7), Inter. = interaction between conditions and cognitions.

We observed a significant effect of the conditions for the ML and AP displacement (RoM) for the speed-related parameters (MLml, MVap, TMV) and for the TUG but no effect of the cognitive loads and no interaction between the conditions and the cognitive loads. Statistically significant mean differences between the conditions with 95% confidence intervals are presented in Figure 2, and complete results and the post hoc analysis results are presented in Supplementary Table S1.

Table 3 summarizes the results for dual-task cost mobility and cognitive for the postural control and TUG. We observe a decrease in motor and cognitive performance during the dual task. This decrease in performance seemed to be more marked for the oral task.

Table 3. Dual-task costs for the different studied parameters. Mean [95% CI].

Parameters	Conditions	Oral	Mental
Balance			
Cognitive	3	4 [−19; 3]%	/
	7	6 [−25; 37]%	/
MVml	3	30 [14; 47]%	6 [−7; 18]%
	7	25 [8; 42]%	0 [−8; 9]%
MVap	3	44 [24; 64]%	9 [−6; 25]%
	7	44 [28; 60]%	1 [−12; 15]%
TMV	3	40 [23; 57]%	8 [−5; 22]%
	7	39 [25; 54]%	2 [−10; 14]%
TUG			
Cognitive	3	4 [−12; 19]%	/
	7	67 [59; 75]%	/
Time (mobility)	3	20 [14; 26]%	17 [11; 23]%
	7	30 [23; 36]%	20 [13; 26]%

DTC, dual-task cost; cogn, cognitive; mob, mobility; O, oral; M, mental; 3, countback 3; 7 countback 7, MVml; mean medio-lateral velocity, MVap; mean antero-posterior velocity, TMV; total mean velocity.

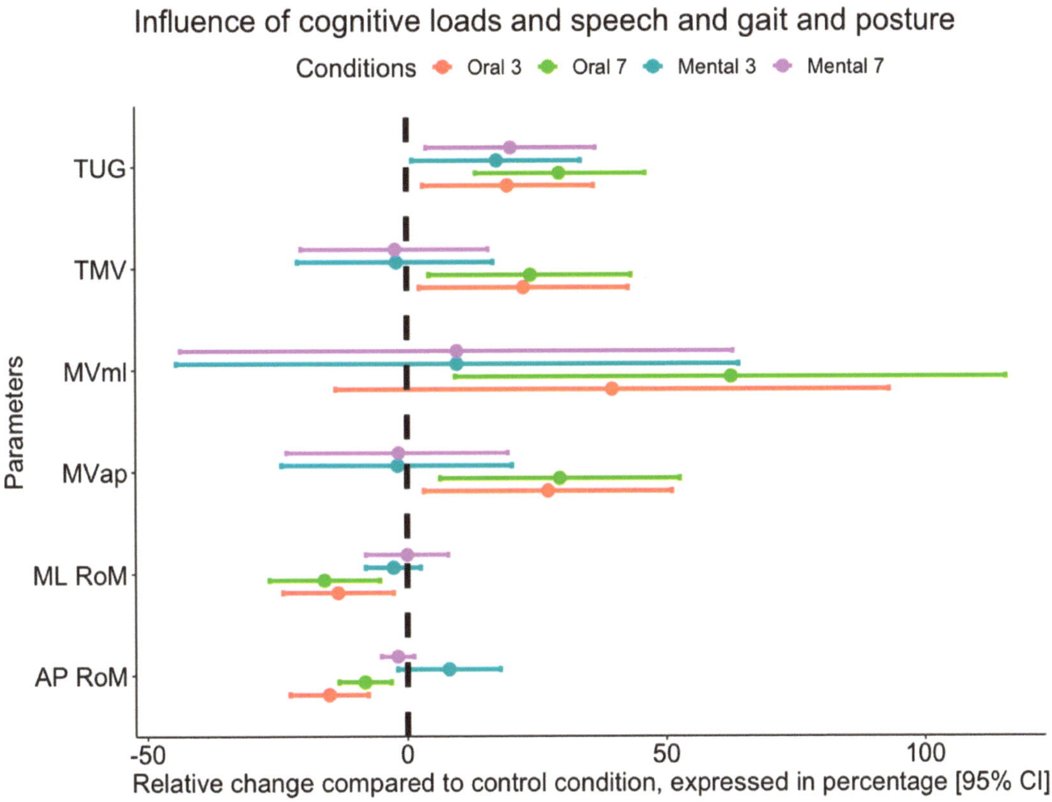

Figure 2. Influence of the different modalities on gait (TUG) and balance-related parameters.

The effect of speech production on oscillation velocity and TUG is presented in Table 4. There was a significant effect of speech production on all oscillation speeds. The TUG only increased for countback 7.

Table 4. Mean difference [95% CI] for speech production effect and cognitive load level on oscillation velocity and TUG. p-value are the results of the comparison with control situation (paired t-test).

Effect of Speech Production	**Diff.**	**p-Value**	**Cognitive Load Level**	**Diff.**	**p-Value**
			Balance		
SP3 MVml	−25 [−45; −3]%	0.0001	CLO MVml	5 [−18; 29]%	0.334
SP3 MVap	−35 [−60; −10]%	0.0002	CLO MVap	−0 [−25; 26]%	0.961
SP3 TMV	−32 [−54; −9]%	0.0001	CLO TMV	1 [−22; 2]%	0.842
SP7 MVml	−25 [−44; −17]%	0.003	CLM MVml	5 [−11; 21]%	0.389
SP7 MVap	−43 [−64; −21]%	3.19×10^{-5}	CLM MVap	8 [−13; 29]%	0.429
SP7 TMV	−37 [−56; −17]%	4.69×10^{-5}	CLM TMV	6 [−12; 25]%	0.411
			TUG		
SP3	3 [−6; 11]%	0.343	CLO	−11 [−26; −13]%	0.00025
SP7	10 [1; 20]%	0.0014	CLM	−3 [−11; −6]%	0.16

SP: effect of speech production, 3: countback 3, 7: countback 7, MVml: mean velocity medio-lateral, MVap; mean velocity antero-posterior, TMV: total mean velocity, CLO: cognitive load level oral task, CLM: cognitive load level mental task.

There was no significant effect of cognitive load level for the MVml (5 [−18; 29]%, p = 0.334), MVap (0 [−25; 26]%, p = 0.961) and TMV (1 [−22; 2]%, p = 0.842) for the oral condition. We observed the same results on MVml (5 [−11; 21]%, p = 0.389), MVap (8 [−13; 29]%, p = 0.429) and TMV (6 [−12; 25]%, p = 0.411) for the mental condition. The cognitive load level affected the TUG only in the oral condition: CLO (−11 [−26; −13]%, p = 0.00025), CLM (−3 [−11; −6]%, p = 0.16).

Finally, to compare the static and dynamic (TUG) aspects, we compared the relative changes (in comparison with the control condition) induced by the different cognitive tasks on balance and TUG. We did not find a correlation between these changes for any of the studied conditions (R = 0.20 for Oral 3, 0.36 for Oral 7, 0.08 for Mental 3 and 0.11 for Mental 7, see Figure 3).

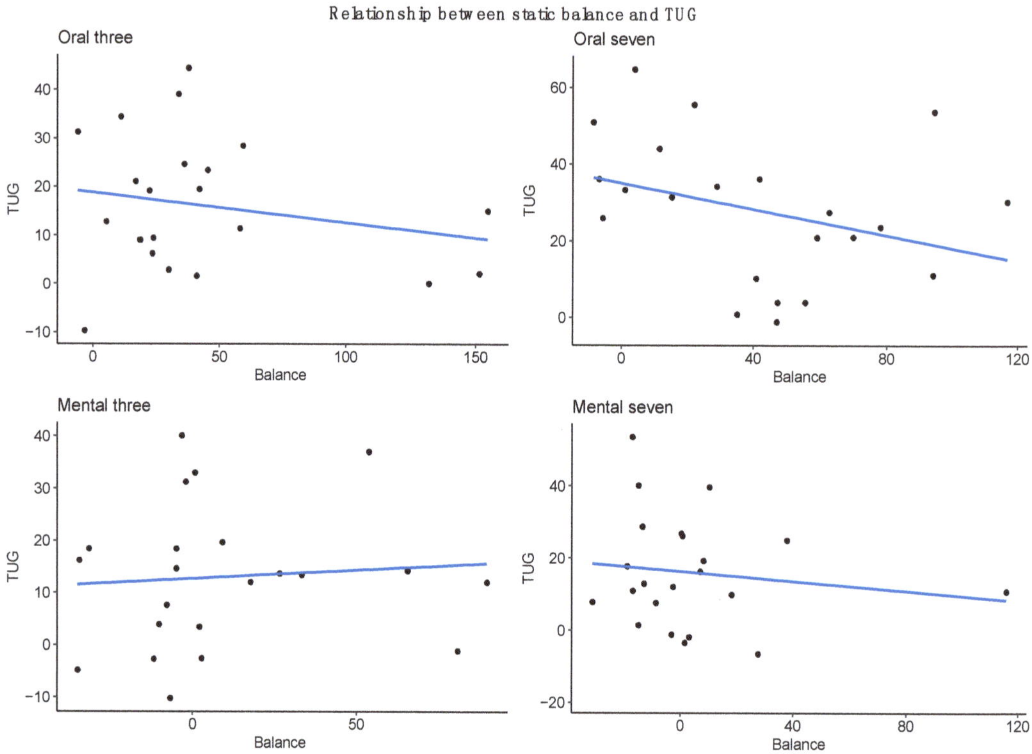

Figure 3. Relative changes (percentage of change relative to the control condition) for the different tasks for TUG and balance.

4. Discussion

The main result of this study is that speech production had a direct influence on gait and posture. First, we evaluated the effect of a cognitive task (countback 3 and 7; oral or mental) on postural control and TUG.

In our research, neither the cognitive task nor its level (cognitive load level) appeared to influence static postural control. Dual tasking is the interference of one task with another when they are performed simultaneously. The effect of a cognitive task depends on its difficulty level. Indeed, a light task decreases postural oscillations [12], whereas a difficult task increases them [1,31]. Our result may indicate that the cognitive task was not sufficiently challenging to produce cognitive-motor interference. The effect of dual tasking on postural velocity illustrates this finding. The impact of dual tasking on sway velocity is comparable across cognitive levels. In spite of this, a recent study demonstrated that, regardless of the level of difficulty, postural oscillation velocities decrease with cognitive load [32]. It should be noted that these studies use different cognitive tasks: arithmetic calculations [32] and executive function on a tablet (i.e., shifting, inhibition, updating) [31]. This disparity in testing can perhaps explain the conflicting results. For dynamic postural control, we used a protocol similar to that of Brustio et al. [16] but we focused on the effect of the dual task on the timed up and go test. The mobility performance (i.e., the dual-task cost) is altered with the cognitive load with a reduction of 20% and 30% for countback 3 and 7, respectively. In contrast to the Brustio study [16], we did not observe any difference in TUG between the levels of cognitive load. However, the dual-task cost on time is 10% higher for countback 7. Other studies have also shown a similar effect of cognitive load level on gait [33]. On the other hand, the cognitive load levels have a negative impact on timed up go in the oral condition. This may be due to a combination of two factors: an increased cognitive load and articulation. It is also known that talking while walking decreases gait speed [16,34]. Subjects seem to prioritise speaking over walking [35]. The difference between countback 3 and 7 may be increased by speaking prioritization.

In this study, we found statistically significant differences between the oral and mental conditions. These results are consistent with those found in the literature on elderly subjects [36].

We found a more significant decrease in postural control with the oral tasks than the mental tasks and control conditions. However, in our study, only the oscillation velocity (i.e., TMV, MVap, MVml) increased, not the COP displacement (DOT, area, AP RoM, ML RoM). The dual task increases the time on TUG; however, there was no difference between the type of tasks (oral vs. mental) and cognitive load (3 vs. 7). Secondly, we analysed the mobility performance (TUG), postural control performance (TMV, MVap, MVml) and cognitive performance during the dual task (dual-task cost). We observed a more significant decrease in postural control performance during the oral task compared to the mental task. An example with the TMV showed a score of 8% for the mental countback 3 and 40% for oral countback 3. As a reminder, the more positive the score, the more the performance is altered. We observed the same thing for cognitive load 7. The TUG evolved similarly but with less marked differences between the oral and mental tasks. The cognitive performance during the dual task in quiet standing was only slightly impaired. This difference was more marked during the TUG. During this dual task, the cognitive performance was more impaired with cognitive load 7 than 3 (67% vs. 4%). Third, to refine our results, we calculated the effect of speech production and cognitive load. Speech production had a significant effect on oscillation velocity. However, this was only observed for countback 7 for the TUG. Nevertheless, these observations further emphasize the importance of the oral task on balance.

The cognitive load level did not affect oscillation velocity but did affect the TUG with countback 7 in the oral condition. These last observations reinforce the effect of speech on postural control.

When comparing our results with the literature, we found that a previous study highlighted that mental tasks could improve postural control [6]; however, when the task is performed orally, there is an increase in postural sway [9,19]. The difficulty of the task also had a negative influence on postural control [1]. In our study, only the oscillation velocity was significantly increased during cognitive tasks in oral conditions. Our results are, therefore, at least partially, in agreement with those of previous studies [9,19,20]. The increase in oscillation velocity can be interpreted as an alteration of balance control [37]. It did not seem to have a difference between the oral tasks (O3 vs. O7) except for mediolateral velocity, nor between the different mental tasks (M3 vs. M7). These observations were corroborated by the effect of speech production. These changes in oscillating velocity during the oral tasks can be attributed to the motor requirements during speech production [19] more than the effect of attention itself. However, other parameters such as the tidal volume changes during speech or mental tasks could also explain this. During a mental task, we found a decrease in tidal volume and a stabilization [6]. Meanwhile, during speech, we found an increase in tidal volume [8], which could induce an increase in oscillations; other parameters such as changes in lung volume could also explain these changes. In this study, we did not assess respiratory parameters and breathing patterns and therefore cannot evaluate these hypotheses. Another hypothesis is that the movements of the jaw will induce modifications in the maintenance of the head. This is important for postural control. Indeed, a stabilization of the cervical spine during a cognitive task (mental) increases postural disturbances [38]. It has been previously demonstrated, for example, that jaw clenching or biting may reduce the postural sway [39]. However, to the authors' best knowledge, there is currently no study assessing the impact of jaw motion on postural control during speech production. This would be an important factor to study and to determine the implications for the clinics. We did not measure the influence of cognitive load on speaking. However, by modifying the duration of speech-related movements or their variability, it can influence static postural control [17]. This is to be also determined in the future.

In addition, the effect of conducting a dual task on cognitive performance was not modified when the subject was in quiet standing. Cognitive performance was maintained despite the increased cognitive load. It has been previously shown that quiet postural standing [40] and speech production [13] can involve the cortex. Therefore, the task performed in this study could be interpreted as a triple task: cognition, speech production and postural control [21]. However, only postural control was altered. As the cognitive performance was maintained, we hypothesize that, given the possible competitive nature of the cognitive tests, the subjects prioritized the cognitive task over the balance control [41]. We did give any prioritization instructions; however, it could be carried out automatically.

For the TUG, we observed an increase in time associated with the cognitive load, which agrees with a previous study [15]. However, there was no difference in task difficulty or oral or mental type in our study. Nevertheless, the effect of speech production was significant for countback 7, and the effect of the level of the cognitive load was significant during the oral task. This tautologically reinforces our observation of the effect of an oral complex cognitive task on motor performance. This decrease in mobility performance can be explained by the association between the high cognitive task and speech production [21,42]. On the other hand, as previously said, large pre-phonatory breaths involve the premotor cortex [13]. The premotor cortex is also involved in anticipatory postural adjustments during stepping leg selection [43]. A lesion in this area will thus lead to gait dysfunctions. In our study we evaluated healthy subjects, we can therefore hypothesize a competition between speech production and gait control occurring in the premotor cortex.

The multi-tasking effect (speech production, motor task and cognition) was more important for a mobility task than for postural control, which is logical when comparing the tasks' complexity (balance control vs. TUG). There was a significant impact of mobility on cognition, unlike quiet standing. During the dual task on cognition, the cognitive performance was more impacted for countback 7 than 3 for TUG, which can be considered a marker of a more pronounced interaction between cognition and mobility in the oral condition.

The absence of correlation found between the changes induced by the cognitive loads for the static balance and the TUG (Figure 2) highlights the importance of analysing these two tests individually to have a more precise evaluation of the cognitive-motor interaction.

While these data are important for determining the choice of test type, they are also important for the rehabilitation of these patients. It is important to train the subjects in dual tasks in order to get situations as close as possible to real life. This dual-task training improves walking ability [44] and balance [45] and is more effective than sequential training [46] in elderly subjects with different impairments. However, the exact modalities are yet to be determined [24]. Our study leads to possible research in this area. This study applies to young but the influence of cognitive on postural sways is the same in young and old adults [31]. Multiple tasks, either static or dynamic, are performed in the course of daily life. Or study highlights the possible need to train dynamic and static postural control in a specific manner. In addition, it would be intriguing to observe the results of training with mental and/or oral tasks. In our research, we did not ask the participant to rank the tasks. It may be of interest to investigate the impact of prioritized training (cognitive, motor) as well as the level of difficulty on the subject's performance improvement [47].

This study has a few limitations, and the results must be interpreted carefully. A first limitation is the absence of tidal volume measurement during the cognitive task. It has been shown that the cognitive task impacts tidal volume and could therefore influence the postural sway [6]. Our observations are, thus, a net result of the oral and cognitive task on postural control without discriminating the effects of the dual task and tidal volume. Another potential limitation is that participants were not given specific instructions on whether they should focus more on the cognitive or motor tasks—this can impact the subjects' performance, since some may decide to focus more on motor or cognitive strategies [41].

Another limitation of the study is that there is no precise objective assessment of COP during or after the functional task (TUG). The analysis of such variables may bring relevant information into future investigations, particularly in assessing the risk of fall in patients with chronic respiratory diseases [48].

Despite these limitations, we highlighted the complex interaction between speech production, motor aspect and complexity of the cognitive task. Therefore, these factors are important to consider when determining the best tests to assess patients suffering from specific diseases. Further studies need to focus on the impact of various pathologies on these outcomes and relationships. The proposed solution is cost-effective, portable and easy to use and could therefore be easily implemented into daily care.

5. Conclusions

In this study, we have shown that the influence of different cognitive tasks on postural control is influenced by the production of speech. During static balance, the oral task seems to significantly alter the balance in healthy subjects more than the similar task performed mentally. However, during a more functional task (TUG), both conditions (oral and mental) had a similar impact on postural control.

This study opens new perspectives for assessing patients with respiratory diseases, cognitive limitations or speaking problems. The results of this study are also of importance for the implementation of specific rehabilitation programs for these patients. Future investigations are needed to confirm our findings and determine the implications of the pathologies in this relationship.

Supplementary Materials: The following supporting information can be downloaded at: https://www.mdpi.com/article/10.3390/brainsci12081018/s1, Supplementary Table S1: Difference [95% CI] and p-value of group comparisons after post-hoc corrections.

Author Contributions: Conceptualization, O.V.H. and B.B.; formal analysis, O.V.H.; funding acquisition, B.B.; investigation, O.V.H.; methodology, O.V.H. and B.B.; supervision, B.B.; visualization, B.B.; writing—original draft, O.V.H.; writing—review and editing, O.V.H., R.P., P.P., A.M.C., S.N., V.F., G.D. and B.B. All authors have read and agreed to the published version of the manuscript.

Funding: This research received no external funding.

Institutional Review Board Statement: This study was approved by the Ethical Committee of Erasme Hospital (B4062021000062).

Informed Consent Statement: Informed consent was obtained from all subjects involved in the study.

Data Availability Statement: The data that support the findings of this study are available on request from the corresponding author (BB).

Conflicts of Interest: The authors declare no conflict of interest.

References

1. Pellecchia, G.L. Postural Sway Increases with Attentional demands of concurrent cognitive task. *Gait Posture* **2003**, *18*, 29–34. [CrossRef]
2. Shumway-Cook, A.; Woollacott, M.H. *Motor Control: Translating Research into Clinical Practice*; Lippincott Williams & Wilkins: Philadelphia, PA, USA, 2007.
3. Kuznetsov, N.A.; Riley, M.A. Effects of breathing on multijoint control of center of mass position during upright stance. *J. Mot. Behav.* **2012**, *44*, 241–253. [CrossRef] [PubMed]
4. Van Hove, O.; Van Muylem, A.; Leduc, D.; Jansen, B.; Feipel, V.; Van Sint Jan, S.; Bonnechère, B. Validation of the Wii Balance Board to assess balance modifications induced by increased respiratory loads in healthy subjects. *Gait Posture* **2019**, *68*, 449–452. [CrossRef]
5. Janssens, L.; Brumagne, S.; Polspoel, K.; Troosters, T.; McConnell, A. The effect of inspiratory muscles fatigue on postural control in people with and without recurrent low back pain. *Spine* **2010**, *35*, 1088–1094. [CrossRef] [PubMed]
6. Hagio, K.; Obata, H.; Nakazawa, K. Effects of breathing movement on the reduction of postural sway during postural-cognitive dual tasking. *PLoS ONE* **2018**, *13*, e0197385. [CrossRef] [PubMed]
7. Caron, O.; Fontanari, P.; Cremieux, J.; Joulia, F. Effects of ventilation on body sway during human standing. *Neurosci. Lett.* **2004**, *366*, 6–9. [CrossRef] [PubMed]
8. Binazzi, B.; Lanini, B.; Bianchi, R.; Romagnoli, I.; Nerini, M.; Gigliotti, F.; Duranti, R.; Milic-Emili, J.; Scano, G. Breathing pattern and kinematics in normal subjects during speech, singing and loud whispering. *Acta Physiol.* **2006**, *186*, 233–246. [CrossRef] [PubMed]
9. Yardley, L.; Gardner, M.; Leadbetter, A.; Lavie, N. Effect of articulatory and mental tasks on postural control. *Neuroreport* **1999**, *10*, 215–219. [CrossRef] [PubMed]
10. Andersson, G.; Hagman, J.; Talianzadeh, R.; Svedberg, A.; Larsen, H.C. Effect of cognitive load on postural control. *Brain Res. Bull.* **2002**, *58*, 135–139. [CrossRef]
11. Vuillerme, N.; Nafati, G. How attentional focus on body sway affects postural control during quiet standing. *Psychol. Res.* **2007**, *71*, 192–200. [CrossRef]
12. Vuillerme, N.; Nougier, V.; Teasdale, N. Effects of a reaction time task on postural control in humans. *Neurosci. Lett.* **2000**, *291*, 77–80. [CrossRef]
13. Tremoureux, L.; Raux, M.; Ranohavimparany, A.; Morélot-Panzini, C.; Pouget, P.; Similowski, T. Electroencephalographic evidence for a respiratory-related cortical activity specific of the preparation of prephonatory breaths. *Respir Physiol. Neurobiol.* **2014**, *204*, 64–70. [CrossRef] [PubMed]
14. Yogev, G.; Hausdorff, J.M.; Giladi, N. The Role of Executive Function and Attention in Gait. *Mov. Disord.* **2008**, *23*, 329–472. [CrossRef]
15. Srygley, J.M.; Mirelman, A.; Herman, T.; Giladi, N.; Hausdorff, J.M. When does walking alter thinking? Age and task associated findings. *Brain Res.* **2009**, *1253*, 92–99. [CrossRef] [PubMed]
16. Brustio, P.R.; Magistro, D.; Zecca, M.; Rabaglietti, E.; Liubicich, M.E. Age-related decrements in dual-task performance: Comparison of different mobility and cognitive tasks. A cross sectional study. *PLoS ONE* **2017**, *12*, e0181698. [CrossRef] [PubMed]
17. MacPherson, M.K. Cognitive Load Affects Speech Motor Performance Differently in Older and Younger Adults. *J. Speech Lang. Hear. Res.* **2019**, *62*, 1258–1277. [CrossRef]
18. García, A.M.; Ibáñez, A. A touch with words: Dynamic synergies between manual actions and language. *Neurosci. Biobehav. Rev.* **2016**, *68*, 59–95. [CrossRef] [PubMed]
19. Dault, M.C.; Yardley, L.; Frank, J.S. Does articulation contribute to modifications of postural control during dual-task paradigms? *Brain Res. Cogn. Brain Res.* **2003**, *16*, 434–440. [CrossRef]
20. Rodriguez, A.D.; McCabe, M.L.; Nocera, J.R.; Reilly, J. Concurrent Word Generation and Motor Performance: Further Evidence for Language-Motor Interaction. *PLoS ONE* **2012**, *7*, e37094. [CrossRef] [PubMed]
21. Armieri, A.; Holmes, J.D.; Spaulding, S.J.; Jenkins, M.E.; Johnson, A.M. Dual task performance in a healthy young adult population: Results from a symmetric manipulation of task complexity and articulation. *Gait Posture* **2009**, *29*, 346–348. [CrossRef] [PubMed]
22. Plummer-D'Amato, P.; Altmann, L.J.P.; Saracino, D.; Fox, E.; Behrman, A.L.; Marsiske, M. Interactions between cognitive tasks and gait after stroke: A dual task study. *Gait Posture* **2008**, *27*, 683–688. [CrossRef] [PubMed]
23. Al-Yahya, E.; Dawes, H.; Smith, L.; Dennis, A.; Howells, K.; Cockburn, J. Cognitive motor interference while walking: A systematic review and meta-analysis. *Neurosci. Biobehav. Rev.* **2011**, *35*, 715–728. [CrossRef] [PubMed]
24. Li, K.Z.H.; Bherer, L.; Mirelman, A.; Maidan, I.; Hausdorff, J.M. Cognitive Involvement in Balance, Gait and Dual-Tasking in Aging: A Focused Review From a Neuroscience of Aging Perspective. *Front. Neurol.* **2018**, *9*, 913. [CrossRef] [PubMed]
25. Strobach, T.; Wendt, M.; Janczyk, M. Editorial: Multitasking: Executive Functioning in Dual-Task and Task Switching Situations. *Front. Psychol.* **2018**, *9*, 108. [CrossRef]
26. Leone, C.; Feys, P.; Moumdjian, L.; D'Amico, E.; Zappia, M.; Patti, F. Cognitive-motor dual-task interference: A systematic review of neural correlates. *Neurosci. Biobehav. Rev.* **2017**, *75*, 348–360. [CrossRef]
27. MacPherson, S.E. Definition: Dual-tasking and multitasking. *Cortex* **2018**, *106*, 313–314. [CrossRef] [PubMed]

28. Bonnechère, B.; Jansen, B.; Omelina, L.; Rooze, M.; Van Sint Jan, S. Interchangeability of the Wii Balance Board for Bipedal Balance Assessment. *JMIR Rehabil. Assist. Technol.* **2015**, *2*, e8. [CrossRef] [PubMed]
29. Clark, R.A.; Mentiplay, B.F.; Pua, Y.-H.; Bower, K.J. Reliability and validity of the Wii Balance Board for assessment of standing balance: A systematic review. *Gait Posture* **2018**, *61*, 40–54. [CrossRef] [PubMed]
30. Podsiadlo, D.; Richardson, S. The timed "Up & Go": A test of basic functional mobility for frail elderly persons. *J. Am. Geriatr. Soc.* **1991**, *39*, 142–148. [CrossRef]
31. Ward, N.; Menta, A.; Ulichney, V.; Raileanu, C.; Wooten, T.; Hussey, E.K.; Marfeo, E. The Specificity of Cognitive-Motor Dual-Task Interference on Balance in Young and Older Adults. *Front. Aging Neurosci.* **2021**, *13*, 804936. [CrossRef]
32. Gursoy, Z.G.; Yilmaz, U.; Celik, H.; Arpinar-Avsar, P.; Kirazci, S. Effect of individualized cognitive and postural task difficulty levels on postural control during dual task condition. *Gait Posture* **2022**, *96*, 1–8. [CrossRef] [PubMed]
33. Hausdorff, J.M.; Schweiger, A.; Herman, T.; Yogev-Seligmann, G.; Giladi, N. Dual-task decrements in gait: Contributing factors among healthy older adults. *J. Gerontol. A Biol. Sci. Med. Sci.* **2008**, *63*, 1335–1343. [CrossRef] [PubMed]
34. Hollman, J.H.; Kovash, F.M.; Kubik, J.J.; Linbo, R.A. Age-related differences in spatiotemporal markers of gait stability during dual task walking. *Gait Posture* **2007**, *26*, 113–119. [CrossRef] [PubMed]
35. Plummer, P.; Altmann, L.; Feld, J.; Zukowski, L.; Najafi, B.; Giuliani, C. Attentional prioritization in dual-task walking: Effects of stroke, environment, and instructed focus. *Gait Posture* **2020**, *79*, 3–9. [CrossRef] [PubMed]
36. Rizzato, A.; Paoli, A.; Andretta, M.; Vidorin, F.; Marcolin, G. Are Static and Dynamic Postural Balance Assessments Two Sides of the Same Coin? A Cross-Sectional Study in the Older Adults. *Front. Physiol.* **2021**, *12*, 681370. [CrossRef] [PubMed]
37. Kwok, B.-C.; Clark, R.A.; Pua, Y.-H. Novel use of the Wii Balance Board to prospectively predict falls in community-dwelling older adults. *Clin. Biomech.* **2015**, *30*, 481–484. [CrossRef]
38. Clavel, L.; Attali, V.; Rivals, I.; Niérat, M.-C.; Laveneziana, P.; Rouch, P.; Similowski, T.; Sandoz, B. Decreased respiratory-related postural perturbations at the cervical level under cognitive load. *Eur. J. Appl. Physiol.* **2020**, *120*, 1063–1074. [CrossRef] [PubMed]
39. Hellmann, D.; Giannakopoulos, N.N.; Blaser, R.; Eberhard, L.; Schindler, H.J. The effect of various jaw motor tasks on body sway. *J. Oral Rehabil.* **2011**, *38*, 729–736. [CrossRef]
40. Varghese, J.P.; Beyer, K.B.; Williams, L.; Miyasike-daSilva, V.; McIlroy, W.E. Standing still: Is there a role for the cortex? *Neurosci. Lett.* **2015**, *590*, 18–23. [CrossRef] [PubMed]
41. Yogev-Seligmann, G.; Rotem-Galili, Y.; Mirelman, A.; Dickstein, R.; Giladi, N.; Hausdorff, J.M. How does explicit prioritization alter walking during dual-task performance? Effects of age and sex on gait speed and variability. *Phys. Ther.* **2010**, *90*, 177–186. [CrossRef] [PubMed]
42. Beauchet, O.; Dubost, V.; Aminian, K.; Gonthier, R.; Kressig, R.W. Dual-task-related gait changes in the elderly: Does the type of cognitive task matter? *J. Mot. Behav.* **2005**, *37*, 259–264.
43. Chang, W.-H.; Tang, P.-F.; Wang, Y.-H.; Lin, K.-H.; Chiu, M.-J.; Chen, S.-H.A. Role of the premotor cortex in leg selection and anticipatory postural adjustments associated with a rapid stepping task in patients with stroke. *Gait Posture* **2010**, *32*, 487–493. [CrossRef] [PubMed]
44. Kuo, H.-T.; Yeh, N.-C.; Yang, Y.-R.; Hsu, W.-C.; Liao, Y.-Y.; Wang, R.-Y. Effects of different dual task training on dual task walking and responding brain activation in older adults with mild cognitive impairment. *Sci. Rep.* **2022**, *12*, 8490. [CrossRef] [PubMed]
45. Akin, H.; Senel, A.; Taskiran, H.; Kaya Mutlu, E. Do motor-cognitive and motor-motor dual task training effect differently balance performance in older adults? *Eur. Geriatr. Med.* **2021**, *12*, 371–378. [CrossRef] [PubMed]
46. Spanò, B.; Lombardi, M.G.; De Tollis, M.; Szczepanska, M.A.; Ricci, C.; Manzo, A.; Giuli, S.; Polidori, L.; Griffini, I.A.; Adriano, F.; et al. Effect of Dual-Task Motor-Cognitive Training in Preventing Falls in Vulnerable Elderly Cerebrovascular Patients: A Pilot Study. *Brain Sci.* **2022**, *12*, 168. [CrossRef] [PubMed]
47. Wollesen, B.; Voelcker-Rehage, C. Training effects on motor–cognitive dual-task performance in older adults. *Eur. Rev. Aging Phys. Act.* **2014**, *11*, 5–24. [CrossRef]
48. Hakamy, A.; Bolton, C.E.; Gibson, J.E.; McKeever, T.M. Risk of fall in patients with COPD. *Thorax* **2018**, *73*, 1079–1080. [CrossRef]

brain sciences

MDPI

Article

Using Interaction between Cognitive and Motor Impairment for Risk Screening of Major Neurocognitive Disorders: Results of the EPIDOS Observational Cohort Study

Olivier Beauchet [1,2,3,4,*], Jacqueline Matskiv [2], Yves Rolland [5], Anne-Marie Schott [6] and Gilles Allali [7]

1 Departments of Medicine and Geriatrics, University of Montreal, Montreal, QC H3W 1W5, Canada
2 Research Centre of the Geriatric University Institute of Montreal, Montreal, QC H3W 1W5, Canada; jacquelinematskiv@gmail.com
3 Department of Medicine, Division of Geriatric Medicine, Sir Mortimer B. Davis Jewish General Hospital and Lady Davis Institute for Medical Research, McGill University, Montreal, QU H3T 1E2, Canada
4 Lee Kong Chian School of Medicine, Nanyang Technological University, Singapore 636921, Singapore
5 Department of Geriatric, Toulouse University Hospital, 31059 Toulouse, France; rolland.y@chu-toulouse.fr
6 HESPER EA 7425, Hospices Civils de Lyon, Pôle de Santé Publique, Université Claude Bernard Lyon 1, 69229 Lyon, France; anne-marie.schott-pethelaz@chu-lyon.fr
7 Leenaards Memory Center, Lausanne University Hospital and University of Lausanne, 1011 Lausanne, Switzerland; gilles.allali@chuv.ch
* Correspondence: olivier.beauchet@umontreal.ca; Tel.: +1-514-340-3540 (ext. 3637); Fax: +1-514-340-4765

Abstract: Background and purpose: Cognitive and motor impairments are risk factors of major neurocognitive disorders (MNCD). Inability to name the date and use of a walking aid and/or history of falls are two items which are surrogate measures of cognitive and motor impairments. This study aims to examine the association of inability to name the date (i.e., cognitive impairment), use of a walking aid and/or history of falls (i.e., motor impairment) and their combination with incident MNCD in community-dwelling older adults. Methods: A total of 709 participants (mean age 79.8 ± 3.7; 100% female) of the EPIDémiologie de l'OStéoporose (EPIDOS) study recruited in Toulouse (France) were selected for this study. EPIDOS is an observational population-based cohort study with a 7-year follow-up period for Toulouse participants. Inability to name the date and use of a walking aid and/or history of falls were collected at baseline. Incident MNCD and their type (i.e., Alzheimer's disease (AD) and non-AD) were diagnosed at the end of the 7-year follow-up. Results: Overall incidence of MNCD was 29.1%. Cox regressions revealed that inability to name the date and its combination with use of a walking aid and/or history of falls was associated with a significant increased incidence of MNCD (hazard ratio (HR) = 1.10 with $p = 0.003$ and HR = 1.81 with $p = 0.011$, respectively) and AD (HR =1.13 with $p = 0.003$ and HR = 2.80 with $p = 0.016$, respectively). Conclusions: Increased incident MNCD was reported when inability to name the date and use of a walking aid and/or history of falls were combined, suggesting that this combination of items may be used for risk screening of MNCD in the older population, especially for incident AD.

Keywords: older adults; epidemiology; cohort study; dementia; screening

check for updates

Citation: Beauchet, O.; Matskiv, J.; Rolland, Y.; Schott, A.-M.; Allali, G. Using Interaction between Cognitive and Motor Impairment for Risk Screening of Major Neurocognitive Disorders: Results of the EPIDOS Observational Cohort Study. *Brain Sci.* **2022**, *12*, 1021. https://doi.org/10.3390/brainsci12081021

Academic Editor: Daniele Corbo

Received: 30 June 2022
Accepted: 25 July 2022
Published: 31 July 2022

1. Introduction

Cognitive impairment is a typical risk factor of major neurocognitive disorders (MNCD) [1,2]. Motor impairment is also a risk factor for MNCD in the aging population [3–7]. For instance, mild parkinsonian signs—which are prevalent in aging—and slow walking speed have been associated with the occurrence of MNCD [5,6]. Both cognitive and motor impairments are independent risk factors of MNCD which may interact. Cognitive impairment is a risk factor of motor impairment and vice versa [1,3–8]. In addition, when they coexist, the risk of MNCD increases significantly [8–10]. The co-occurrence of slow walking speed and subjective cognitive complaint (SCC) in individuals free of MNCD defines motoric cognitive risk syndrome (MCR) [8]. MCR is a pre-MNCD stage and poses a greater risk of MNCD than each MCR component considered individually [8,9], suggesting a synergistic effect which may be used to screen MNCD in the older population [10].

Frailty is also associated with increased risk of MNCD [11]. Cognitive frailty is a clinical syndrome which combines physical and cognitive impairment [11,12]. Individuals with cognitive frailty have a higher risk of MNCD than those with physical frailty alone [11–13]. "Emergency Room Evaluation and Recommendations" (ER^2) is a clinical tool which screens frailty and its related risk of adverse outcomes in older emergency department users [14]. Inability to name the date and use of a walking aid and/or history of falls are two ER^2 items which are surrogate measures of cognitive and motor impairments. We hypothesized that, like the MCR components, interaction between these two ER^2 items could be associated with an increased risk of MNCD. This study thus aims to examine the association of inability to name the date (i.e., cognitive impairment) and use of a walking aid and/or history of falls (i.e., motor impairment) and their combination with incident MNCD in community-dwelling older adults.

2. Material and Methods

2.1. Design

The "EPIDémiologie de l'OStéoporose" (EPIDOS) database was used for the present study [15]. EPIDOS is an observational population-based cohort study designed to examine risk factors for hip fracture in older French women. The participants selected for the present study were recruited in Toulouse (city in Southern France). They had an additional 3-year follow-up after the initial 4-year EPIDOS follow-up period, which included a final full cognitive assessment, which was performed at the University Hospital of Toulouse or at the participant's home.

2.2. Population

The initial set of EPIDOS participants was composed of 7598 women, aged 75 and over, living in communities in five French cities (Amiens, Lyon, Montpellier, Paris and Toulouse). A total of 1462 (19.2%) participants were recruited in Toulouse. We excluded from this subset of participants those with a suspicion of MNCD at baseline using the threshold value of $\geq$3 incorrect answers on the Short Portable Mental Status Questionnaire (SPMSQ), and those without information on their cognitive status (i.e., no MNCD versus MNCD and its etiology coded as Alzheimer's Disease (AD) vs. non-AD) at the end of follow-up [16]. A total of 709 (48.5% of the Toulouse EPIDOS participants) participants were finally selected for the present study.

2.3. Baseline Assessment

Age, living in residence, high education level (i.e., high school level or higher completed), living alone, frequency of contact with someone over the past week, number of drugs taken daily, measured weight (in kg) and height (in cm), regular physical activity (i.e., $\geq$ one hour a week during the past month), use of a walking aid, history of falls in the past 6 months and inability to name the date were recorded at baseline assessment using a standardized face-to-face physical examination. Age was stratified into two groups using the threshold value $\geq$ 85. Body mass index (BMI) was calculated. Overweight and/or obesity were defined as a BMI $\geq$ 25 kg/m^2. Polypharmacy was defined as $\geq$5 drugs taken daily. Social isolation was defined as living alone and no contact with someone over the past week.

2.4. Definition of MNCD

At the end of the 7-year follow-up, a face-to-face cognitive assessment including the SPMSQ [10], the Mini Mental State Examination [17] and the Grober and Buschke test (i.e., Free and Cued Selective Reminding Test) was performed [18]. Data collected were analyzed by a geriatrician and a neurologist in a double-blind manner to determine the cognitive status of participants. DSM-IV criteria were used for the diagnosis of MNCD [13,14]. AD diagnosis was made using the criteria of the NINCDS-ADRDA Work Group [19–22]. Participants who satisfied DSM-IV criteria but not NINCDS-ADRDA criteria were classified with a diagnosis of non-AD. Participants were separated in four groups: no MNCD, all categories of MNCD, AD and non-AD.

2.5. Standard Protocol Approval and Patient Consents

The Research Ethics Boards (REB) of Toulouse University Hospital approved the EPIDOS protocol (protocol code EPIODS (@ and 1992/01/05). Written informed consent for research was obtained for all recruited EPIDOS participants.

3. Statistics

The participants' baseline characteristics were described using means, standard deviation (SD), percentages and confidence intervals. Cox regressions were performed to examine the association of inability to name the date, use of a walking aid and/or history of falls and their combination (independent variables; separated model for each variable) with incident MNCD (dependent variable; separated model for each type of MNCD). All models are adjusted by age, place of living, education level, abnormal body mass index (i.e., $\geq$25 kg/m^2), regular physical activity, polypharmacy and social isolation; p-values < 0.05 were considered statistically significant. All statistics were performed using SPSS (version 28.0; SPSS, Inc., Chicago, IL, USA).

4. Results

Table 1 shows the baseline characteristics of participants. The incidence of MNCD was 29.1%. AD was more incident than non-AD (15.5% vs. 13.5%). Cox regressions showed that inability to name the date and its combination with use of a walking aid and/or history of falls were significantly associated with an increased incidence of MNCD (hazard ratio (HR) = 1.10 with p = 0.003 and HR = 1.81 with p = 0.011) and AD (HR =1.13 with p = 0.003 and HR = 2.80 with p = 0.016) (Table 2). No significant association was found with incident non-AD. Use of a walking aid and/or history of falls was not associated with incident MNCD, including its subtypes.

Table 1. Participants' baseline characteristics and incident major neurocognitive disorders (n = 709).

Characteristics	Value	[95% CI]
Age (year)		
Mean ± SD	79.8 ± 3.7	[79.5; 80.1]
Age ≥ 85, n (%)	69 (9.7)	[7.5; 11.9]
Living in residence, n (%)	77 (10.9)	[8.6; 13.3]
Social isolation *, n (%)	276 (38.9)	[35.2; 42.4]
High education level †, n (%)	299 (42.2)	[38.4; 45.7]
Number of drugs taken daily		
Mean ± SD	5.0 ± 2.9	[4.8; 5.3]
Polypharmacy ‡	388 (54.7)	[51.0; 58.4]
Body mass index (kg/m^2)		
Mean ± SD	25.0 ± 3.9	[24.7; 25.3]
Overweight/Obese ¶	324 (45.7)	[42.01; 49.3]
Regular physical activity #	297 (41.9)	[38.5; 45.8]
Use of walking aid and/or history of fall in the past 6 months	227 (32.0)	[29.5; 34.6]
Inability to name day's date	147 (20.7)	[18.1; 23.7]
Incident major neurocognitive disorders, n (%)		
All categories	206 (29.1)	[25.5; 32.2]
Non-Alzheimer's disease	96 (13.5)	[11.0; 16.0]
Alzheimer's disease	110 (15.5)	[12.7; 18.0]

SD: standard deviation; CI: confidence interval; *: living alone and no contact with someone over the past week; †: high school and greater; ‡: number of drugs taken daily ≥ 5; ¶: value ≥ 25 kg/m^2; #: at least one recreational physical (walking, gymnastics, cycling, swimming or gardening) activity for at least one hour a week for the past month or more.

Table 2. Cox regressions showing the association of inability to name day's date, use of walking aid and/or history of falls and their combination (independent variable; separated model for each variable) and incident major neurocognitive disorders (all categories, non-Alzheimer's disease, Alzheimer's disease; dependent variable; separated model for each category) in EPIDOS participants (n = 709).

	Major Neurocognitive Disorders								
	All Categories			Non-Alzheimer's Disease			Alzheimer's Disease		
	HR	[95% CI]	p-Value	HR	[95% CI]	p-Value	HR	[95% CI]	p-Value
Inability to name the day's date	1.10	[1.03; 1.17]	**0.003**	1.06	[0.97; 1.16]	0.226	1.13	[1.05; 1.23]	**0.003**
Use of walking aid and/or history of falls	1.01	[0.96; 1.08]	0.646	0.98	[0.90; 1.07]	0.679	1.04	[0.96; 1.13]	0.316
Inability to name the day's date AND use of walking aid and/or history of falls	1.81	[1.15; 2.85]	**0.011**	1.53	[0.76; 3.09]	0.235	2.08	[1.14; 3.78]	**0.016**

HR: hazard ratio; CI: confidence interval; all models are adjusted by age, place of living, social isolation, education level, body mass index $\geq$ 25 kg/m^2, regular physical activity and polypharmacy; p-value significance (i.e., <0.05) indicated in bold.

5. Discussion

The findings show that an increased incidence of MNCD and AD were associated with inability to name the date alone and its combination with use of a walking aid and/or history of falls in EPIDOS participants. The greatest incidence of MNCD was reported when inability to name the date was combined with use of a walking aid and/or history of falls and AD.

We found that inability to name the date, but not use of a walking aid and/or history of falls, was associated with incident MNCD and AD. This result is consistent with a previous study which examined the association of MCR and its components (i.e., slow walking speed and SCC) with incident MNCD [6]. In this former study, the cognitive component of MCR (i.e., subjective cognitive complaint) and not slow walking speed was associated with incident MNCD in older community dwellers. An explanation of this specific association may be due to the fact that the onset of MNCD is characterized by cognitive impairment [1]. Inability to name the date may be assimilated as an objective cognitive impairment. In our study, selected participants were free of MNCD at baseline and only those with temporal impairment had significant risk of incident MNCD. Although specific motor impairment related to gait disorders or mild parkinsonian signs may predict cognitive decline, the association between cognitive impairment and incident MNCD is stronger than between motor impairment and incident MNCD [7–9]. Finally, there are more mixed results regarding the association between motor impairment and incident MNCD compared to cognitive impairment [7–10].

Our findings also revealed that the association between both items (i.e., inability to name the date and use of a walking aid and/or history of falls) and incident MNCD was significant and greater when combined, compared to the inability to name the date by itself. Again, this result is consistent with previous results reported on MCR [9,10]. Indeed, we report that the magnitude of this risk is two-fold when compared to the cognitive impairment item alone. This result highlights an interaction between cognitive impairment and motor impairment, which may be used to screen individuals at risk of MNCD and AD. Interestingly, inability to name the date alone or combined with use of a walking aid and/or history of falls did not predict non-AD. This contrast with the prediction of AD may be related to the heterogeneity of the patients included in the non-AD dementia group. Indeed, vascular dementia, Lewy bodies dementia, or other neurodegenerative conditions affecting the oldest old (i.e., PART, LATE) rely on different neuropathogenic mechanisms.

The 7-year duration of the prospective, observational follow-up and the sample size are the main strengths of the present study, but some limitations emerged. First, even if EPIDOS's design was appropriate for the objective of our study, examining an association between the ER^2 items and incident MNCD was not initially planned. Second, we selected only EPIDOS participants recruited in Toulouse, only about half of which were included in the present study, which may have introduced selection bias and impacted outcomes. Third, Cox models were adjusted for participants' baseline characteristics, but residual confounders may still be present and modify the association between the ER^2 items and incident MNCD. For instance, chronic morbidities may influence both cognitive and motor impairments, and thus their association with incident dementia [11]. We tried to control for the effects of comorbidities by adjusting for polypharmacy, which is a surrogate measure of accumulation of morbidities [23]. Finally, the generalization of the study findings does not apply to males, as EPIDOS included only women.

6. Conclusions

Increased incident MNCD was reported when inability to name the date and use of a walking aid and/or history of falls were combined, suggesting that these items may be used for risk screening of MNCD in the older population. Both items are easy to collect at the level of an older population, which creates new opportunities for MNCD risk identification and the preventive care of its modifiable risk factors [24,25].

Author Contributions: Conceived and designed the experiments: O.B. and G.A. Cohort data collection: A.-M.S. and Y.R. Analyzed and interpreted the data: O.B. and G.A. Contributed reagents, materials, analysis tools or data: O.B. Writing of the manuscript: O.B., J.M. and G.A. Revision of manuscript: A.-M.S. and Y.R. All authors have read and agreed to the published version of the manuscript.

Funding: The French Ministry of Health fi-nancially supported the study. Dr. Beauchet and Dr. Allali were supported by the National Institute of Health/National Institute on Aging grants PO1 AG03949 and R01AG057548-01A1.

Institutional Review Board Statement: The study was conducted in accordance with the Declaration of Helsinki, and approved by the Research Ethics Boards (REB) of Toulouse University Hospital (protocol code EPIODS(@ and 1992/01/05).

Informed Consent Statement: Informed consent was obtained from all subjects involved.

Data Availability Statement: Access to the EPIDOS Database can be obtained by contacting Olivier Beauchet via Olivier.beauchet@umontreal.ca.

Acknowledgments: The main investigators of the EPIDOS study were G. Breart, P. Dargent-Molina, P.J. Meunier, A.M. Schott, D. Hans, and P.D. Delmas, and the co-investigators were C. Baudoin and J.L. Sebert (Amiens), M.C. Chapuy and A.M. Schott (Lyon), F. Favier and C. Marcelli (Montpellier), C.J. Menkes, C. Cormier, and E. Hausherr (Paris) and H. Grandjean and C. Ribot (Toulouse).

Conflicts of Interest: The authors declare no conflict of interest. The French Ministry of Health financially supported the study. Beauchet and Allali were supported by the National Institute of Health/National Institute on Aging grants PO1 AG03949 and R01AG057548-01A1. The sponsors had no role in designing and conducting the study, nor in the collection, management, analysis and interpretation of the data, nor in the preparation, review or approval of the manuscript.

References

1. Mendonça, M.D.; Alves, L.; Bugalho, P. From Subjective Cognitive Complaints to Dementia: Who is at Risk?: A Systematic Review. *Am. J. Alzheimers Dis. Other Demen.* **2016**, *31*, 105–114. [CrossRef]
2. Hugo, J.; Ganguli, M. Dementia and Cognitive Impairment: Epidemiology, Diagnosis, and Treatment. *Clin. Geriatr. Med.* **2014**, *30*, 421–442. [CrossRef] [PubMed]
3. Kueper, J.K.; Speechley, M.; Lingum, N.R.; Montero-Odasso, M. Motor function and incident dementia: A systematic review and meta-analysis. *Age Ageing* **2017**, *46*, 729–738. [CrossRef] [PubMed]
4. Mitchell, S.L.; Rockwood, K. The association between parkinsonism, Alzheimer's disease, and mortality: A comprehensive approach. *J. Am. Geriatr. Soc.* **2000**, *48*, 422–425. [CrossRef]
5. Mahoney, J.R.; Verghese, J.; Holtzer, R.; Allali, G. The evolution of mild parkinsonian signs in aging. *J. Neurol.* **2014**, *261*, 1922–1928. [CrossRef] [PubMed]

6. Beauchet, O.; Annweiler, C.; Callisaya, M.L.; De Cock, A.M.; Helbostad, J.L.; Kressig, R.W.; Srikanth, V.; Steinmetz, J.P.; Blumen, H.M.; Verghese, J.; et al. Poor Gait Performance and Prediction of Dementia: Results From a Meta-Analysis. *J. Am. Med. Dir. Assoc.* **2016**, *17*, 482–490. [CrossRef] [PubMed]
7. Quan, M.; Xun, P.; Chen, C.; Wen, J.; Wang, Y.; Wang, R.; Chen, P.; He, K. Walking Pace and the Risk of Cognitive Decline and Dementia in Elderly Populations: A Meta-analysis of Prospective Cohort Studies. *J. Gerontol. A Biol. Sci. Med. Sci.* **2017**, *72*, 266–270. [CrossRef]
8. Verghese, J.; Wang, C.; Lipton, R.B.; Holtzer, R. Motoric cognitive risk syndrome and the risk of dementia. *Gerontol. Ser. A Biomed. Sci. Med. Sci.* **2013**, *68*, 412–418. [CrossRef]
9. Mullin, D.S.; Cockburn, A.; Welstead, M.; Luciano, M.; Russ, T.C.; Muniz-Terrera, G. Mechanisms of motoric cognitive risk-Hypotheses based on a systematic review and meta-analysis of longitudinal cohort studies of older adults. *Alzheimers Dement.* **2022**. *ahead of print*. [CrossRef]
10. Beauchet, O.; Sekhon, H.; Cooper-Brown, L.; Launay, C.P.; Gaudreau, P.; Morais, J.A.; Allali, G. Motoric cognitive risk syndrome and incident dementia in older adults from the Québec NuAge cohort. *Age Ageing* **2021**, *50*, 969–973. [CrossRef]
11. Chu, W.; Chang, S.F.; Ho, H.Y. Adverse Health Effects of Frailty: Systematic Review and Meta-Analysis of Middle-Aged and Older Adults With Implications for Evidence-Based Practice. *Worldviews Evid. Based Nurs.* **2021**, *18*, 282–289. [CrossRef] [PubMed]
12. Blodgett, J.; Theou, O.; Kirkland, S.; Andreou, P.; Rockwood, K. Frailty in NHANES: Comparing the frailty index and phenotype. *Arch. Gerontol. Geriatr.* **2015**, *60*, 464–470. [CrossRef] [PubMed]
13. De Vries, N.M.; Staal, J.B.; Van Ravensberg, C.D.; Hobbelen, J.S.; Olde Rikkert, M.G.; Nijhuis-van der Sanden, M.W. Outcome instruments to measure frailty: A systematic review. *Ageing Res. Rev.* **2011**, *10*, 104–114. [CrossRef] [PubMed]
14. Launay, C.P.; Galery, K.; Vilcocq, C.; Afilalo, M.; Beauchet, O. Risk for short-term undesirable outcomes in older emergency department users: Results of the ER2 observational cohort study. *PLoS ONE* **2021**, *16*, e0249882. [CrossRef]
15. Dargent-Molina, P.; Favier, F.; Grandjean, H.; Baudoin, C.; Schott, A.M.; Hausherr, E.; Meunier, P.J.; Bréart, G. Fall-related factors and risk of hip fracture: The EPIDOS prospective study. *Lancet* **1996**, *348*, 145–149. [CrossRef]
16. Pfeiffer, E. A short portable mental status questionnaire for the assessment of organic brain deficit in elderly patients. *J. Am. Geriatr. Soc.* **1975**, *23*, 433–441. [CrossRef] [PubMed]
17. Folstein, M.; Folstein, S.; McHugh, P. Mini Mental State. A practical method for grading the cognitive state of patients for the clinician. *J. Psychiatr. Res.* **1975**, *12*, 189–198. [CrossRef]
18. Grober, E.; Buschke, H.; Crystal, H.; Bang, S.; Dresner, R. Screening for dementia by memory testing. *Neurology* **1988**, *38*, 900. [CrossRef] [PubMed]
19. American Psychiatric Association. *Diagnostic and Statistical Manual of Mental Disorders*, 4th ed.; American Psychiatric Association: Washington, DC, USA, 1994.
20. Tucker, G.J. The diagnosis of delirium and DSM-IV. *Dement. Geriatr. Cogn. Disord.* **1999**, *10*, 359–363. [CrossRef]
21. McKhann, G.; Drachman, D.; Folstein, M.; Katzman, R.; Price, D.; Stadlan, E.M. Clinical diagnosis of Alzheimer's disease: Report of the NINCDS-ADRDA Work Group under the auspices of Department of Health and Human Services Task Force on Alzheimer's Disease. *Neurology* **1984**, *34*, 939–944. [CrossRef]
22. Hogervorst, E.; Barnetson, L.; Jbst, K.A.; Nagy, Z.; Combrinck, M.; Smith, A.D. Diagnosing dementia: Interrater reliability assessment and accuracy of the NINCDS/ADRDA criteria versus CERAD histopathological criteria for Alzheimer's disease. *Dement. Geriatr. Cogn. Disord.* **2000**, *11*, 107–113. [CrossRef] [PubMed]
23. De Decker, L.; Launay, C.; Annweiler, C.; Kabeshova, A.; Beauchet, O. Number of drug classes taken per day may be used to assess morbidity burden in older inpatients: A pilot cross-sectional study. *J. Am. Geriatr. Soc.* **2013**, *61*, 1224–1225. [CrossRef] [PubMed]
24. Livingston, G.; Huntley, J.; Sommerlad, A.; Ames, D.; Ballard, C.; Banerjee, S.; Brayne, C.; Burns, A.; Cohen-Mansfield, J.; Cooper, C.; et al. Dementia prevention, intervention, and care: 2020 report of the Lancet Commission. *Lancet* **2020**, *396*, 413–446. [CrossRef]
25. Curran, E.; Chong, T.W.H.; Godbee, K.; Abraham, C.; Lautenschlager, N.T.; Palmer, V.J. General population perspectives of dementia risk reduction and the implications for intervention: A systematic review and thematic synthesis of qualitative evidence. *PLoS ONE* **2021**, *16*, e0257540. [CrossRef] [PubMed]

brain sciences

MDPI

Systematic Review

Cognitive and Academic Outcomes of Fundamental Motor Skill and Physical Activity Interventions Designed for Children with Special Educational Needs: A Systematic Review

Pinja Jylänki [1,*], Theo Mbay [1], Anni Byman [1], Airi Hakkarainen [1], Arja Sääkslahti [2] and Pirjo Aunio [1]

[1] Faculty of Educational Sciences, University of Helsinki, 00170 Helsinki, Finland; theo.mbay@helsinki.fi (T.M.); anni.byman@helsinki.fi (A.B.); airi.hakkarainen@helsinki.fi (A.H.); pirjo.aunio@helsinki.fi (P.A.)

[2] Faculty of Sport and Health Sciences, University of Jyväskylä, 40600 Jyväskylä, Finland; arja.saakslahti@jyu.fi

* Correspondence: pinja.jylanki@helsinki.fi

Abstract: This systematic review aimed to investigate the methodological quality and the effects of fundamental motor skills and physical activity interventions on cognitive and academic skills in 3- to 7-year-old children with special educational needs. The review was reported in accordance with the Preferred Reporting Items for Systematic Reviews and Meta-Analyses (PRISMA 2020) statement. A literature search was carried out in April 2020 (updated in January 2022) using seven electronic databases, including ERIC, Scopus, Web of Science, PsycINFO, CINAHL, PubMed, and SPORTDiscus. The methodological quality of the studies was assessed with Effective Public Health Practice Project (EPHPP) Quality Assessment Tool. Cohen's *d* effect sizes and post-hoc power analyses were conducted for the included studies. Altogether 22 studies (1883 children) met the inclusion criteria, representing children at-risk for learning difficulties, due to family background ($n_{studies}$ = 8), children with learning difficulties ($n_{studies}$ = 7), learning disabilities ($n_{studies}$ = 5), and physical disabilities ($n_{studies}$ = 2). Two of the included 22 studies displayed strong, one moderate, and 19 studies weak methodological quality. The intervention effects appeared to be somewhat dependent on the severity of the learning difficulty; in cognitive and language skills, the effects were largest in children at-risk due to family background, whereas in executive functions the effects were largest in children with learning disabilities. However, due to the vast heterogeneity of the included studies, and a rather low methodological quality, it is challenging to summarize the findings in a generalizable manner. Thus, additional high-quality research is required to determine the effectiveness of the interventions.

Keywords: academic skills; cognition; early intervention; motor skills; physical activity; special educational needs; systematic review

Citation: Jylänki, P.; Mbay, T.; Byman, A.; Hakkarainen, A.; Sääkslahti, A.; Aunio, P. Cognitive and Academic Outcomes of Fundamental Motor Skill and Physical Activity Interventions Designed for Children with Special Educational Needs: A Systematic Review. *Brain Sci.* **2022**, *12*, 1001. https://doi.org/10.3390/brainsci12081001

Academic Editors: Daniele Corbo and Fiorenzo Moscatelli

Received: 27 June 2022
Accepted: 26 July 2022
Published: 28 July 2022

1. Introduction

Children's cognitive (e.g., executive functions) and academic skills (e.g., early numeracy and literacy skills) start to develop during the early years [1,2] which provide important grounds for later development [3]. During these years, in particular, the development of cognitive and academic skills is highly interrelated [4]. Thus, early childhood education has an important role in children's development, especially for children with special educational needs (SEN) [3] whose later academic success is at risk [5]. Children with SEN are not a homogeneous group, but rather include a wide range of children with various types and extents of learning difficulties or disabilities [4,6]; stemming, for instance, from biological, neurobiological, intellectual, genetic, or environmental factors [7]. While the challenges of children with SEN differ widely, in general, a requirement for customized special education is observed [4]. Early childhood education provides a valuable environment for the implementation of effective interventions to support the learning of children with SEN, and thereby minimize or prevent the impact of learning difficulties, disabilities, or at-risk conditions on children's development and future capabilities [4,8].

Fundamental motor skills (FMS, i.e., balance, and manipulative and locomotor movement skills) [9] and physical activity (PA; i.e., bodily movements that increase energy expenditure) [10] have been found to be related to cognitive and academic skills in children [11,12]. The close relationship between FMS and cognitive skills has been explained by co-activation in the same brain areas (i.e., cerebellum, prefrontal cortex, and basal ganglia) [13]. In addition, studies have shown that the relationship between PA and cognitive skills may be mediated through improved executive functions, especially inhibition [12,14], and that particularly high-intensity PA may affect cognitive skills through changes in the brain, via increased cerebral blood volume, and other physiological changes, such as increased neurogenesis [11,14,15].

In recent decades, studies examining the effects of FMS and PA interventions on cognitive skills have increased rapidly [12]. A recent systematic review demonstrated the positive effects of FMS and PA interventions on preschoolers' cognitive and academic skills in typically developing children [16]. However, the effects, as well as the quality of FMS and PA interventions on children with SEN have not been previously analyzed. Considering that children with SEN are at risk for developing more severe problems in their academic skills during later years [17], it is highly important to investigate the most effective evidence-based practices for supporting cognitive and academic learning at an early age [18]. As FMS and PA are associated with cognitive and academic skills already in early childhood [13], and FMS and PA interventions have been found to improve typically developing children's cognitive and academic skills [16], it is plausible that FMS and PA interventions similarly support the learning of children with SEN. Thus, the aim of the present systematic review was to investigate the methodological quality and the effects of FMS and PA interventions on cognitive and academic skills in children aged 3–7 years-old with SEN. Since children with SEN include a wide range of children with various types and extents of learning difficulties or disabilities [4,6], the intervention effects were examined in groups based on the assumed severity of the children's learning difficulties.

2. Materials and Methods

The present systematic review was reported in accordance with the Preferred Reporting Items for Systematic Reviews and Meta-Analyses (PRISMA 2020) statement, which is designed for systematic reviews that evaluate the effects of intervention studies [19]. The review protocol was not pre-registered. The systematic review was conducted as follows:

Step 1: A literature search, including abstract rating and full-text screening, based on the pre-determined eligibility criteria, was carried out in April 2020 by two authors (P.J. and T.M.)

Step 2: In January 2022, an updated literature search was carried out for studies published between April 2020 and January 2022 by two authors (T.M. and A.B.) following the aforementioned protocol.

Step 3: Methodological quality of the studies was assessed with the Effective Public Health Practice Project Quality Assessment Tool for Quantitative Studies (EPHPP).

Step 4: Cohen's *d* effect sizes were calculated, and post hoc power analyses were conducted to determine the statistical power of the included studies.

2.1. Eligibility Criteria

Peer-reviewed intervention studies investigating the effects of FMS or PA interventions on cognitive or academic skills in preschoolers with SEN published in English, were eligible for the present systematic review. Specific eligibility criteria are reported according to the PICO framework [20]:

Population: Published peer-reviewed intervention studies that were published in English and included children aged 3–7-years old with SEN. For instance, children at risk for learning difficulties due to family background (e.g., low socioeconomic status (SES)), children with learning difficulties (e.g., high risk of attention deficit hyperactivity disorder (ADHD)), learning disabilities (e.g., autism spectrum disorder), and physical disabilities (e.g., cerebral palsy) were included.

Intervention: All intervention studies with only FMS and/or PA practices or a program that combined FMS and/or PA with cognitive or academic skill practices (e.g., children counting balls while playing with them or collecting items in a particular order) [21] were included. For study designs, all intervention designs except case studies were included. Considering that the present study focused on children with a wide range of SEN (e.g., children with cerebral palsy) it may prove difficult to find a comparable control group, and thus, this was not required.

Comparator: The business-as-usual control groups were used in the analysis as a comparator. Studies that used an active control group were analyzed as separate interventions.

Outcome: The effects of the intervention had to be assessed via cognitive or academic outcomes, such as measures of cognitive, language, and numeracy skills or executive functions.

The systematic literature search was carried out on 16 April 2020, by two authors, P.J. and T.M., using seven electronic databases, including ERIC, Scopus, Web of Science, PsycINFO, CINAHL, PubMed, and SPORTDiscus. Since there was more than one year from the previous literature search, an updated search, following the same methods, was carried out on 18 January 2022 by the authors T.M. and A.B. Search terms were designed in accordance with the PICO framework [20] and consisted of the following: "early education" OR child * AND motor * OR "physical activity" AND intervention OR program * OR treatment OR training OR instruction AND cognit * OR academic *. When possible, additional search filters were used to exclude studies that investigated children older than seven years, and in the updated literature search only articles published from 2020 onwards were examined.

2.2. Study Selection

Three authors performed the article selection according to the predetermined eligibility criteria. The articles were initially screened based on the abstract. In terms of inclusion, the abstracts were coded as "yes", "maybe" or "no". The inter-rater agreement was determined during the abstract rating processes by calculating Cohen's weighted kappa. In the first literature search, both authors rated the first 40% (n = 2266) of the abstracts independently, after which the inter-rater agreement was 0.718 and the remaining abstracts were divided between the authors. In the updated literature search, both authors rated all of the abstracts (n = 2198) independently with an inter-rater agreement of 0.771. In both cases, the inter-rater agreement could be considered as good [22]. Following the abstract screening, all of the eligible articles underwent full-text screening, where the authors independently decided whether to "exclude", "include" or "maybe" include each article.

2.3. Methodological Quality

The methodological quality of the eligible studies was assessed with the Effective Public Health Practice Project Quality Assessment Tool for Quantitative Studies (EPHPP) [23,24]. The tool is suitable for evaluating the quality of a variety of study designs (e.g., randomized controlled trials (RCT) and pre-post designs (PPD)) [25] and has been used previously in systematic reviews in this particular field [16,26]. The inter-rater agreement has been shown to be more consistent with the EPHPP tool compared to the Cochrane Collaboration Risk of Bias Tool [25], for instance. Three authors rated each study procedure as "strong", "moderate" or "weak". Final ratings were formed based on six sections (selection bias, study design, confounders, blinding, data collection methods, and withdrawals and drop-outs) with the following criteria: studies with no weak ratings and at least four strong ratings were considered as "strong"; studies with less than four strong ratings and one weak rating were considered as "moderate"; studies with two or more weak ratings were considered as "weak". For more detailed criteria see the Quality Assessment Tool for Quantitative Studies Dictionary [23]. Any disagreements were solved in consensus meetings with all authors.

2.4. Data Extraction

Data were extracted from the eligible studies independently by three authors. Extracted data included the geographical location, study design, sample size, children's age, gender, and reason for SEN, cognitive and academic outcomes, intervention exposure, intervention details (only FMS and/or PA interventions and combined interventions), control conditions, and data for effect size calculations. If missing data was encountered, the corresponding author was contacted in order to receive the required information.

2.5. Effect Size Calculations

Cohen's d effect sizes [27] were calculated to allow for the quantification and comparison of the effects across the studies. Effect sizes were calculated for the studies that demonstrated significant effects and provided sufficient information (i.e., pre- and post-scores, as well as the associated standard deviations or standard errors). If a study demonstrated significant effects for multiple outcomes, all of them were included. Cohen's *d* effect sizes of <0.2, 0.2, 0.5, and 0.8, correspond to trivial, small, medium, and large effects, respectively [27].

Between-group effects were calculated in accordance with the following;

$$ES_{(d)} = \frac{\left(M_{post,\,E} - M_{pre,\,E}\right) - \left(M_{post,\,C} - M_{pre,\,C}\right)}{SDpooled_{pre}}$$

where

$$SDpooled_{pre} = \sqrt{\frac{(n_E - 1)\ SD_{pre2,\,E} + (n_C - 1)\ SD_{pre2,\,C}}{(n_E + n_C - 2)}}$$

And within-group effects were calculated as:

$$ES_{(d)} = \frac{\left(M_{post} - M_{pre}\right)}{SD_{pre}}$$

$ES_{(d)}$ = Cohen's *d* effect size
M_{post} = mean post-score
M_{pre} = mean pre-score
E = experimental group
C = control group
SD_{pooled} = pooled standard deviation
n = sample size

2.6. Power Analyses

Power calculations were carried out with G*power 3.1.9.6 [28]. If a study had multiple groups, the power calculations were conducted on a sub-group basis in order to determine the power of specific group comparisons. Type 1 error probability (α) was computed as 0.05, corresponding to a significance level of 5%. A medium effect size (0.5) was used as the reference point to establish observed power for each outcome and a type 2 error probability (β) of 0.2, corresponding to a power of 0.8 ($1 - \beta$), or 80%, was selected as the cut-off point for adequate power [29].

3. Results

3.1. Search Results

The stages of the systematic selection of the studies are presented in detail in Figure 1. In the updated literature search, a total of 3211 articles were found, which became 2198 articles after removal of duplicates. Of these, 2128 articles were excluded due to not meeting the eligibility criteria and the remaining 70 articles underwent full-text screening. Finally, 2 and 20 articles from the updated and the previous literature search (i.e., studies which were identified in the previous systematic review [16] but excluded since the review focused on typically developing children), respectively, were included.

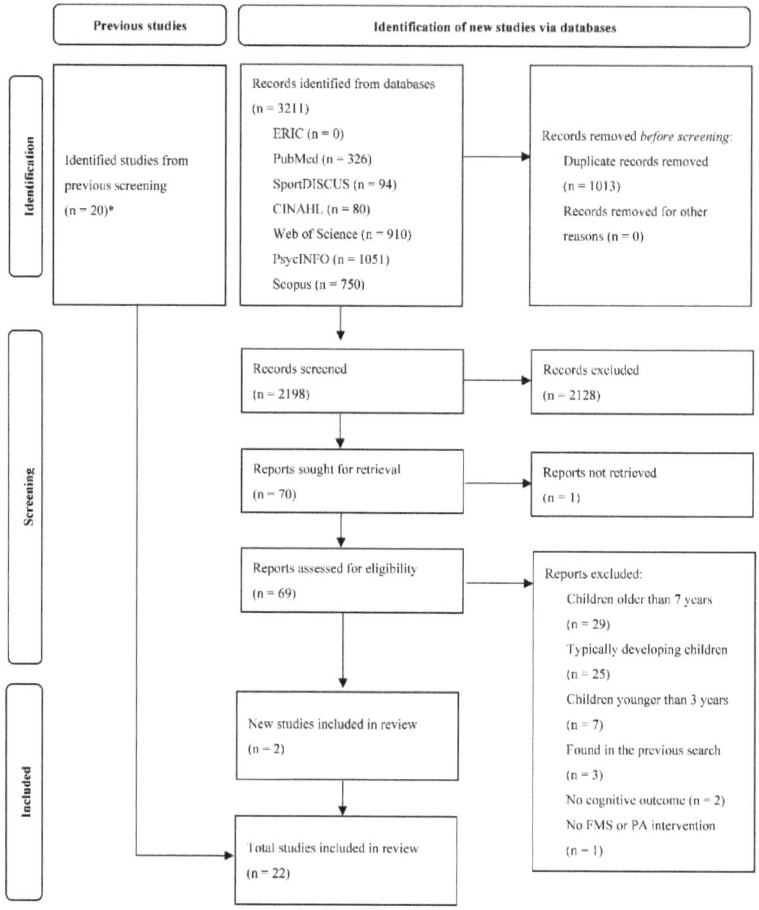

Figure 1. PRISMA flow diagram of the stages associated with the systematic selection of studies. * Studies were identified in the previous systematic review [16], but excluded since the review focused on typically developing children.

3.2. Study Characteristics and Population

Study characteristics are presented in detail in Supplementary Table S1. The included 22 studies represented 1883 children with various types and extents of SEN. In order to compare the intervention effects, children were divided into four groups based on the assumed severity of the learning difficulty. Thereafter, the following groups were formed: children at-risk for learning difficulties due to family background (n_{studies} = 8; e.g., low SES) [30], children with learning difficulties (n_{studies} = 7; e.g., high risk of ADHD) [31], learning disabilities (n_{studies} = 5; e.g., autism spectrum disorder) [32], physical disabilities (n_{studies} = 2; e.g., cerebral palsy) [33]. The mean ages of the participants ranged from 3.8 [34] to 7.4 years [35], and all of the studies included both boys and girls, apart from two studies that only included the former [36,37]. In terms of geographical location, the included studies were conducted in ten countries representing North America, Europe, Asia, and Africa, and were published between 1972 [38] and 2021 [35].

3.3. Intervention Characteristics

In total, 22 studies with 25 intervention programs were included in the present review. Three of the studies [38–40] included two intervention programs that met the eligibility criteria, and thus, were analyzed separately. A total of 14 intervention programs focused on FMS only interventions [30,31,38,39,41–44], PA only interventions [35,40], or FMS and PA only interventions [32,45], while 11 programs combined FMS [3,21,33,36,37,46,47], PA [34,48], or FMS and PA [40,49] with cognitive or academic skill practices. Intervention duration ranged from five weeks [35] to one academic year [3,36–38,41,45]. The duration of sessions ranged from 10 min [43] to two and a half hours [49], and sessions were held once a week [21,42,46] to two times every preschool day [34,48]. Outcome measures were divided into five categories based on the provided descriptions: cognitive skills (e.g., Miller assessment for preschoolers) [36,37], executive functions (e.g., Childhood executive functioning inventory) [32], academic skills (e.g., Comprehensive test of basic skills) [39], language skills (e.g., Assessment of children's language comprehension) [49], and numeracy (e.g., Counting and number recognition) [43].

3.4. Methodological Quality

The methodological quality was determined based on the following factors: study design, selection bias, confounders, blinding, data collection methods, and withdrawals and drop-outs [23]. Only two of the included 22 studies (9%) demonstrated strong methodological quality, while one study (5%) had moderate quality, and 19 studies (86%) were considered methodologically weak. The rating for each section, as well as the overall quality, of the studies is presented in Table 1.

Of the 22 included studies, 15 were controlled clinical trials (CCTs, i.e., quasi-experimental designs and RCTs that did not report the randomization process), three studies were RCTs, and the remaining four studies were PPDs. While only one study [30] referred the participants through randomization, in eight studies [3,36–39,41,43], the participants were referred from a source (e.g., preschool) in a systematic manner, and, thus, the participants were considered only somewhat likely to be representative of the target population. Important confounders (i.e., participants' age, gender, health status, or pre-intervention score) were observed in four studies [21,36,37,40]. Of these, the confounders were 80–100% controlled in three studies [21,36,40], while in seven studies [3,32,34,41,43,45,48], no important confounders were observed between the groups. While most of the studies [3,21,30,33,34,37–39,41,42,44,46–49] did not report the outcome assessors' blinding, six of the studies [31,35,36,40,43,45] reported that the outcome assessors were not aware, and only one study [32] reported that the outcome assessors were aware of the intervention or exposure status of the participants. Data collection methods were shown to be valid in seven studies [30,32,33,39,40,44,45] and of these, only three studies [30,40,44] reported that the data collection methods demonstrated good reliability in that particular data. Thus, only three studies [30,40,44] demonstrated strong quality in terms of data collection methods. Only five of the studies [21,32,44,45,48] reported both withdrawals and drop-outs in terms of numbers and reasons.

Of the included studies, ten (45%) were found to be underpowered to detect a medium effect size [31,34,36,38–40,43,44,46,48], while seven (32%) were confirmed to be adequately powered [3,21,30,32,35,41,45]; for the remaining five (23%) post hoc power could not be determined (i.e., within-group designs without required information) [33,37,42,47,49]. Only three (14%) of the included studies reported the conducting of a priori power analysis [32,36,37], and six studies stated small sample size as a limitation of the study [31,34,35,42,43,48]. It should be noted, that while underpowered to detect a medium effect size, in one study [36], the authors conducted a priori power calculations with a large (0.74) estimated effect size, based on a pilot study, for which the study was adequately powered.

Table 1. Methodological quality of the included studies.

Authors and Year	Selection Bias	Study Design	Confoun-ders	Blinding	Data Collection Methods	Withdrawals and Drop-Outs	Overall Quality Scores
Bala et al., 2013 [41]	moderate	strong	strong	moderate	weak	weak	weak
Berrol, 1984 [39]	moderate	strong	weak	moderate	moderate	weak	weak
Chevalier et al., 2017 [31]	moderate	strong	weak	moderate	weak	weak	weak
Coleman & Andersson, 1978 [49]	weak	moderate	weak	moderate	weak	weak	weak
Connor-Kuntz & Dummer, 1996 [40]	moderate	strong	strong	moderate	strong	weak	moderate
Devesa et al., 2011 [33]	weak	moderate	NA	moderate	moderate	weak	weak
Draper et al., 2012 [21]	weak	strong	weak	moderate	weak	moderate	weak
Fisher & Turner, 1972 [38]	moderate	strong	weak	moderate	weak	weak	weak
Flippin et al., 2021 [35]	weak	moderate	NA	strong	weak	weak	weak
Golos et al., 2011 [36]	moderate	strong	strong	strong	strong	strong	strong
Golos et al., 2013 [37]	weak	weak	weak	moderate	strong	weak	weak
Hendry & Kerr, 1983 [46]	weak	strong	weak	moderate	weak	weak	weak
Iwanaga et al., 2014 [42]	weak	moderate	weak	moderate	moderate	NA	weak
Kirk et al., 2014 [34]	weak	strong	strong	moderate	weak	weak	weak
Kirk & Kirk, 2016 [48]	weak	strong	strong	moderate	weak	strong	weak
Lam et al., 2019 [3]	moderate	strong	strong	moderate	weak	weak	weak
Mische Lawson et al., 2012 [43]	moderate	strong	weak	strong	weak	weak	weak
Moore et al., 1984 [44]	weak	strong	weak	moderate	strong	strong	weak
Mulvey et al., 2018 [30]	strong	strong	weak	moderate	strong	weak	weak
Puder et al., 2011 [45]	strong	strong	strong	strong	moderate	strong	strong
Wang et al., 2020 [32]	weak	strong	strong	moderate	moderate	weak	weak
Zawadzka et al., 2012 [47]	weak	moderate	weak	moderate	weak	weak	weak

Note. Some modifications were made to the EPHPP tool to solve misunderstandings between the raters. *Study design*: Studies that used quasi-experimental design were coded as CCT. *Confounders*: The confounders of interest included age, gender, health status, and pre-intervention score. *Blinding*: In question 2 "Were the study participants aware of the research question?" we chose to code "no" if there was no mention that participants were aware of the research question. This decision was made based on the young age of the participants. *Data collection methods*: The outcome of interest (cognitive or academic measurement) was evaluated. Methods were coded to be "valid" if the validity was mentioned in the article or if there was a citation to a test manual or another article where the validity was reported. Some well-known methods were seen as valid methods without a separate mention (e.g., Wechsler Intelligence Scale for Children or Bayley Scales of Infant and Toddler Development). Methods were coded as "reliable" only if the reliability was measured and reported in that specific data set. *Withdrawals and drop-outs*: In question 1, "Were withdrawals and drop-outs reported in terms of numbers and/or reasons per group?", if both numbers and reasons were reported it was coded as "yes", otherwise "no" was selected. Withdrawals and drop-outs were considered as children that did not finish the intervention, i.e., not missing data.

3.5. Effect Sizes

Individual effect sizes for each outcome and sub-group within the included studies are reported in Table 2. The effect sizes were presented in four groups based on the assumed severity of the participants learning difficulty:

Children at-risk for learning difficulties due to family background. In total, eight of the included studies (one with two separate interventions) [38] investigated the effects of FMS and PA interventions in children with low SES [21,30,34,35,38,44,48], while one was carried out with immigrant children [45]. Two studies assessed cognitive skills as an outcome; one with an FMS only intervention [38] and one with a combined FMS intervention [21]. Both studies demonstrated a beneficial effect of the intervention. The effect was large (d = 3.0) for the latter, while an effect size could not be calculated for the former due to the lack of required data. Language skills were assessed in two studies [34,48], both of which demonstrated large beneficial effects of a combined PA intervention (d = 0.78–1.57 $\bar{x}$ 1.18). Three of the identified studies included executive functions as an outcome; one demonstrated a small beneficial effect of an FMS only intervention (d = 0.48) [30]; one found a significant benefit of a PA only intervention, but an effect size could not be calculated due to the lack of required data [35]; while one did not observe significant effects of an FMS/PA only intervention [45]. Finally, two studies investigated the effects of FMS only interventions on academic skills [38,44]; of which one demonstrated beneficial effects [38]; however, an effect size could not be calculated due to the lack of required data. The null-finding was underpowered to detect a medium effect [44].

Children with learning difficulties. In total, three studies assessed the effects of FMS and PA interventions in children at-risk for learning difficulties with low SES backgrounds [36,37,43], two studies on children with learning and perceptual-motor difficulties [39,46], one study on children with delays in language development [49], and one study on children at high risk for ADHD [31]. Three studies assessed cognitive skills as an outcome; two with combined FMS interventions [36,37]; and one with two separate FMS only interventions [39]. Both combined FMS interventions found large beneficial effects on cognitive skills ($d = 0.78 - 1.87\ \bar{x}\ 1.33$); while in two FMS only interventions the effects were assessed on both cognitive and academic skills and no significant effects were found [39]. The null-findings [39] were underpowered to detect a medium effect. Three studies assessed language skills as an outcome [43,46,49]. One study with combined FMS and PA intervention [49] and one study with FMS only intervention [43] reported a beneficial effect on language skills; however, effect sizes could not be calculated due to the lack of required data. The combined FMS intervention observed no significant benefits on language skills [46]. The null-finding [46] was underpowered to detect a medium effect. Finally, one study [31] investigated the effects of an FMS only intervention on executive functions and demonstrated a large beneficial effect ($d = 1.48$).

Children with learning disabilities. Three of the included studies investigated the effects of FMS and PA interventions in children with autism spectrum disorder [32,42,47]. One study with two separate interventions involved children with significant delays in cognition, social, motor, speech, or language development [40] and one study was on children with global developmental delay, autism spectrum disorder, or speech development delay [3]. Two studies assessed the effects on cognitive skills with combined FMS interventions [3,47], and both demonstrated beneficial effects; one study with large effects ($d = 1.15$) [47], while the other demonstrated a medium beneficial effect ($d = 0.52$) [3]. Three studies (one with two interventions) [40] assessed the effects on language skills [3,40,42]. A trivial effect was found with the FMS and PA only intervention ($d = 0.07$) [40], whereas the effect was small with the combined FMS and PA intervention ($d = 0.34$) [40] demonstrating significantly greater benefits than the FMS and PA only intervention ($d = 0.27$) [40]. Medium beneficial effects were found with combined FMS intervention ($d = 0.57$) [3]. For an FMS only intervention, while reporting beneficial effects, the effect size could not be calculated [42]. One study assessed the effects of an FMS and PA only intervention on executive functions and demonstrated large beneficial effects ($d = 1.40$) [32]. For academic skills, a trivial beneficial effect was found with the combined FMS and PA intervention ($d = 0.10$) [40] and a small effect with the FMS and PA only intervention ($d = 0.30$) [40]; with no significant differences between the groups.

Children with physical disabilities. Two studies assessed the effects of FMS and PA interventions on children with physical disabilities; with one of the studies including children with cerebral palsy and growth hormone deficiency [33]; and one including children who had below average physical development at birth [41]. Both studies assessed the effects on cognitive skills and no significant effects were found, either with combined FMS intervention [33] or with an FMS only intervention [41]. Of these, the former study [33] was underpowered to detect a medium effect.

3.6. Methodological Quality and Effect Sizes

Methodological quality and effect sizes are presented in Table 3. Large effects were found in children's cognitive skills [21,36,47], executive functions [31,32], and language skills [48]. Of these six studies, only one (17%) [36] had a strong methodological quality, while five (83%) [21,31,32,47,48] displayed a weak methodological quality. In addition, only one study that found large effects [32] used outcome measures that were shown to be valid, while three studies [21,36,48] used outcomes that were shown to be reliable. Two of these studies [31,47] used outcome measures that were neither shown to be valid nor reliable. The studies that received a strong rating in terms of data collection methods demonstrated small effects in two studies [30,40], and trivial effects in one study [40]. Five studies reported that the intervention effects were significant [35,38,42,43], or children's skills improved during the intervention [49]; however, effect sizes could not be calculated due to limited data availability.

Table 2. Individual effect sizes for each outcome and subgroup within the included studies.

Reference	Outcome	Sub-Group within Study	Effect Size (d)	Sufficient Power to Detect a Medium Effect?
Children at Risk for Learning Difficulties due to Family Background				
Draper et al., 2012 [21]	Cognitive skills: Herbst Early Childhood Development Criteria test	Intervention cf. Control	ns.	yes
	Cognitive skills: Herbst Early Childhood Development Criteria test	Within group analysis	3.00	
Fisher & Turner, 1972a; 1972b [38]	Cognitive skills: Slosson Intelligence Test	Intervention (experimental 1 and 2) cf. Control	sign *	no
	Academic skills: Metropolitan Readiness Test	Intervention (experimental 1 and 2) cf. Control	sign *	
Flippin et al., 2021 [35]	Executive functions, sustained attention: on-task behavior	Intervention period cf. Control period	sign. *	yes
Kirk et al., 2014 [34]	Language skills: Pre-school Literacy Individual Growth and Development Indicators, alliteration	Intervention cf. Control	0.78	no
	Language skills: Pre-school Literacy Individual Growth and Development Indicators, picture naming	Intervention cf. Control	0.21	
Kirk & Kirk, 2016 [48]	Lanugage skills: Pre-school Literacy Individual Growth and Development Indicators, alliteration	Intervention cf. Control	0.38	no
	Language skills: Pre-school Literacy Individual Growth and Development Indicators, rhyming	Intervention cf. Control	1.57	
Moore et al., 1984 [44]	Academic skills: The Tests of Basic Experience (TOBE) Level K General Concepts Test	Intervention cf. Control	ns.	no
Mulvey et al., 2018 [30]	Executive functions: Head, Toes, Knees, SKIP -task	Intervention cf. Control	0.48	yes
Puder et al., 2011 [45]	Executive functions, attention: Konzentrations-Hand-lungsverfahren für Vorschulkinder	Intervention cf. Control	ns.	yes
	Executive functions, spatial working memory: subtest taken from the Intelligence and Development Scales	Intervention cf. Control	ns.	
Children with learning difficulties				
Berrol, 1984a [39]	Academic skills: Comprehensive Test of Basic Skills	Intervention (Dance/movement therapy) cf. Control	ns.	no
	Executive functions, sustained attention: the Children's Checking Test (CCT)	Intervention (Dance/movement therapy) cf. Control	ns.	
Berrol, 1984b [39]	Academic skills: Comprehensive Test of Basic Skills	Intervention (Sensory integration activity) cf. Control	ns.	no
	Executive functions, sustained attention: the Children's Checking Test (CCT)	Intervention (Sensory integration activity) cf. Control	ns.	
Chevalier et al., 2017 [31]	Executive functions, inhibition, Animal Stroop Test	Within group analysis	1.20	no
	Executive functions, inhibition, Animal Stroop Test	Intervention cf. Control	ns.	
	Executive functions, attention: Conners' Kiddie Continuous Performance Test	Intervention cf. Control	ns.	
		Within group analysis	ns.	
	Executive functions, selective attention: Neuropsychological Assessment Battery (NEPSY).	Intervention cf. Control	1.48	
Coleman & Andersson, 1978 [49]	Language skills: Language recognition inventory	Within group analysis (experimental 1)	improved	n/a
		Within group analysis (experimental 2)	improved	
	Language skills: Assessment of Children's Language Comprehension	Within group analysis (experimental 1)	improved	
		Within group analysis (experimental 2)	improved	
Golos et al., 2011 [36]	Cognitive skills: Miller Assessment for Preschoolers, complex skills subset	Intervention cf. Control (at-risk or with developmental delays)	1.87	no
	Cognitive skills: Miller Assessment for Preschoolers, non-verbal abilities subset	Intervention cf. Control (at-risk or with developmental delays)	ns.	
Golos et al., 2013 [37]	Cognitive skills: Miller Assessment for Preschoolers, complex skills subset	Within group analysis (2-year group)	ns.	n/a
	Cognitive skills: Miller Assessment for Preschoolers, non-verbal abilities subset	Within group analysis (2-year group)	0.78	
Hendry & Kerr, 1983 [46]	Language skills: grouping of items by shape, size, family name and placing picture cards in a logical story sequence	Intervention cf. Control	ns.	no
Mische Lawson et al., 2012 [43]	Language skills: recognition of alphabets, short words, geometric shapes, and incomplete pictures	Intervention cf. Control	ns.	
	Language skills, grade report: shape recognition	Intervention cf. Control	sign. *	no
	Language skills, grade report: letter recognition	Intervention cf. Control	ns.	
	Language skills, grade report: writing	Intervention cf. Control	ns.	
	Language skills, grade report: color recognition	Intervention cf. Control	ns.	
	Numeracy, grade report: counting	Intervention cf. Control	ns.	
	Numeracy, grade report: number recognition	Intervention cf. Control	ns.	

Table 2. *Cont.*

Reference	Outcome	Sub-Group within Study	Effect Size (*d*)	Sufficient Power to Detect a Medium Effect?
Children with learning disabilities				
Connor-Kuntz & Dummer, 1996a [40]	Academic skills: school readiness composite	Within group analysis (combined, developmentally delayed)	0.1	no
	Langugage skills: Bracken Basic Concept Scale, the direction/position subscale	Within group analysis (combined, developmentally delayed)	0.34	
Connor-Kuntz & Dummer, 1996a; 1996b [40]	Academic skills: school readiness composite	Combined intervention cf. Control	ns.	
	Langugage skills: Bracken Basic Concept Scale, the direction/position subscale	Combined intervention cf. Control (developmentally delayed)	0.27	
Connor-Kuntz & Dummer, 1996b [40]	Academic skills: school readiness composite	Within group analysis (control, developmentally delayed)	0.3	no
	Langugage skills: Bracken Basic Concept Scale, the direction/position subscale	Within group analysis (control, developmentally delayed)	0.07	
Iwanaga et al., 2014 [42]	Language skills: Japanese Miller Assessment for Preschoolers, verbal subset	Within group analysis (individual sensory integration)	ns.	n/a
	Language sklils: Japanese Miller Assessment for Preschoolers, non-verbal subset	Within group analysis (individual sensory integration)	sign. *	
Lam et al., 2019 [3]	Cognitive skills: Cognitive subtest of the Developmental Assessment Chart Revised (DAC-R)	Intervention cf. Control	0.52	yes
	Language skills, verbal comprehension: Reynell Developmental Language Scales (RDLS) Cantonese Version	Intervention cf. Control	0.40	
	Language skills, expressive language: Reynell Developmental Language Scales (RDLS) Cantonese Version	Intervention cf. Control	0.57	
Wang et al., 2020 [32]	Executive functions, working memory: Childhood Executive Functioning Inventory	Intervention cf. Control	0.96	yes
	Executive functions, inhibition: Childhood Executive Functioning Inventory	Intervention cf. Control	1.1	
	Executive functions, regulation: Childhood Executive Functioning Inventory	Intervention cf. Control	1.4	
Zawadzka et al., 2012 [47]	Cognitive skills: Behaviour Observation Scale adapted for children, cognitive subset	Within group analysis	1.15	n/a
Children with physical disabilities				
Bala et al., 2013 [41]	Cognitive skills: Raven's Matrices	Intervention cf. Control (below average development at birth)	ns.	yes
Devesa et al., 2011 [33]	Cognitive skills: The Battelle Developmental Inventory Screening Test, cognitive subset	Within group analysis (pre-treatment period)	ns.	n/a

* sign = significant effects were reported but effect sizes could not be calculated due to limited data availability. improved = beneficial effects were reported with no description of statistical analyses. n/a = not applicable; power analyses could not be conducted for within-group analyses. ns = nonsignifican differences.

Table 3. Summary of the relationship between methodological quality and intervention effects.

Intervention	Outcome	Not Significant	Significant, but Effect Sizes Could not be Calculated *	Effect Size (d)			
				Trivial	Small	Medium	Large
FMS	Executive functions	*Berrol, 1984a* [c] [39]; *Berrol, 1984b* [c] [39]; *Chevalier et al. 2017* [c,b] [31]			*Mulvey et al. 2018* [c] [30]		*Chevalier et al. 2017* [b,c] [31]
	Language skills	*Mische Lawson et al. 2012* [c] [43]; *Iwanga et al. 2014* [b] [42]	*Mische Lawson et al. 2012* [c] [43]; *Iwanga et al. 2014* [b] [42]				
	Cognitive skills	*Bala et al. 2013* [c] [41]	*Fisher & Turner, 1972* [a,c] [38]				
	Numeracy	*Mische Lawson et al. 2012* [c] [43]					
	Academic skills	*Moore et al. 1984* [c] [44]; *Berrol, 1984a* [c] [39]; *Berrol, 1984b* [c] [39]	*Fisher & Turner, 1972* [a,c] [38]				
PA	Executive functions		*Flippin et al. 2021* [c] [35]				
FMS & PA	Executive functions	**Puder et al. 2011** [c] [45]					*Wang et al. 2020* [c] [32]
	Language skills			Connor-Kuntz & Dummer 1996b [b] [40]			
	Academic skills				Connor-Kuntz & Dummer 1996b [a] [40]		
FMS combined	Language skills	*Hendry & Kerr, 1983* [c] [46]			*Lam et al. 2019* [c] [3]	*Lam et al. 2019* [c] [3]	
	Cognitive skills	*Draper et al. 2012* [c] [21]; **Golos et al. 2011** [c] [36]; *Golos et al. 2013* [b] [37]; *Devesa et al. 2011* [33] [b]				*Golos et al. 2013* [b] [37]; *Lam et al. 2019* [c] [3]	*Draper et al. 2012* [b] [21]; **Golos et al. 2011** [c] [36]; *Zawadska et al. 2012* [b] [47]
PA combined	Language skills				*Kirk et al. 2014* [c] [34]; *Kirk & Kirk, 2016* [c] [48]	*Kirk et al. 2014* [c] [34]	*Kirk & Kirk, 2016* [c] [48]
FMS & PA combined	Academic skills	Connor-Kuntz & Dummer, 1996a [d] [40]		Connor-Kuntz & Dummer 1996a [b] [40]			
	Language skills		*Coleman & Andersson, 1978* [b,e] [49]		Connor-Kuntz & Dummer 1996a [b,d] [40]		

* Significant effects were reported but effect sizes (*d*) could not be calculated due to limited data availability. [a] Two interventions within one study [38] analyzed together. [b] Within group analysis. [c] Intervention compared to control. [d] Combined intervention [40] compared to FMS & PA only intervention [40]. [e] Beneficial effects were reported with no description of statistical analyses. Methodological quality based on the EPHPP: *weak*, moderate, **strong**.

4. Discussion

The present systematic review aimed to investigate the methodological quality and the effects of FMS and PA interventions on cognitive and academic skills in preschool-aged children with SEN. The results demonstrated that only 9% of the included 22 studies had strong methodological quality, while 86% of the studies were rated as methodologically weak. The most often used outcome measures were cognitive and language skills and the largest effect sizes were found for cognitive skills, executive functions, and language skills. The intervention effects appeared to be somewhat dependent on the severity of the difficulty; in cognitive and language skills, the intervention effects were largest in children with minor learning difficulties (i.e., children at-risk due to family background), whereas in executive functions the intervention effects were largest in children with more severe difficulties (i.e., children with learning disabilities). However, due to the vast heterogeneity of the included studies, and rather low methodological quality, it is challenging to summarize the findings in a generalizable manner.

The finding that most of the included studies were methodologically weak is in line with the findings from a previous systematic review that investigated the effects of FMS and PA interventions on cognitive and academic skills in typically developing preschoolers [16]. The low ratings were mostly a result of inadequate reporting practices, especially in participant selection processes, confounders, blinding, data collection methods, and withdrawals [23]. Inadequate reporting practices are a common limitation in other educational interventions as well [50], and, thus, the use of reporting guidelines is highly recommended in the future.

It is recognized that difficulties exist in recruiting adequate sample sizes in children with SEN, and, thus, a large portion of the included studies were underpowered. Nonetheless, the limitations and potential risks of conducting underpowered studies cannot be dismissed (i.e., studies may result in substantially inflated effects or lead to false negative findings) [51]. Thus, results from underpowered studies may lead to erroneous conclusions as per the efficacy of studied interventions, which can subsequently lead to misguided decision making.

When considering the efficacy of PA and/or FMS interventions in children with SEN, the benefits appear to be somewhat dependent on the severity of the difficulty. Indeed, while large improvements in language skills were found for children at-risk due to family background, the effects were trivial-to-medium in children with learning disabilities. Importantly, the intervention effects observed in children at-risk due to family background were comparable to the effects of children without at-risk conditions in the previous review [16]. In cognitive skills, while medium-to-large improvements were demonstrated among all children with SEN, except for children with physical disabilities, a similar trend was observed. Indeed, the effects were progressively smaller in magnitude with increasing severity of the difficulty. It should be noted, however, that the improvements in cognitive skills—regardless of the severity of the difficulty (apart from children with physical disabilities)—were comparable to the ones experienced by typically developing children [16]. These findings indicate that it is easier to support children with more minor difficulties with FMS and/or PA interventions. Indeed, children that are at-risk due to family background, usually lack the opportunities to develop their cognitive [8] and language skills [52] in their home environment, and, thus, they lag behind their average-performing peers. With the right kind of early education support, these children have the possibility to develop their skills, which can have a huge effect on their later success during formal schooling [5,8].

The improvements in executive functions also appeared to be contingent on the severity of the children's difficulties. In contrast, however, here the effects were large in children with learning difficulties and disabilities, whereas the improvements were small in children at-risk due to family background. In accordance, children with learning difficulties and learning disabilities improved executive functions to a greater extent than typically developing children [16]; which might be reflective of a greater potential to develop executive functions in children with a lower level of executive functions. In line with our findings, studies have demonstrated larger beneficial effects in older children with ADHD in comparison to their typically developing counterparts [53].

Notably, only one of the studies assessed numeracy as an outcome, which was further limited to only a few dimensions of numeracy (i.e., counting and number recognition). Thus, in addition to cognitive and language skills, more studies investigating the effects of FMS and PA interventions on numeracy in children with SEN are required.

In terms of the intervention type, evidence was found for the efficacy of combined interventions for cognitive skills and FMS and/or PA only interventions for executive functions. Due to an insufficient number of studies, comparison between intervention types was possible only for language skills, and in line with our previous findings in typically developing children [16], combined interventions appeared more effective than FMS and PA only interventions. It should be noted, however, that in the combined interventions, the outcome was typically related to the intervention content; thus, these differences might simply stem from the direct practice of the assessed outcome. Finally, the comparison between FMS and PA only interventions could not be done due to the small number of studies and the vast heterogeneity of the participants in the included studies.

Study Limitations and Strengths

One of the strengths of the present systematic review was that both FMS and PA, as well as combined FMS and PA, interventions were included. In addition, while it is increasingly common for systematic reviews to only include RCTs [54], we included all study designs apart from case studies. This is important, as the use of RCT designs in children with SEN is largely impossible and, thus, remains scarce [55]. Furthermore, the present effect size calculations allowed the quantification and comparison of the intervention effects between the studies. However, some limitations of the present study should be addressed. Namely, only studies that were published in English were included, and some populations were vastly underrepresented, as only two studies assessed children with physical disabilities; making the generalization of the findings unreasonable.

5. Conclusions

These results indicate that FMS and PA interventions may be beneficial in the support of cognitive and academic skills in children with SEN. The intervention effects appear to be somewhat dependent on the severity of the difficulty; in cognitive and language skills the intervention effects were largest in children with minor difficulties (i.e., children at-risk due to family background), whereas in executive functions the intervention effects appeared to be largest in children with more severe difficulties (i.e., children with learning disabilities). Moreover, in line with the findings from typically developing children, combined interventions appeared to be more effective compared to FMS and PA only interventions. However, the results should be treated with caution as most of the studies had low methodological quality and displayed vast heterogeneity. More studies including combined interventions as well as FMS and/or PA only interventions in children with SEN are required to confirm the present findings. Finally, adherence to reporting guidelines and the inclusion of a priori power analyses are strongly encouraged for future studies.

Supplementary Materials: The following supporting information can be downloaded at: https://www.mdpi.com/article/10.3390/brainsci12081001/s1, Table S1: Intervention characteristics of the included studies.

Author Contributions: Conceptualization, P.J., A.S. and P.A.; methodology, P.J., T.M., A.H., A.S. and P.A.; software, not applicable; validation, P.J., T.M., A.B., A.H., A.S. and P.A.; formal analysis, P.J., T.M. and A.B.; investigation, not applicable; resources, not applicable; data curation, not applicable; writing—original draft preparation, P.J., T.M., A.B., A.S. and P.A.; writing—review and editing, P.J., T.M., A.B., A.S. and P.A.; visualization, P.J., T.M., A.B.; supervision, A.S. and P.A.; project administration, P.A.; funding acquisition, A.S. and P.A. All authors have read and agreed to the published version of the manuscript.

Funding: This research was funded by the Ministry of Education and Culture [grant number 48825]; and the Finnish Cultural Foundation [Huhtamäen rahasto 2019, no grant number available].

Institutional Review Board Statement: Not applicable.

Acknowledgments: Open access funding provided by University of Helsinki.

Conflicts of Interest: The authors declare no conflict of interest. The funders had no role in the design of the study; in the collection, analyses, or interpretation of data; in the writing of the manuscript; or in the decision to publish the results.

References

1. Duncan, G.J.; Dowsett, C.J.; Claessens, A.; Magnuson, K.; Huston, A.C.; Klebanov, P.; Pagani, L.S.; Feinstein, L.; Engel, M.; Brooks-Gunn, J.; et al. School readiness and later achievement. *Dev. Psychol.* **2007**, *43*, 1428–1446. [CrossRef]
2. Luna, B. Developmental Changes in Cognitive Control through Adolescence. *Adv. Child Dev. Behav.* **2009**, *37*, 233–278. [CrossRef]
3. Lam, S.-F.; Tsang, N.; Keung, Y.-C.; Tong, Y.; Mok, F.; Chiu, A.; Lai, Y.; Yuen, L.; Soh, D. A comprehensive service delivery model for preschoolers with special educational needs: Its characteristics and effectiveness. *Res. Dev. Disabil.* **2018**, *85*, 20–30. [CrossRef]
4. Wilson, R.A. *Special Educational Needs in the Early Years*, 2nd ed.; Routledge Falmer: London, UK, 2003.
5. Lenkeit, J.; Hartmann, A.; Ehlert, A.; Knigge, M.; Spörer, N. Effects of special educational needs and socioeconomic status on academic achievement. Separate or confounded? *Int. J. Educ. Res.* **2022**, *113*, 101957. [CrossRef]
6. Koutrouba, K.; Vamvakari, M.; Steliou, M. Factors correlated with teachers' attitudes towards the inclusion of students with special educational needs in Cyprus. *Eur. J. Spéc. Needs Educ.* **2006**, *21*, 381–394. [CrossRef]
7. Swanson, H.L.; Harris, K.R.; Graham, S. *Handbook of Learning Disabilities*, 2nd ed.; Guilford Press: New York, NY, USA, 2013.
8. Burger, K. How does early childhood care and education affect cognitive development? An international review of the effects of early interventions for children from different social backgrounds. *Early Child. Res. Q.* **2010**, *25*, 140–165. [CrossRef]
9. Gallahue, D.L.; Ozmun, J.C. *Understanding Motor Development: Infants, Children, Adolescents, Adults*; McGraw-Hill: New York, NY, USA, 2002.
10. Caspersen, C.J.; Powell, K.E.; Christenson, G.M. Physical activity, exercise, and physical fitness: Definitions and distinctions for health-related research. *Public Health Rep.* **1985**, *100*, 126–131.
11. Hillman, C.H.; Biggan, J.R. A Review of Childhood Physical Activity, Brain, and Cognition: Perspectives on the Future. *Pediatr. Exerc. Sci.* **2017**, *29*, 170–176. [CrossRef]
12. Pesce, C.; Vazou, S.; Benzing, V.; Álvarez-Bueno, C.; Anzeneder, S.; Mavilidi, M.F.; Leone, L.; Schmidt, M. Effects of chronic physical activity on cognition across the lifespan: A systematic meta-review of randomized controlled trials and realist synthesis of contextualized mechanisms. *Int. Rev. Sport Exerc. Psychol.* **2021**, 1–39. [CrossRef]
13. Diamond, A. Close Interrelation of Motor Development and Cognitive Development and of the Cerebellum and Prefrontal Cortex. *Child Dev.* **2000**, *71*, 44–56. [CrossRef]
14. Best, J.R. Effects of physical activity on children's executive function: Contributions of experimental research on aerobic exercise. *Dev. Rev.* **2010**, *30*, 331–351. [CrossRef]
15. Carson, V.; Hunter, S.; Kuzik, N.; Wiebe, S.A.; Spence, J.C.; Friedman, A.; Tremblay, M.S.; Slater, L.; Hinkley, T. Systematic review of physical activity and cognitive development in early childhood. *J. Sci. Med. Sport* **2015**, *19*, 573–578. [CrossRef]
16. Jylänki, P.; Mbay, T.; Hakkarainen, A.; Sääkslahti, A.; Aunio, P. The effects of motor skill and physical activity interventions on preschoolers' cognitive and academic skills: A systematic review. *Prev. Med.* **2022**, *155*, 106948. [CrossRef]
17. Kirk, S.A.; Gallagher, J.J.; Anastasiow, N.J. *Educating Exceptional Children*, 9th ed.; Houghton Mifflin: Boston, MA, USA, 2000.
18. Kavale, K.A. Quantitative research synthesis: Meta-analysis of research on meeting special educational needs. In *The Sage Handbook of Special Education*; Florian, L., Ed.; Sage: London, UK, 2007; pp. 207–221.
19. Page, M.J.; McKenzie, J.E.; Bossuyt, P.M.; Boutron, I.; Hoffmann, T.C.; Mulrow, C.D.; Shamseer, L.; Tetzlaff, J.M.; Akl, E.A.; Brennan, S.E.; et al. The PRISMA 2020 statement: An updated guideline for reporting systematic reviews. *BMJ* **2021**, *88*, 105906. [CrossRef]
20. Melnyk, B.M.; Fineout-Overholt, E. *Evidence-Based Practice in Nursing & Healthcare: A Guide to Best Practice*; Lippincott Williams & Wilkins: Philadelphia, PA, USA, 2005; p. 29.
21. Draper, C.E.; Achmat, M.; Forbes, J.; Lambert, E. Impact of a community-based programme for motor development on gross motor skills and cognitive function in preschool children from disadvantaged settings. *Early Child Dev. Care* **2012**, *182*, 137–152. [CrossRef]
22. Byrt, T. How Good Is That Agreement? *Epidemiology* **1996**, *7*, 561. [CrossRef]
23. National Collaborating Centre for Methods and Tools. *Quality Assessment Tool for Quantitative Studies*; McMaster University: Hamilton, ON, USA, 2008.
24. Thomas, B.; Ciliska, D.; Dobbins, M.; Micucci, S. A Process for Systematically Reviewing the Literature: Providing the Research Evidence for Public Health Nursing Interventions. *Worldviews Evid. Based Nurs.* **2004**, *1*, 176–184. [CrossRef]

25. Armijo-Olivo, S.; Stiles, C.R.; Hagen, N.A.; Biondo, P.D.; Cummings, G.G. Assessment of study quality for systematic reviews: A comparison of the Cochrane Collaboration Risk of Bias Tool and the Effective Public Health Practice Project Quality Assessment Tool: Methodological research. *J. Eval. Clin. Pract.* **2010**, *18*, 12–18. [CrossRef]
26. Norris, E.; Shelton, N.; Dunsmuir, S.; Duke-Williams, O.; Stamatakis, E. Physically active lessons as physical activity and educational interventions: A systematic review of methods and results. *Prev. Med.* **2015**, *72*, 116–125. [CrossRef]
27. Cohen, J. *Statistical Power Analysis for the Behavioral Sciences. Chapter 1: The Concepts of Power Analyses*, 2nd ed.; Lawrence Erlbaum Associates: Mahwah, NJ, USA, 1988.
28. Faul, F.; Erdfelder, E.; Lang, A.-G.; Buchner, A. G*Power 3: A flexible statistical power analysis program for the social, behavioral, and biomedical sciences. *Behav. Res. Methods* **2007**, *39*, 175–191. [CrossRef]
29. Schulz, K.F.; Grimes, A.D. Sample size calculations in randomised trials: Mandatory and mystical. *Lancet* **2005**, *365*, 1348–1353. [CrossRef]
30. Mulvey, K.L.; Taunton, S.; Pennell, A.; Brian, A. Head, Toes, Knees, SKIP! Improving Preschool Children's Executive Function Through a Motor Competence Intervention. *J. Sport Exerc. Psychol.* **2018**, *40*, 233–239. [CrossRef] [PubMed]
31. Chevalier, N.; Parent, V.; Rouillard, M.; Simard, F.; Guay, M.-C.; Verret, C. The Impact of a Motor-Cognitive Remediation Program on Attentional Functions of Preschoolers with ADHD Symptoms. *J. Atten. Disord.* **2012**, *21*, 1121–1129. [CrossRef] [PubMed]
32. Wang, J.-G.; Cai, K.-L.; Liu, Z.-M.; Herold, F.; Zou, L.; Zhu, L.-N.; Xiong, X.; Chen, A.-G. Effects of Mini-Basketball Training Program on Executive Functions and Core Symptoms among Preschool Children with Autism Spectrum Disorders. *Brain Sci.* **2020**, *10*, 263. [CrossRef]
33. Devesa, J.; Alonso, B.; Casteleiro, N.; Couto, P.; Castañón, B.; Zas, E.; Reimunde, P. Effects of recombinant growth hormone (GH) replacement and psychomotor and cognitive stimulation in the neurodevelopment of GH-deficient (GHD) children with cerebral palsy: A pilot study. *Ther. Clin. Risk Manag.* **2011**, *7*, 199–206. [CrossRef]
34. Kirk, S.M.; Vizcarra, C.R.; Looney, E.C.; Kirk, E.P. Using Physical Activity to Teach Academic Content: A Study of the Effects on Literacy in Head Start Preschoolers. *Day Care Early Educ.* **2013**, *42*, 181–189. [CrossRef]
35. Flippin, M.; Clapham, E.D.; Tutwiler, M.S. Effects of using a variety of kinesthetic classroom equipment on elementary students' on-task behaviour: A pilot study. *Learn. Environ. Res.* **2020**, *24*, 137–151. [CrossRef]
36. Golos, A.; Sarid, M.; Weill, M.; Weintraub, N. Efficacy of an Early Intervention Program for At-Risk Preschool Boys: A Two-Group Control Study. *Am. J. Occup. Ther.* **2011**, *65*, 400–408. [CrossRef]
37. Golos, A.; Sarid, M.; Weill, M.; Weintraub, N. The Influence of Early Intervention Length on the Participation of Low Socioeconomic At-Risk Preschool Boys: A Two-Group Control Study. *J. Occup. Ther. Sch. Early Interv.* **2013**, *6*, 188–202. [CrossRef]
38. Fisher, M.D.; Turner, R.V. The Effects of a Perceptual-Motor Training Program Upon the Academic Readiness of Culturally Disadvantaged Kindergarten Children. *J. Negro Educ.* **1972**, *41*, 142. [CrossRef]
39. Berrol, C. The effects of two movement therapy approaches on selected academic, physical and socio-behavioral measures of first grade children with learning and perceptual-motor problems. *Am. J. Dance Ther.* **1984**, *7*, 32–48. [CrossRef]
40. Connor-Kuntz, F.J.; Dummer, G.M. Teaching across the Curriculum: Language-Enriched Physical Education for Preschool Children. *Adapt. Phys. Act. Q.* **1996**, *13*, 302–315. [CrossRef]
41. Bala, G.; Krneta, Z.; Drid, P. The effects of kinesiological activity on motor and intellectual functioning of children in relation to their physical constitution at birth. *Coll. Antropol.* **2013**, *37*, 161–169.
42. Iwanaga, R.; Honda, S.; Nakane, H.; Tanaka, K.; Toeda, H.; Tanaka, G. Pilot Study: Efficacy of Sensory Integration Therapy for Japanese Children with High-Functioning Autism Spectrum Disorder. *Occup. Ther. Int.* **2013**, *21*, 4–11. [CrossRef]
43. Mische Lawson, L.A.; Cox, J.; Blackwell, A.L. Yoga as a Classroom Intervention for Preschoolers. *J. Occup. Ther. Sch. Early Interv.* **2012**, *5*, 126–137. [CrossRef]
44. Moore, J.B.; Guy, L.M.; Reeve, T.G. Effects of the Capon Perceptual-Motor Program on Motor Ability, Self-Concept, and Academic Readiness. *Percept. Mot. Ski.* **1984**, *58*, 711–714. [CrossRef]
45. Puder, J.J.; Marques-Vidal, P.; Schindler, C.; Zahner, L.; Niederer, I.; Burgi, F.; Ebenegger, V.; Nydegger, A.; Kriemler, S. Effect of multidimensional lifestyle intervention on fitness and adiposity in predominantly migrant preschool children (Ballabeina): Cluster randomised controlled trial. *BMJ* **2011**, *343*, d6195. [CrossRef]
46. Hendry, J.; Kerr, R. Communication through Physical Activity for Learning Disabled Children. *Percept. Mot. Ski.* **1983**, *56*, 155–158. [CrossRef]
47. Zawadzka, D.; Zawadzka, P.; Bugaj, R.; Curyło, M.; Gumułka, A.; Bukowska, A. Evaluation of effectiveness of the therapeutic program combining elements of occupational therapy and Veronica Sherborne's Developmental Movement approach in autistic children. A preliminary study. *Pol. J. Physiother.* **2012**, *12*, 405–416. [CrossRef]
48. Kirk, S.M. Sixty Minutes of Physical Activity per Day Included Within Preschool Academic Lessons Improves Early Literacy. *J. Sch. Health* **2016**, *86*, 155–163. [CrossRef]
49. Coleman, R.O.; Anderson, D.E. Enhancement of Language Comprehension in Developmentally Delayed Children. *Lang. Speech Hear. Serv. Sch.* **1978**, *9*, 241–253. [CrossRef]
50. Albarqouni, L.; Glasziou, P.; Hoffmann, T. Completeness of the reporting of evidence-based practice educational interventions: A review. *Med Educ.* **2017**, *52*, 161–170. [CrossRef]
51. La Caze, A.; Duffull, S. Estimating risk from underpowered, but statistically significant, studies: Was APPROVe on TARGET? *J. Clin. Pharm. Ther.* **2010**, *36*, 637–641. [CrossRef]

52. Hoff, E. How social contexts support and shape language development. *Dev. Rev.* **2006**, *26*, 55–88. [CrossRef]
53. Wass, S.; Scerif, G.; Johnson, M. Training attentional control and working memory—Is younger, better? *Dev. Rev.* **2012**, *32*, 360–387. [CrossRef]
54. Zeng, N.; Ayyub, M.; Sun, H.; Wen, X.; Xiang, P.; Gao, Z. Effects of Physical Activity on Motor Skills and Cognitive Development in Early Childhood: A Systematic Review. *BioMed Res. Int.* **2017**, *2017*, 1–13. [CrossRef]
55. Odom, S.L. Education of Students with Disabilities, Science, and Randomized Controlled Trials. *Res. Pract. Pers. Sev. Disabil.* **2021**, *46*, 132–145. [CrossRef]

brain sciences

Article

Factors Associated with Fear of Falling in Individuals with Different Types of Mild Cognitive Impairment

Pei-Hao Chen [1,2,3,†], Ya-Yuan Yang [4,5,†], Ying-Yi Liao [6], Shih-Jung Cheng [1,2,7], Pei-Ning Wang [8,9,10] and Fang-Yu Cheng [4,*]

1 Department of Neurology, MacKay Memorial Hospital, Taipei 104, Taiwan; a7662888@gmail.com (P.-H.C.); csjneuro@gmail.com (S.-J.C.)
2 Department of Medicine, MacKay Medical College, New Taipei City 252, Taiwan
3 Graduate Institute of Mechanical and Electrical Engineering, National Taipei University of Technology, Taipei 106, Taiwan
4 Institute of Long-Term Care, MacKay Medical College, New Taipei City 252, Taiwan; nina40137@gmail.com
5 Kaifeng Minquan Day Care Center, Taipei 104, Taiwan
6 Department of Gerontological Health Care, National Taipei University of Nursing and Health Sciences, Taipei 112, Taiwan; ianliao1209@gmail.com
7 Department of Physical Therapy and Assistive Technology, National Yang Ming Chiao Tung University, Taipei 112, Taiwan
8 Department of Neurology, Neurological Institute, Taipei Veterans General Hospital, Taipei 112, Taiwan; linda2860@gmail.com
9 Department of Neurology, School of Medicine, National Yang Ming Chiao Tung University, Taipei 112, Taiwan
10 Brain Research Center, National Yang Ming Chiao Tung University, Taipei 112, Taiwan
* Correspondence: fycheng@mmc.edu.tw; Tel.: +886-226-360-303
† These authors contributed equally to this work.

Citation: Chen, P.-H.; Yang, Y.-Y.; Liao, Y.-Y.; Cheng, S.-J.; Wang, P.-N.; Cheng, F.-Y. Factors Associated with Fear of Falling in Individuals with Different Types of Mild Cognitive Impairment. *Brain Sci.* **2022**, *12*, 990. https://doi.org/10.3390/brainsci12080990

Academic Editor: Daniele Corbo

Received: 5 July 2022
Accepted: 25 July 2022
Published: 26 July 2022

Abstract: Mild cognitive impairment (MCI) is considered an intermediate state between normal aging and early dementia. Fear of falling (FOF) could be considered a risk indicator for falls and quality of life in individuals with MCI. Our objective was to explore factors associated with FOF in those with MCI due to Alzheimer's disease (AD-MCI) and mild cognitive impairment in Parkinson's disease (PD-MCI). Seventy-one participants were separated into two groups, AD-MCI (n = 37) and PD-MCI (n = 34), based on the disease diagnosis. FOF was assessed using the Activities-specific Balance Confidence scale. The neuropsychological assessment and gait assessment were also measured. FOF was significantly correlated with global cognitive function, attention and working memory, executive function, Tinetti assessment scale scores, gait speed, and stride length in the AD-MCI group. Moreover, attention and working memory were the most important factors contributing to FOF. In the PD-MCI group, FOF was significantly correlated with gait speed, and time up and go subtask performance. Furthermore, turn-to-walk was the most important factor contributing to FOF. We noted that FOF in different types of MCI was determined by different factors. Therapies that aim to lower FOF in AD-MCI and PD-MCI populations may address attention and working memory and turn-to-walk, respectively.

Keywords: mild cognitive impairment; fear of falling; Alzheimer disease; Parkinson disease

1. Introduction

Falls among community-dwelling older adults are common events in daily life, and they can lead to disability, hospitalization and even death [1]. Older adults, especially those at risk of falls, may also exhibit peculiar psychological features, such as fear of falls, linked to the experience of negative emotions related to falls [2]. FOF is defined as cautious concern with falling that leads to an individual losing confidence and avoiding activities associated with daily life [3]. Having a FOF was an independent risk factor for falling among persons older than 65 years of age [3]. Therefore, it is important to investigate the factors associated with FOF to incorporate these factors in prevention and rehabilitation programs. Several factors may contribute to FOF among aging adults. Age, sex, history of falls, balance and gait performance and depression are significantly associated with FOF [4]. The completion time in the timed up and go task was also shown to be significantly correlated with FOF [4].

Cognitive impairment has been identified as a risk factor for falls in aging [5]. Mild cognitive impairment (MCI) is considered a clinical stage between the expected cognitive decline in normal aging and the more serious decline in early dementia [6]. Falls are more prevalent in older adults with MCI than in age-matched healthy subjects [7]. A previous study noted that older adults with MCI reported FOF more often than patients with mild Alzheimer's disease and older adults with healthy cognition [2]. FOF not only leads to an increased risk for falling but also causes restriction and avoidance of activities that eventually result in degenerated physical and mental status [8]. Identifying factors related to FOF in MCI patients may be useful in developing multicomponent strategies to decrease FOF and improve quality of life. However, there is no study investigating the factors associated with FOF in older adults with MCI.

There are many types of MCI. Clinical presentations have shown that they can be amnestic, and they can involve a single nonmemory domain or involve multiple cognitive domains [9]. Each of these clinical presentations may have multiple etiologies, such as degenerative, vascular, metabolic, traumatic, and psychiatric etiologies [9]. Based on etiopathology, the most common neurodegenerative disorders associated with MCI are Alzheimer's disease and Parkinson's disease. Patients who are diagnosed with mild cognitive impairment due to AD (AD-MCI) usually have amnestic MCI and positive biomarkers for both Aβ and neuronal injury [10]. On the other hand, based on the guidelines proposed by the International Parkinson and Movement Disorders Society, the definition of mild cognitive impairment in PD (PD-MCI) refers to patients with a diagnosis of Parkinson's disease, whose cognitive abilities decline and are not caused by other comorbidities or diseases [11]. These two types of MCI have different clinical manifestations because of the different underlying diseases. For example, memory is the most commonly impacted domain among patients with AD-MCI [10]; in contrast, both nonamnestic and amnestic domains of cognition can be affected in persons with PD-MCI [12]. The factors that relate to FOF in these two types of MCI populations may be different. However, there is limited knowledge regarding the contributing factors. Hence, this study aimed to determine factors associated with FOF among people with different types of MCI (i.e., AD-MCI and PD-MCI).

2. Materials and Methods

2.1. Participants and Study Design

This was an observational, cross-sectional design, and quantitative study. We recruited 71 older adults with MCI from the neurological outpatient clinics of a medical center in Northern Taiwan. We included older adults (age 60 years and older) who were able to walk 10 m independently and to meet the inclusion criteria of MCI (subjective cognitive complaints, a global clinical dementia rating of 0.5, and a clinical dementia rating sum of boxes of 0.5–4.0 [13]). We excluded individuals with dementia, positive psychiatric history or unstable medical conditions and those taking any medications causing cognitive complaints during the past 3 months. A total of 100 participants provided written informed consent prior to enrollment. The study procedures were approved by the ethics committee of the institution (number: 18MMHIS005e). The inclusion process for this study is depicted in Figure 1 (n = 71). The participants' age, sex, education, body mass index, history of fall (have fallen in the last year), and history of metabolic disease were obtained from patient interviews and medical charts. We administered the Barthel Index to explore participants' self-care activities, for example, transferring, bathing, and toileting [14], and the Instrumental Activities of Daily Living scale to assess participants' ability to perform task such as cooking, using a telephone, laundry, and handling finances [15].

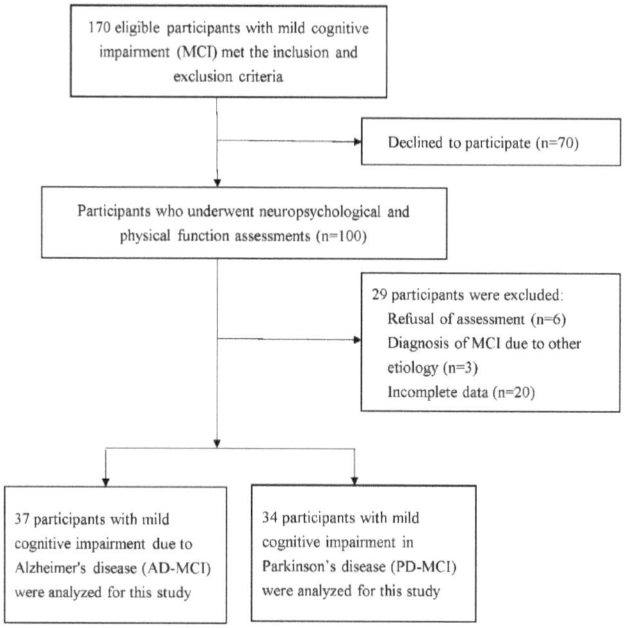

Figure 1. Flow chart showing the process of selecting subjects in this study.

2.2. *Defining the MCI Etiologies*

The diagnosis for the MCI etiology of each participant was relied on reviewing clinical chart, brain imaging data, neuropsychological tests, and biochemical results. Diagnosis was made for AD-MCI according to the National Institute on Aging and the Alzheimer's Association workgroup consensus criteria [10]. AD-MCI was diagnosed if the patient met the following criteria: (a) meets the core clinical criteria for MCI, and (b) has positive biomarkers for both Aβ and neuronal injury. For the diagnosis of PD-MCI, we used the Movement Disorder Society Task Force diagnostic criteria and classified patients with either clinically established Parkinson's disease or clinically probable Parkinson's disease [11]. The diagnosis of PD-MCI was made if participants met the following criteria: (a) meets the core clinical criteria for MCI, and (b) diagnosis of Parkinson's disease as based on the UK PD Brain Bank Criteria [16]. Patients with MCI due to cerebrovascular disease, other etiologies, or at least two etiologies were excluded.

2.3. *Assessment of Fear of Falling*

We used the Activities-specific Balance Confidence (ABC) scale to measure FOF. The ABC scale is a 16-item self-report measure of balance confidence in performing activities without losing balance [17]. This scale was shown to have good validity [18] and excellent test-retest reliability [17] in measuring FOF in community-dwelling older adults. The participants are asked to rate their confidence in performing various activities without losing their balance or becoming unsteady, where 0 is "no confidence" and 100 is "completely confident". The overall score is calculated by adding the item scores and dividing the total by 16.

2.4. *Outcome Measures*

Participants included in the study underwent comprehensive neuropsychological testing and physical activity testing.

The neuropsychological assessment included the Mini-Mental State Examination, parts A and B of the Trail Making Test, the category fluency test, the forward and backward digit recall tests, the California Verbal Learning Test, the Judgment of Line Orientation test-Short Form, the Boston Naming Test, the Geriatric Depression Scale-15, and the General Anxiety Disorder-7 questionnaire. The Mini-Mental State Examination is a widely used and valid test of global cognitive function among elderly individuals [19]. The parts A and B of the Trail Making Test was designed to measure visual attention and executive function [20]. The category fluency test is one of the most commonly used measures of executive function and language [21]. The forward and backward digit recall tests are neuropsychological tests of short-term verbal language and executive function [22]. The California Verbal Learning Test is a comprehensive, detailed assessment of episodic memory in older adults [23]. The short form of the Judgment of Line Orientation test is a commonly used measure of visuospatial perception [24]. The Boston Naming Test is a valid assessment tool for the measure of confrontation naming in individuals with language impairments caused by stroke, Alzheimer's disease and dementia [25]. The Geriatric Depression Scale-15 is a widely used assessment tool to evaluate the prevalence of depressive symptoms in the older adults [26]. The General Anxiety Disorder-7 questionnaire was designed to screen for anxiety or to measure its severity [27].

Physical activity was assessed using the Tinetti assessment scale, straight walking performance, and the timed up and go test. The Tinetti assessment scale is a very good indicator of the fall risk of an older adult and evaluates balance ability and gait performance [28]. The G-WALK® (BTS Bioengineering Corp., Quincy, MA, USA) system, which is a wearable system for the functional analysis of movement, was used for straight walking performance and the TUG test. This system comprises a portable inertial sensor placed in the area of the S1–S2 vertebrae to record specific movements. The validity and reliability of this system to evaluate movement performance has been well established [29]. In the straight walking test, the participants were instructed to walk straight for 10 m at a usual speed three times. In the TUG test, the participants were asked to rise from a 45 cm-high chair, walk 3 m, turn around, return, and sit down three times. The total TUG duration was divided into four subtasks: time to stand (sit-to-stand), time to turn around in the midway (turn-to-walk), time to turn around to reach the chair (turn-to-sit), and time to sit down in the chair (stand-to-sit). The average of the three trials in each test was used for data analysis.

2.5. Statistics

Statistical analyses were conducted using SPSS 22.0 (SPSS, Inc., Chicago, IL, USA). Descriptive data were reported in terms of means, SDs or numbers. Independent t-tests (continuous variables) or chi-square tests (categorical variables) were used to compare the between-group differences. In this study, the correlations were first established, and the factors significantly correlating with ABC scores were further analyzed using multivariable linear regression models. Covariates in multivariable models included age, sex, fall history, education, and comorbidities (e.g., hypertension, diabetes mellitus and cardiovascular disease). Spearman's rank correlation analysis was used to examine correlations between ABC scores and cognitive performance and gait performance. The level of significance was set at a p value of less than 0.05.

3. Results

3.1. Baseline Demographic Data and Comparisons between AD-MCI and PD-MCI

Seventy-one participants (male: 38; female: 33) participated in this study. Table 1 presents the characteristics and performance of the participants in the AD-MCI (N = 37) and PD-MCI (N = 34) groups. The mean ages of the participants in the AD-MCI and PD-MCI groups were 72.4 $\pm$ 8.8 and 68.4 $\pm$ 10.2 years, respectively. There were no statistically significant differences in the basic characteristics (sex, age, education, onset duration, body mass index, functional status and falls history), cognitive functions, depression, anxiety, straight walking performance, or performance in two timed up and go subtasks (sit-to-stand and stand-to-sit) between the two groups. However, the participants with PD-MCI had significantly lower ABC scale scores ($p = 0.002$), which indicated less confidence in performing activities without losing balance, than did the AD-MCI participants. In addition, compared to the patients with AD-MCI, those with PD-MCI had significantly lower Tinetti gait and balance scale scores (Tinetti gait, $p = 0.018$; Tinetti balance, $p = 0.019$) and longer completion times on the two timed up and go turning subtasks (turn-to-walk, $p = 0.007$; turn-to-sit, $p = 0.022$).

Table 1. Comparison of participants' characteristics between the AD-MCI and PD-MCI groups ($n = 71$).

Variable	Total ($n = 71$)	AD-MCI (N = 37)	PD-MCI (N = 34)	p-Value
Gender (male/female)	38/33	20/17	18/16	0.936
Age (years)	70.46 ± 9.64	72.41 ± 8.84	68.35 ± 10.16	0.077
Education (years)	7.85 ± 3.95	7.03 ± 4.11	8.74 ± 3.61	0.068
Onset duration (years)	1.83 ± 1.80	2.11 ± 1.91	1.53 ± 1.54	0.167
Falls history (n)	28	11	17	0.081
BMI	23.93 ± 3.42	23.86 ± 3.6	24.00 ± 3.27	0.861
Barthel Index	98.21 ± 7.42	99.59 ± 1.82	96.67 ± 10.51	0.124
IADL	21.54 ± 4.76	22.27 ± 4.37	20.73 ± 5.11	0.178
Hypertension (n)	42	26	16	0.047
Diabetes mellitus (n)	20	12	8	0.405
Cardiovascular disease (n)	25	19	6	0.003
Global cognitive function				
MMSE	25.32 ± 3.57	24.65 ± 4.10	26.06 ± 2.79	0.096
Episodic memory				
CVLT-SF	18.79 ± 5.40	18.70 ± 5.36	18.88 ± 5.51	0.890
Visuospatial performance				
Judgment of line orientation	13.51 ± 3.56	13.25 ± 3.23	13.79 ± 3.91	0.534
Attention and working memory				
Forward digits	7.24 ± 1.53	7.43 ± 1.56	7.03 ± 1.49	0.269
Backward digits	3.97 ± 1.52	4.03 ± 1.67	3.91 ± 1.38	0.753
Executive function				
Category fluency test	12.45 ± 3.79	12.57 ± 3.87	12.32 ± 3.76	0.789
Trail making test A (s)	25.60 ± 15.57	24.51 ± 15.22	26.81 ± 16.09	0.542
Trail making test B (s)	67.00 ± 36.49	63.67 ± 36.67	70.65 ± 36.50	0.431
Language				
Boston naming test	22.39 ± 5.24	21.78 ± 5.48	23.06 ± 4.97	0.309
Depression				
GDS-15	3.07 ± 3.34	2.76 ± 3.11	3.42 ± 3.60	0.408
Anxiety				
GAD-7	2.43 ± 3.67	2.00 ± 3.66	2.91 ± 3.68	0.304
Fear of falling				
ABC scale	82.40 ± 22.60	90.28 ± 16.10	73.58 ± 25.62	0.002
Balance and gait performance				
Tinetti gait	14.91 ± 2.52	15.62 ± 1.21	14.12 ± 3.30	0.018
Tinetti balance	11.44 ± 1.52	11.86 ± 0.67	10.97 ± 2.01	0.019
Straight walking performance				
Gait speed (m/s)	0.84 ± 0.22	0.87 ± 0.22	0.82 ± 0.21	0.302
Stride length (cm)	63.01 ± 19.39	63.36 ± 17.38	62.63 ± 21.57	0.876
Cadence (steps/min)	97.86 ± 14.18	97.60 ± 11.39	98.13 ± 16.81	0.878
TUG subtasks				
Sit-to-stand (s)	1.83 ± 0.57	1.74 ± 0.58	1.91 ± 0.55	0.209
Turn-to-walk (s)	2.65 ± 1.34	2.23 ± 0.80	3.11 ± 1.64	0.007
Turn-to-sit (s)	2.50 ± 1.53	2.09 ± 0.93	2.96 ± 1.91	0.022
Stand-to-sit (s)	2.45 ± 0.89	2.40 ± 0.83	2.51 ± 0.97	0.593

Abbreviation: AD-MCI, mild cognitive impairment due to Alzheimer's disease. PD-MCI, mild cognitive impairment in Parkinson's disease. BMI, body mass index. MMSE, Mini-Mental State Examination. IADL, Instrumental Activities of Daily Living. GDS-15, Geriatric Depression Scale-15. GAD-7, General Anxiety Disorder-7 questionnaire. ABC, Activities-specific Balance Confidence Scale.

3.2. Factors Determining FOF

The correlation coefficients between cognitive performance and ABC scores are shown in Table 2. In the AD-MCI group, the ABC scores were positively correlated with Mini-Mental State Examination, forward digits recall test and backward digits recall test scores ($r = 0.414$ to 0.516, $p < 0.05$), meaning that greater balance confidence was associated with better global cognition, attention, and working memory performance. The Trail Making Test B scores were negatively correlated with the ABC scores ($r = -0.409$, $p < 0.05$), meaning that less balance confidence was associated with worse visual attention and executive function. In the PD-MCI group, there were no significant differences between cognitive functions and ABC scores.

Table 2. Correlation coefficients between cognitive performance and ABC scores.

Group		AD-MCI Group (N = 37)		PD-MCI Group (N = 34)	
		ABC Score		ABC Score	
Domain	**Outcome Measures**	**r**	***p*-Value**	**r**	***p*-Value**
Cognition	MMSE	0.431 *	0.008	0.244	0.165
	CVLT-SF	0.315	0.058	0.221	0.209
	Judgment of line orientation	0.217	0.203	0.011	0.952
	Forward digits	0.516 *	0.001	0.297	0.088
	Backward digits	0.414 *	0.012	0.117	0.510
	Category fluency test	0.155	0.360	0.279	0.109
	Trail making test A	−0.290	0.082	−0.319	0.070
	Trail making test B	−0.409 *	0.013	−0.221	0.217
	Boston naming test	0.234	0.164	0.252	0.150
Depression	GDS-15	−0.250	0.136	−0.056	0.754
Anxiety	GAD-7	−0.218	0.195	−0.173	0.329

Abbreviation: AD-MCI, mild cognitive impairment due to Alzheimer's disease. PD-MCI, mild cognitive impairment in Parkinson's disease. MMSE, Mini-Mental State Examination. CVLT-SF, California verbal language test-short form. GAD-7, General Anxiety Disorder-7 questionnaire. GDS-15, Geriatric Depression Scale-15. ABC, Activities-specific Balance Confidence Scale. * $p < 0.05$.

The correlation coefficients between balance and gait performance and ABC scores are shown in Table 3. The gait speed, stride length, cadence, and Tinetti assessment scale scores were positively correlated with the ABC scores in the AD-MCI group ($r = 0.336$ to 0.507, $p < 0.05$), meaning that greater balance confidence was associated with better balance and gait performance. In the PD-MCI group, gait speed was positively correlated with ABC scores ($r = 0.385$, $p < 0.05$), meaning that greater balance confidence was associated with better gait speed. Performance on the timed up and go subtasks was negatively correlated with the ABC scores ($r = -0.503$ to -0.663), $p < 0.05$), meaning that less balance confidence was associated with worse functional mobility.

Based on the regression models (Table 4), after adjusted for age, sex, fall history, education, and comorbidities, attention and working memory were the most important factors in determining ABC scores in the AD-MCI group ($F = 10.162$, effect size $f2 = 0.328$, power = 0.92, $p = 0.003$). On the other hand, turn-to-walk time was the most important factor in determining ABC scores in the PD-MCI group ($F = 35.292$, effect size $f2 = 1.179$, power = 0.99, $p < 0.001$).

Table 3. Correlation coefficients between gait performance and ABC scores.

Group		AD-MCI Group (N = 37)		PD-MCI Group (N = 34)	
		ABC Score		ABC Score	
Domain	**Outcome Measures**	**r**	***p*-Value**	**r**	***p*-Value**
Gait performance	Speed (m/s)	0.507 *	0.002	0.385 *	0.027
	Stride Length	0.439 *	0.007	0.321	0.065
	Cadence step	0.349 *	0.037	0.119	0.503
Tinetti assessment scale	Tinetti Posture	0.482 *	0.003	0.165	0.352
	Tinetti Gait	0.336 *	0.042	0.317	0.068
TUG subtasks	Turn-to-walk	−0.316	0.056	−0.663 **	<0.001
	Turn-to-sit	−0.302	0.069	−0.464 *	0.006
	Sit-to-stand	−0.292	0.079	−0.503 *	0.002
	Stand-to-sit	−0.138	0.415	−0.229	0.192

Abbreviation: AD-MCI, mild cognitive impairment due to Alzheimer's disease. PD-MCI, mild cognitive impairment in Parkinson's disease. ABC, Activities-specific Balance Confidence Scale. TUG, timed up and go test. * $p < 0.05$; ** $p < 0.005$.

Table 4. Summary of linear regression analyses in the AD-MCI and PD-MCI groups.

AD-MCI Group			PD-MCI Group		
Variable	**β**	**Coefficient**	**Variable**	**β**	**Coefficient**
Age	−0.338	−0.344	Age	−0.113	−0.015
Sex	−0.198	−0.224	Sex	−0.089	−0.132
Fall history	−0.056	−0.063	Fall history	−0.191	−0.277
Education level	0.168	0.192	Education level	0.113	0.167
Hypertension	−0.037	−0.042	Hypertension	0.019	0.028
Diabetes mellitus	0.001	0.001	Diabetes mellitus	0.197	0.291
Cardiovascular disease	−0.286	−0.328	Cardiovascular disease	−0.020	−0.029
MMSE	0.176	0.188	Speed	0.068	0.080
Forward digits	0.497		Turn-to-walk	−0.735	
Backward digits	0.049	0.047	Turn-to-sit	−0.021	−0.014
Trail making test B	−0.115	−0.127	Sit-to-stand	−0.136	−0.173
Speed	0.268	0.297	R square	0.541	
Stride Length	0.389	0.443	Adjusted R square	0.525	
Cadence step	−0.047	−0.052	*p* value	<0.001	
Tinetti Posture	0.071	0.081			
R square	0.247				
Adjusted R square	0.223				
p value	0.003				

Abbreviations: AD-MCI, mild cognitive impairment due to Alzheimer's disease. PD-MCI, mild cognitive impairment in Parkinson's disease. MMSE, Mini-Mental State Examination. Models: adjusted for age, sex, fall history, education, and comorbidities (e.g., hypertension, diabetes mellitus and cardiovascular disease).

4. Discussion

This study investigated the factors associated with FOF in older people with different types of MCI (i.e., AD-MCI and PD-MCI). The results indicated that attention and working memory were the most important factors in determining ABC scores in the AD-MCI group, whereas turn-to-walk was the most important factor in determining ABC scores in the PD-MCI group.

A previous study indicated that older adults with cognitive impairments have greater FOF [30], and the prevalence of FOF in those with MCI was significantly higher than that in cognitively healthy older adults. Interestingly, we found different results in the two MCI groups. In the PD-MCI group, the ABC scale score was significantly lower than that in the AD-MCI group, which indicated that the individuals with PD-MCI had less confidence/greater FOF in performing activities than those with AD-MCI. This might be due to poor balance and gait performance in the PD-MCI group. However, the ABC scale score in the AD-MCI group was not only significantly higher than that in the PD-MCI group but also significantly higher than that in age-matched controls (90.28 vs. 79.89 from Huang et al. [31], one-sample t-test, $p < 0.001$). This result indicated that the participants with AD-MCI had more confidence/less FOF in conducting activities of daily living that required transferring, bending, reaching, or walking than cognitively healthy elderly participants. However, the prevalence of falls in older individuals with MCI and Alzheimer's disease is higher than that in cognitively healthy individuals [2]. Do they truly have confidence in engaging in daily activities, or do they lack an awareness of the risk of falling? Henry et al. found that patients with mild dementia had difficulties differentiating high- from low-danger situations, and this difficulty was related to more general cognitive decline [32]. A cross-sectional study reported that cognitive impairment was significantly associated with the absence of FOF in community-living older individuals, which indicated that worse cognitive function may inhibit FOF [33]. Thus, the participants with AD-MCI in the present study may have impaired threat perception, which led to lower FOF. Lack of FOF might lead to subsequent fall incidents [34]. Future studies can further investigate this interesting and important point.

In the AD-MCI group, global cognitive function, executive function, attention and working memory were significantly correlated with ABC scores. The regression model showed that attention and working memory were the most important factors in determining FOF. There have been inconsistent findings on the relationship between FOF and cognition in the aging population. Shirooka et al. indicated that cognitive impairment, especially executive function, is related with the absence of FOF in older adults with frailty [33]. A one-year longitudinal study of 406 community-dwelling older adults reported that the presence of subjective memory complaints at baseline was related to FOF [35]. Another longitudinal study of 4931 middle-age and older adults showed that FOF was associated with a greater odds of decreased Montreal Cognitive Assessment but not Mini-Mental State Examination scores [36]. A 3 year prospective study of 4280 older adults with normal cognition noted that the older adults who were very afraid of falling were 1.45 times more likely to cognitive decline than those who were not afraid of falling [37]. These inconsistent results demonstrated that cognitive impairment may play different roles in determining FOF in different aging populations. In our study, we found that the most important factors in determining FOF in AD-MCI, but not PD-MCI, were attention and working memory. Previous study indicated that the cognitive worsening over time is associated with a worse motor performance at baseline [38]. Attention and working memory are crucial for postural and gait control, and adding a concurrent secondary task while walking or maintaining balance requires even more attention, especially in older adults [39]. Impaired attention and working memory may have a negative effect on postural control, which leads to FOF. The finding of present study appears in line with a broad psychogeriatric perspective, regarding the peculiar involvement of psychological factors along cognitive-physical trajectories of aging.

In the PD-MCI group, the turn-to-walk measure was the most important factor in determining ABC scores. A cross-sectional study of 104 people with Parkinson's disease reported that the strongest contributing factor to FOF was walking difficulties [40]. Haertner et al. indicated that FOF was associated with a longer turning duration in individuals with Parkinson's disease, which is in line with our result [41]. A novel finding in this study was that even in the presence of cognitive degradation, turning performance remained the most important factor associated with FOF in individuals with Parkinson's disease. Parkinson's disease is a neurodegenerative disease that leads to motor dysfunctions, including walking difficulties. With the progression of the disease, people with Parkinson's disease show impairments in cognitive functions. Impaired cognitive function has been suggested to be related to FOF [42]. However, our regression models indicated that turning performance, not cognitive function, was the most important factor contributing to FOF in persons with Parkinson's disease. Turning is a complex movement that requires interlimb coordination, posture and gait control, and continuous movement of the center of gravity [43]. Turning difficulties are a common problem in Parkinson's disease patients due to impairments in balance control and freezing of gait that can lead to falls [44]. When compared with healthy subjects, people with Parkinson's disease need more time to turn, have a narrower base of support, and usually present freezing of gait during turning periods [45]. These disturbances not only influence the quality of their turning performance but can also cause FOF.

FOF has been shown to be a major barrier to performing daily activities and to engaging in exercise, which may cause social isolation and functional decline in the older adults. Thus, there is a need to understand the factors contributing to the FOF to efficiently address this in clinical practice and research. Based on our results, attention and working memory were the most important factors in determining FOF in the AD-MCI group, and the turn-to-walk ability was the most important factor in determining FOF in the PD-MCI group. We suggest that training programs that focus on improving cognitive function and turning performance may decrease FOF in individuals with AD-MCI and PD-MCI, respectively. Parry et al. reported that cognitive-behavioural therapy probably reduces FOF and depression in community-living older adults with undue FOF [46]. A meta-analysis of thirty trials noted that exercise intervention may have a small to moderate positive effect immediately after intervention in community-dwelling older individuals [47]. Therefore, future studies could design new cognitive and exercise programs based on our results and explore their effects on FOF in aging population.

There were several limitations should be pointed out in this study. Firstly, the relatively small sample size may limit the strength of the results. Future studies should have a larger sample size to confirm our results. Moreover, we used a cross-sectional study design, which limited our exploration of the predictive factors of FOF in these two groups. To better understand the cause-effect relationship between contributing factors and FOF in older adults with high risk of cognitive impairment, a large cohort study is encouraged. Furthermore, we did not measure a parameter for the severity of motor impairment such as Unified Parkinson's Disease Rating Scale (MDS-UPDRS) score in PD-MCI group. The turn-to-walk parameter was the most important parameter explaining FOF, which may be explained by motor impairment. Treatment effects of cognitive training and turning performance training to improve FOF in MCI populations are warranted and encouraged for future studies.

5. Conclusions

The present study found that FOF in individuals with different types of MCI was influenced by different factors. Attention and working memory were the most important contributing factors in the AD-MCI group, whereas the turn-to-walk ability was the most important contributing factor in the PD-MCI group. Thus, therapies that aim to lower FOF in AD-MCI and PD-MCI populations may address attention and working memory and turn-to-walk performance, respectively.

Author Contributions: Conceptualization, P.-H.C. and Y.-Y.Y.; methodology, Y.-Y.Y.; validation, Y.-Y.L.; resources, S.-J.C. and P.-N.W.; data curation, P.-H.C.; writing—original draft preparation, Y.-Y.Y. and P.-H.C.; writing—review and editing, F.-Y.C.; funding acquisition, P.-H.C. and F.-Y.C. All authors have read and agreed to the published version of the manuscript.

Funding: This research was funded by grants from the MacKay Medical College (MMC-RD-109-1B-09) and MacKay Memorial Hospital (MMH-MM-10811). The APC was funded by Ministry of Science and Technology (MOST 110-2314-B-715-006-MY3).

Institutional Review Board Statement: The study was conducted in accordance with the Declaration of Helsinki, and approved by the institutional review board of Mackay Memorial Hospital (number: 18MMHIS005e).

Informed Consent Statement: Informed consent was obtained from all subjects involved in the study.

Data Availability Statement: The datasets used and/or analysed during the current study are available from the corresponding author on reasonable request.

Acknowledgments: We would like to thank all the participants, as well as the research assistants Yi-Hsuan Ho, Wen-Chun Wu, and Lu-Shan Lee, for participating in the assessment.

Conflicts of Interest: The authors declare no competing interest.

References

1. Katz, R.; Shah, P. The patient who falls: Challenges for families, clinicians, and communities. *JAMA* **2010**, *303*, 273–274. [CrossRef] [PubMed]
2. Borges Sde, M.; Radanovic, M.; Forlenza, O.V. Fear of falling and falls in older adults with mild cognitive impairment and Alzheimer's disease. *Neuropsychol. Dev. Cogn. B Aging Neuropsychol. Cogn.* **2015**, *22*, 312–321. [CrossRef] [PubMed]
3. Gazibara, T.; Kurtagic, I.; Kisic-Tepavcevic, D.; Nurkovic, S.; Kovacevic, N.; Gazibara, T.; Pekmezovic, T. Falls, risk factors and fear of falling among persons older than 65 years of age. *Psychogeriatrics* **2017**, *17*, 215–223. [CrossRef] [PubMed]
4. Hoang, O.T.; Jullamate, P.; Piphatvanitcha, N.; Rosenberg, E. Factors related to fear of falling among community-dwelling older adults. *J. Clin. Nurs.* **2017**, *26*, 68–76. [CrossRef]
5. Holtzer, R.; Friedman, R.; Lipton, R.B.; Katz, M.; Xue, X.; Verghese, J. The relationship between specific cognitive functions and falls in aging. *Neuropsychology* **2007**, *21*, 540–548. [CrossRef] [PubMed]
6. Petersen, R.C.; Doody, R.; Kurz, A.; Mohs, R.C.; Morris, J.C.; Rabins, P.V.; Ritchie, K.; Rossor, M.; Thal, L.; Winblad, B. Current concepts in mild cognitive impairment. *Arch. Neurol.* **2001**, *58*, 1985–1992. [CrossRef]
7. Allali, G.; Launay, C.P.; Blumen, H.M.; Callisaya, M.L.; De Cock, A.M.; Kressig, R.W.; Srikanth, V.; Steinmetz, J.P.; Verghese, J.; Beauchet, O. Falls, Cognitive Impairment, and Gait Performance: Results from the GOOD Initiative. *J. Am. Med. Dir. Assoc.* **2017**, *18*, 335–340. [CrossRef]
8. Scheffer, A.C.; Schuurmans, M.J.; van Dijk, N.; van der Hooft, T.; de Rooij, S.E. Fear of falling: Measurement strategy, prevalence, risk factors and consequences among older persons. *Age Ageing* **2008**, *37*, 19–24. [CrossRef]
9. Winblad, B.; Palmer, K.; Kivipelto, M.; Jelic, V.; Fratiglioni, L.; Wahlund, L.O.; Nordberg, A.; Bäckman, L.; Albert, M.; Almkvist, O.; et al. Mild cognitive impairment–beyond controversies, towards a consensus: Report of the International Working Group on Mild Cognitive Impairment. *J. Intern. Med.* **2004**, *256*, 240–246. [CrossRef]
10. Albert, M.S.; DeKosky, S.T.; Dickson, D.; Dubois, B.; Feldman, H.H.; Fox, N.C.; Gamst, A.; Holtzman, D.M.; Jagust, W.J.; Petersen, R.C.; et al. The diagnosis of mild cognitive impairment due to Alzheimer's disease: Recommendations from the National Institute on Aging-Alzheimer's Association workgroups on diagnostic guidelines for Alzheimer's disease. *Alzheimers Dement.* **2011**, *7*, 270–279. [CrossRef]
11. Litvan, I.; Goldman, J.G.; Tröster, A.I.; Schmand, B.A.; Weintraub, D.; Petersen, R.C.; Mollenhauer, B.; Adler, C.H.; Marder, K.; Williams-Gray, C.H.; et al. Diagnostic criteria for mild cognitive impairment in Parkinson's disease: Movement Disorder Society Task Force guidelines. *Mov. Disord.* **2012**, *27*, 349–356. [CrossRef] [PubMed]
12. Palavra, N.C.; Naismith, S.L.; Lewis, S.J. Mild cognitive impairment in Parkinson's disease: A review of current concepts. *Neurol. Res. Int.* **2013**, *2013*, 576091. [CrossRef] [PubMed]
13. O'Bryant, S.E.; Lacritz, L.H.; Hall, J.; Waring, S.C.; Chan, W.; Khodr, Z.G.; Massman, P.J.; Hobson, V.; Cullum, C.M. Validation of the New Interpretive Guidelines for the Clinical Dementia Rating Scale Sum of Boxes Score in the National Alzheimer's Coordinating Center Database. *Arch. Neurol.* **2010**, *67*, 746–749. [CrossRef] [PubMed]
14. Chen, C.-F.; Chwo, M.-J.; Wen, C.-H.; Yang, C.-C. A Three Month Follow-up Study of Hemorrhagic Stroke Patients Receiving Rehabilitation Therapy as Assessed by the Barthel Index. *Taiwan J. Fam. Med.* **2007**, *17*, 250–263.
15. Chuang, Y.-F.; Sung, H.-C.; Lin, L.-P. Health Status and Health Resource Utilization among Older People in Rural Mining Area. *Tzu Chi Nurs. J.* **2005**, *4*, 34–41.
16. Gibb, W.R.; Lees, A.J. The relevance of the Lewy body to the pathogenesis of idiopathic Parkinson's disease. *J. Neurol. Neurosurg. Psychiatry* **1988**, *51*, 745–752. [CrossRef]
17. Powell, L.E.; Myers, A.M. The Activities-specific Balance Confidence (ABC) Scale. *J. Gerontol. Ser. A Biol. Sci. Med. Sci.* **1995**, *50*, M28–M34. [CrossRef]
18. Talley, K.M.; Wyman, J.F.; Gross, C.R. Psychometric properties of the activities-specific balance confidence scale and the survey of activities and fear of falling in older women. *J. Am. Geriatr. Soc.* **2008**, *56*, 328–333. [CrossRef]
19. Fountoulakis, K.N.; Tsolaki, M.; Chantzi, H.; Kazis, A. Mini Mental State Examination (MMSE): A validation study in Greece. *Am. J. Alzheimer's Dis. Other Dement.* **2000**, *15*, 342–345. [CrossRef]
20. Arnett, J.A.; Labovitz, S.S. Effect of physical layout in performance of the Trail Making Test. *Psychol. Assess.* **1995**, *7*, 220–221. [CrossRef]
21. Brucki, S.M.; Rocha, M.S. Category fluency test: Effects of age, gender and education on total scores, clustering and switching in Brazilian Portuguese-speaking subjects. *Braz. J. Med. Biol. Res.* **2004**, *37*, 1771–1777. [CrossRef] [PubMed]
22. Richardson, J.T. Measures of short-term memory: A historical review. *Cortex* **2007**, *43*, 635–650. [CrossRef]

23. Chang, C.C.; Kramer, J.H.; Lin, K.N.; Chang, W.N.; Wang, Y.L.; Huang, C.W.; Lin, Y.T.; Chen, C.; Wang, P.N. Validating the Chinese version of the Verbal Learning Test for screening Alzheimer's disease. *J. Int. Neuropsychol. Soc.* **2010**, *16*, 244–251. [CrossRef] [PubMed]
24. Calamia, M.; Markon, K.; Denburg, N.L.; Tranel, D. Developing a short form of Benton's Judgment of Line Orientation Test: An item response theory approach. *Clin. Neuropsychol.* **2011**, *25*, 670–684. [CrossRef]
25. del Toro, C.M.; Bislick, L.P.; Comer, M.; Velozo, C.; Romero, S.; Gonzalez Rothi, L.J.; Kendall, D.L. Development of a short form of the Boston naming test for individuals with aphasia. *J. Speech Lang. Hear. Res.* **2011**, *54*, 1089–1100. [CrossRef]
26. Liu, C.Y.; Wang, S.J.; Teng, E.L.; Fuh, J.L.; Lin, C.C.; Lin, K.N.; Chen, H.M.; Lin, C.H.; Wang, P.N.; Yang, Y.Y.; et al. Depressive disorders among older residents in a Chinese rural community. *Psychol. Med.* **1997**, *27*, 943–949. [CrossRef] [PubMed]
27. Spitzer, R.L.; Kroenke, K.; Williams, J.B.; Löwe, B. A brief measure for assessing generalized anxiety disorder: The GAD-7. *Arch. Intern. Med.* **2006**, *166*, 1092–1097. [CrossRef]
28. Tinetti, M.E. Performance-oriented assessment of mobility problems in elderly patients. *J. Am. Geriatr. Soc.* **1986**, *34*, 119–126. [CrossRef]
29. De Ridder, R.; Lebleu, J.; Willems, T.; De Blaiser, C.; Detrembleur, C.; Roosen, P. Concurrent Validity of a Commercial Wireless Trunk Triaxial Accelerometer System for Gait Analysis. *J. Sport Rehabil.* **2019**, *28*, jsr.2018-0295. [CrossRef]
30. Tavares, G.M.S.; Pacheco, B.P.; Gottlieb, M.G.V.; Müller, D.V.K.; Santos, G.M. Interaction between cognitive status, fear of falling, and balance in elderly persons. *Clinics* **2020**, *75*, e1612. [CrossRef] [PubMed]
31. Huang, T.-T.; Wang, W.-S. Comparison of three established measures of fear of falling in community-dwelling older adults: Psychometric testing. *Int. J. Nurs. Stud.* **2009**, *46*, 1313–1319. [CrossRef]
32. Henry, J.D.; Thompson, C.; Ruffman, T.; Leslie, F.; Withall, A.; Sachdev, P.; Brodaty, H. Threat perception in mild cognitive impairment and early dementia. *J. Gerontol. Ser. B Psychol. Sci. Soc. Sci.* **2009**, *64*, 603–607. [CrossRef]
33. Shirooka, H.; Nishiguchi, S.; Fukutani, N.; Tashiro, Y.; Nozaki, Y.; Hirata, H.; Yamaguchi, M.; Tasaka, S.; Matsushita, T.; Matsubara, K.; et al. Cognitive impairment is associated with the absence of fear of falling in community-dwelling frail older adults. *Geriatr. Gerontol. Int.* **2017**, *17*, 232–238. [CrossRef] [PubMed]
34. Muir, S.W.; Gopaul, K.; Montero Odasso, M.M. The role of cognitive impairment in fall risk among older adults: A systematic review and meta-analysis. *Age Ageing* **2012**, *41*, 299–308. [CrossRef]
35. Sakurai, R.; Suzuki, H.; Ogawa, S.; Kawai, H.; Yoshida, H.; Hirano, H.; Ihara, K.; Obuchi, S.; Fujiwara, Y. Fear of falling, but not gait impairment, predicts subjective memory complaints in cognitively intact older adults. *Geriatr. Gerontol. Int.* **2017**, *17*, 1125–1131. [CrossRef]
36. Peeters, G.; Leahy, S.; Kennelly, S.; Kenny, R.A. Is Fear of Falling Associated with Decline in Global Cognitive Functioning in Older Adults: Findings from the Irish Longitudinal Study on Ageing. *J. Am. Med. Dir. Assoc.* **2018**, *19*, 248–254.e3. [CrossRef]
37. Noh, H.M.; Roh, Y.K.; Song, H.J.; Park, Y.S. Severe Fear of Falling Is Associated with Cognitive Decline in Older Adults: A 3-Year Prospective Study. *J. Am. Med. Dir. Assoc.* **2019**, *20*, 1540–1547. [CrossRef] [PubMed]
38. Quattropani, M.C.; Sardella, A.; Morgante, F.; Ricciardi, L.; Alibrandi, A.; Lenzo, V.; Catalano, A.; Squadrito, G.; Basile, G. Impact of cognitive Reserve and Premorbid IQ on Cognitive and Functional Status in Older Outpatients. *Brain Sci.* **2021**, *11*, 824. [CrossRef] [PubMed]
39. Woollacott, M.; Shumway-Cook, A. Attention and the control of posture and gait: A review of an emerging area of research. *Gait Posture* **2002**, *16*, 1–14. [CrossRef]
40. Lindholm, B.; Hagell, P.; Hansson, O.; Nilsson, M.H. Factors associated with fear of falling in people with Parkinson's disease. *BMC Neurol.* **2014**, *14*, 19. [CrossRef] [PubMed]
41. Haertner, L.; Elshehabi, M.; Zaunbrecher, L.; Pham, M.H.; Maetzler, C.; van Uem, J.M.T.; Hobert, M.A.; Hucker, S.; Nussbaum, S.; Berg, D.; et al. Effect of Fear of Falling on Turning Performance in Parkinson's Disease in the Lab and at Home. *Front. Aging Neurosci.* **2018**, *10*, 78. [CrossRef] [PubMed]
42. Uemura, K.; Shimada, H.; Makizako, H.; Doi, T.; Tsutsumimoto, K.; Yoshida, D.; Anan, Y.; Ito, T.; Lee, S.; Park, H.; et al. Effects of mild and global cognitive impairment on the prevalence of fear of falling in community-dwelling older adults. *Maturitas* **2014**, *78*, 62–66. [CrossRef]
43. Cheng, F.Y.; Yang, Y.R.; Wang, C.J.; Wu, Y.R.; Cheng, S.J.; Wang, H.C.; Wang, R.Y. Factors influencing turning and its relationship with falls in individuals with Parkinson's disease. *PLoS ONE* **2014**, *9*, e93572. [CrossRef] [PubMed]
44. Nutt, J.G.; Bloem, B.R.; Giladi, N.; Hallett, M.; Horak, F.B.; Nieuwboer, A. Freezing of gait: Moving forward on a mysterious clinical phenomenon. *Lancet Neurol.* **2011**, *10*, 734–744. [CrossRef]
45. Visser, J.E.; Voermans, N.C.; Oude Nijhuis, L.B.; van der Eijk, M.; Nijk, R.; Munneke, M.; Bloem, B.R. Quantification of trunk rotations during turning and walking in Parkinson's disease. *Clin. Neurophysiol.* **2007**, *118*, 1602–1606. [CrossRef]
46. Parry, S.W.; Bamford, C.; Deary, V.; Finch, T.L.; Gray, J.; MacDonald, C.; McMeekin, P.; Sabin, N.J.; Steen, I.N.; Whitney, S.L.; et al. Cognitive-behavioural therapy-based intervention to reduce fear of falling in older people: Therapy development and randomised controlled trial—The Strategies for Increasing Independence, Confidence and Energy (STRIDE) study. *Health Technol. Assess* **2016**, *20*, 1–206. [CrossRef]
47. Kumar, A.; Delbaere, K.; Zijlstra, G.A.; Carpenter, H.; Iliffe, S.; Masud, T.; Skelton, D.; Morris, R.; Kendrick, D. Exercise for reducing fear of falling in older people living in the community: Cochrane systematic review and meta-analysis. *Age Ageing* **2016**, *45*, 345–352. [CrossRef]

brain sciences

MDPI

Article

The Luria-Nebraska Neuropsychological Battery Neuromotor Tasks: From Conventional to Image-Derived Measures

Daniele Corbo [1,*], Donatella Placidi [2], Roberto Gasparotti [1], Robert Wright [3], Donald R. Smith [4], Roberto G. Lucchini [2,5], Megan K. Horton [3] and Elena Colicino [3]

1 Neuroradiology Unit, Department of Medical and Surgical Specialties, Radiological Sciences and Public Health, University of Brescia and ASST Spedali Civili Hospital, 25121 Brescia, Italy; roberto.gasparotti@unibs.it

2 Occupational Medicine, Department of Medical and Surgical Specialties, Radiological Sciences and Public Health, University of Brescia, 25121 Brescia, Italy; donatella.placidi@unibs.it (D.P.); rlucchin@fiu.edu (R.G.L.)

3 Department of Environmental Medicine and Public Health, Icahn School of Medicine at Mount Sinai, New York, NY 10029, USA; robert.wright@mssm.edu (R.W.); megan.horton@mssm.edu (M.K.H.); elena.colicino@mssm.edu (E.C.)

4 Department of Microbiology and Environmental Toxicology, University of California, Santa Cruz, CA 95064, USA; drsmith@ucsc.edu

5 Environmental Health Sciences, School of Public Health, Florida International University, Miami, FL 33199, USA

* Correspondence: daniele.corbo@unibs.it

Abstract: Background: Sensorimotor difficulties significantly interfere with daily activities, and when undiagnosed in early life, they may increase the risk of later life cognitive and mental health disorders. Subtests from the Luria-Nebraska Neuropsychological Battery (LNNB) discriminate sensorimotor impairments predictive of sensorimotor dysfunction. However, scoring the LNNB sensorimotor assessment is highly subjective and time consuming, impeding the use of this task in epidemiologic studies. Aim: To train and validate a novel automated and image-derived scoring approach to the LNNB neuro-motor tasks for use in adolescents and young adults. Methods: We selected 46 adolescents (19.6 +/− 2.3 years, 48% male) enrolled in the prospective Public Health Impact of Metal Exposure (PHIME) study. We visually recorded the administration of five conventional sensorimotor LNNB tasks and developed automated scoring alternatives using a novel mathematical approach combining optic flow fields from recorded image sequences on a frame-by-frame basis. We then compared the conventional and image-derived LNNB task scores using Pearson's correlations. Finally, we provided the accuracy of the novel scoring approach with Receiver Operating Characteristic (ROC) curves and the area under the ROC curves (AUC). Results: Image-derived LNNB task scores strongly correlated with conventional scores, which were assessed and confirmed by multiple administrators to limit subjectivity (Pearson's correlation $\geq$ 0.70). The novel image-derived scoring approach discriminated participants with low motility (<mean population levels) with a specificity ranging from 70% to 83%, with 70% sensitivity. Conclusions: The novel image-derived LNNB task scores may contribute to the timely assessment of sensorimotor abilities and delays, and may also be effectively used in telemedicine.

Keywords: Luria neuromotor test; sensorimotor impairments; image-derived scoring

Citation: Corbo, D.; Placidi, D.; Gasparotti, R.; Wright, R.; Smith, D.R.; Lucchini, R.G.; Horton, M.K.; Colicino, E. The Luria-Nebraska Neuropsychological Battery Neuromotor Tasks: From Conventional to Image-Derived Measures. *Brain Sci.* **2022**, *12*, 757. https://doi.org/10.3390/brainsci12060757

Academic Editor: Benjamin Straube

Received: 21 April 2022
Accepted: 7 June 2022
Published: 8 June 2022

Publisher's Note: MDPI stays neutral with regard to jurisdictional claims in published maps and institutional affiliations.

1. Introduction

Sensorimotor dysfunction is a pervasive developmental disorder with a prevalence of up to 19% among school-aged children and adolescents [1,2]. Sensorimotor difficulties significantly interfere with daily activities and academic performance, and, if untreated, may lead to an increased risk of cognitive and mental health disorders later in life [1,2]. Childhood sensorimotor dysfunction is defined by decrements in fine and/or gross motor skills, with poor motor performance that is usually slower, less accurate, and more variable than that of an unaffected individual [3]. It is also associated with reduced learning ability, verbal and visuospatial memory problems, and cognitive impairments [1,2,4]. Sensorimotor dysfunction is often a comorbidity with other developmental disorders, such as attention deficit and hyperactivity disorder. Cognitive impairments in children with sensorimotor conditions persist into adolescence and adulthood, and may even extend beyond their motor difficulties [1,5]. Due to the common root and interactions between sensorimotor and cognitive disorders, sensorimotor dysfunction is considered a condition of the central nervous system [1,2,6]. Early detection of abnormal sensorimotor function can lead to timely interventions able to mitigate the adverse long-term consequences [5,6].

Several different tests and exams are designed to detect childhood and adolescence sensorimotor impairment [7–9]. The Luria-Nebraska Neuropsychological Battery (LNNB) stands out for its evaluation of the presence of neuro-motor impairments and identification of focal brain abnormalities that may account for these impairments in children and adolescents [10].

The standard LNNB consists of 11 clinical scales assessing major areas of neuropsychological functioning (motor functions, rhythm, tactile functions, visual functions, receptive speech, expressive speech, writing, reading, arithmetic, memory, intellectual processes), two sensorimotor scales (left hemisphere, right hemisphere), and three summary scales (pathognomonic, profile elevation, impairment) [10]. Although LNNB can capture early sensorimotor impairments, it is not widely used in epidemiological and clinical studies largely because it is both highly subjective and time consuming [10–12]. Indeed, while the administration of the LNNB is standardized, the scoring and thus the interpretation of the LNNB data are time-consuming duties and require the judgment of a trained administrator who is responsible for evaluating the appropriateness of the movement and counting the participant's movements during a specific time frame [10–12].

To address the subjectivity and timeliness of the LNNB in epidemiologic studies, we developed an automated scoring algorithm using a novel mathematical approach. We applied this approach to LNNB data collected from a subset of adolescents enrolled in the ongoing Public Health Impact of Metal Exposure (PHIME) study. Specifically, we trained and validated five image-derived sensorimotor LNNB task scores able to assess the adolescent's movement by combining optic flow fields from recorded image sequences on a frame-by-frame basis. We then validated the accuracy of those novel image-derived LNNB task scores comparing them with conventional LNNB scores. To limit the subjectivity of the conventional LNNB scores, multiple administrators evaluated the appropriateness and counted the participant's movements. This novel automatic image-derived sensorimotor assessment may be implemented during telemedicine services, thus leading to more personalized treatments and earlier interventions in subjects at a higher risk of developing sensorimotor impairment.

2. Methods

Study Population: To test our novel scoring algorithm, we selected 46 participants (ages 15–23; female = 52%) from the ongoing PHIME cohort consisting of 208 children and adolescents, who were enrolled between 2007 and 2014, and followed up between 2017 and 2021. The overall objective of the study was to examine associations between metal exposure and neurodevelopmental outcomes [11]. Participants were at first identified through the local school districts and enrolled with the following inclusion criteria: born in the respective area to a family who resided in the area for at least a generation, lived in the study area since birth [11]. Exclusion criteria included: known hand or finger motor deficits, visual deficits not adequately corrected, and any history of neurological, metabolic, hepatic, or endocrine diseases [11]. Participants were also excluded if they had a history of receiving parenteral nutrition that may cause overload of environmental chemicals (i.e., manganese), or they were taking prescription psychoactive drugs or had known psychiatric disturbances [11]. All participants were followed over adolescence and, during the follow-up visit, 46 of them underwent a LNNB test [11]. The 46 participants were randomly selected and received a sensorimotor LNNB evaluation from two trained administrators, who assessed the appropriateness and counted the participant's movements. Both administrators agreed on the correctness and the number of participant's movements.

Ethics: Written informed consent was obtained from parents and children. Study protocols were approved by the institutional review board at the Ethical Committee of the Public Health Agency of Brescia and Mount Sinai. Extensive data on exposure assessment in these areas were published previously [11].

The Luria-Nebraska Neuropsychological Battery: The LNNB is a battery of motor-neuron tests used in combination with other neurodevelopmental exams to identify childhood or adolescence motor-cognitive deficits or delays [10–12]. Subjects were tested with the LNNB according to the instructions provided in the test manual [13]. Briefly, participants were told to complete the movement/task for 10 s. This motor coordination exam consisted of a standardized test battery of five tasks [10,11]. The five movements/tasks include the dominant hand clench (Task 1), the finger–thumb touching with the dominant hand (Task 2), the non-dominant hand clench (Task 3), the finger–thumb touching with non-dominant hand (Task 4), and alternative hand clench (Task 5) [11]. The sum of the frequency of each of the five task yields a final score that reflects a motor score [11]. The test administrator kept track of time and the frequency of tasks completed in the 10 s period.

Analytical plan: Our novel scoring method leverages the image-derived LNNB task scores with the following steps:

Visual record of all actions. A trained administrator visually recorded the participant's hand movements via an iPad Pro camera (Camera module iSight with8 megapixel) and stored the video in a local hard drive. All participants were seated at a determined distance (1.5 m) from the camera.

Normalization. To remove variability given by sex, age, and recording time, the hand movements in each image were normalized to a pre-defined number of voxels, resolution, and duration using MATLAB scripts.

Compute optic flow. The local motion vectors or optic flow fields were identified from an image sequence on a frame-by-frame basis. Optical flow is an image analysis technique used to detect motion in video sequences [14]. In each image sequence, the optical flow generates a vector for each image pixel representing the apparent motion in the corresponding sampling period. The movement represented by the optical flow is considered an apparent movement because it is a projection of the real three-dimensional (3D) image on a two-dimensional (2D) plane. In addition, previous optical flow computation techniques analyze the brightness variations of each pixel, thus making it impossible to distinguish between true and apparent motion.

We computed the local motion vector for each pixel in the image on a frame-by-frame basis as previously done [15]. The instantaneous full velocity at x, y, z pixel locations that results from a 3D camera rotation, $\omega = (\omega_x, \omega_y, \omega_z)^T$ with ω_p the angular velocity around the p-axis (with p = x, y and z), can be well approximated by $v(x) = B(x)\omega$ [16], where the B(x) matrix can be defined as:

$$B_{(x)} = \begin{bmatrix} xy/f & (-f - x^2)/f & y \\ (f + y^2)/f & -xy/f & -x \end{bmatrix}$$

where f is the focal length of the camera.

The quality of the velocity estimates can be greatly influenced by camera shocks and vibrations. Using this algorithm, we removed this unstable component of the camera motion and stabilized the image sequence, maximizing the temporal local velocity constancy over the entire short sequence. This method, which is embedded within a phase-based optic flow algorithm, was previously tested on both synthetic and complex real-world sequences [17]. The temporal evolution of contours of constant phase can yield a good approximation to the local velocity field [18]. We used this approach because of its computational efficiency as it involves linear systems and simple transformations, the result of which can be computed without time-consuming re-filtering.

Motion quality. We estimated the motional quality of each image-derived LNNB task score as the mean motion over the whole recorded video, as previously done [19]. We analyzed the local motion energy (i.e., speed) of the participant's hand movements in each recorded video as a function of time by averaging speed over pixels [19]. We then calculated mean local motion speed temporal profiles. This resulting measure was expressed in degree/s and provided information about the quantity of motion performed by the hand of the tested subject [19].

Correlation between conventional and image-derived LNNB scores and accuracy evaluation of the image-derived LNNB scores. We estimated the association between conventional and image-derived LNNB task scores with Pearson correlation coefficients. To determine the accuracy of the image-derived LNNB score, we first classified participants with low motility using both conventional and image-derived LNNB scores, and we considered low motility when participant's LNNB scores were below their corresponding mean population levels. We then evaluated the accuracy of the image-derived LNNB scores using the Receiver Operating Characteristic (ROC) curve and the area under the ROC curve (AUC).

Code availability: MATLAB code is publicly available on https://github.com/danielecorbo/Luria/blob/c8cf999e92f5367b60d69e0a6915ca5347c43d6e/motion_calculation.m (accessed on 14 April 2022).

3. Results

PHIME population description. PHIME participants were 19.6 (+/−2.3) years of age and were equally divided into males (48%) and females (52%). The majority of adolescents did not smoke (67%) and did not report any alcohol consumption (57%) (Table 1). All completed the five LNNB tasks (Table 2). The tasks performed by the non-dominant hand (i.e., Task 2 and Task 5) showed lower scores (lower motility) and higher variability (less precision in the movement) compared to the same tasks with the dominant hand (Task 1 and Task 4, respectively) (Table 2).

Table 1. Sociodemographic characteristics of selected participants from the Public Health Impact of Metal Exposure (PHIME) Study.

Characteristics	N (%) or Mean (SD)
Age (years)	19.6 (2.3)
Sex (male)	22 (48%)
Maternal education	
≤High School	38 (83%)
>High School	3 (6%)
NA	5 (11%)
Self-reported cigarette smoking	
No	31 (67%)
Yes	7 (15%)
NA	8 (17%)
Self-reported alcohol consumption	
No	26 (57%)
Yes	12 (26%)
NA	8 (17%)

%: Percentage, SD: Standard Deviation; NA: Not Available.

The image-derived LNNB task scores. Visual recordings of the five sensorimotor LNNB tasks completed by the 46 participants led to 240 visual recordings used to create the novel image-derived LNNB task scores. From each video, we obtained: (a) conventional LNNB task scores from two trained administrators, who both counted the number of performed actions by visual inspection, and agreed on the correctness of the participants' movements and the number of actions; and (b) the novel predicted image-derived

LNNB task scores computed via MATLAB. Distributions and descriptive statistics of both conventional and image-derived LNNB tasks are provided in Table 2 and Figure S1.

Table 2. Conventional and image-derived Luria-Nebraska Neuropsychological Battery (LNNB) test scores among 46 PHIME participants.

Conventional	Mean	Standard Deviation	CV	1st Quartile	2nd Quartile	3rd Quartile	Minimum	Maximum
Task 1	23.48	7.54	2.42	18	22	29	6	37
Task 2	23.15	7.69	2.56	19	21.5	29	5	40
Task 3	17.5	6.31	2.28	13	15.5	23.75	8	31
Task 4	9.67	2.86	0.85	8	10	11	4	17
Task 5	9.13	2.96	0.96	7	9	11	2	16
Image-derived	**Mean**	**Standard Deviation**	**CV**	**1st Quartile**	**2nd Quartile**	**3rd Quartile**	**Minimum**	**Maximum**
Task 1	0.17	0.11	0.07	0.08	0.13	0.24	0.02	0.44
Task 2	0.18	0.11	0.06	0.1	0.17	0.22	0.02	0.43
Task 3	0.14	0.08	0.04	0.09	0.12	0.2	0.02	0.31
Task 4	0.1	0.05	0.03	0.07	0.1	0.13	0.01	0.24
Task 5	0.1	0.05	0.03	0.06	0.09	0.14	0.02	0.24

CV: Coefficient of Variation. Task 1: the dominant hand clench, Task2: the finger–thumb touching with the dominant hand, Task 3: the non-dominant hand clench, Task 4: the finger–thumb touching with non-dominant hand, Task 5 alternative hand clench.

We then evaluated the correlation between conventional and image-derived LNNB task scores. All tasks showed a strong linear relationship between conventional and image-derived LNNB task scores (Figure 1, Table S1). Specifically, the Pearson correlation coefficients between the conventional and image-derived LNNB task scores ranged between 0.70 and 0.74, with all coefficients statistically significant different from 0 (Figure 1, Table S1).

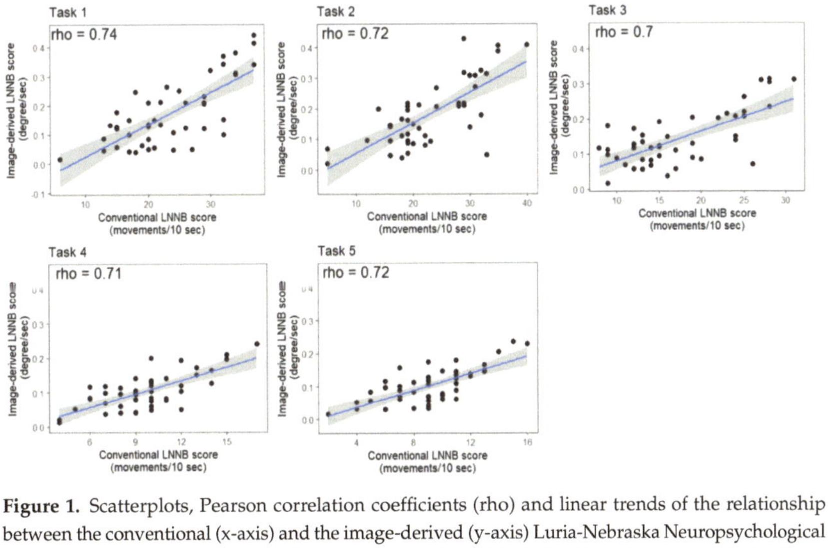

Figure 1. Scatterplots, Pearson correlation coefficients (rho) and linear trends of the relationship between the conventional (x-axis) and the image-derived (y-axis) Luria-Nebraska Neuropsychological Battery (LNNB) task scores. Each dot represents an individual, the black line represents the linear trend of the association between conventional and the image-derived LNNB task scores, and the grey shade indicates the 95% Confidence Interval of the linear trend. Task 1: the dominant hand clench, Task 2: the finger–thumb touching with the dominant hand, Task 3: the non-dominant hand clench, Task 4: the finger–thumb touching with non-dominant hand, Task 5: alternative hand clench.

In this subset and using the conventional LNNB task scores (from Task 1 to Task 5, respectively), 25, 26, 27, 22, and 26 participants showed low motility, i.e., their conventional LNNB scores were below the corresponding mean population levels. With the novel estimated image-derived LNNB tasks and setting a sensitivity level of 70%, we discriminated participants with low levels of motility with a specificity of 71%, 83%, 71%, 63%, 67% for Task 1 to Task 5, respectively. The accuracy of those models ranged from 70% for Task 4 to 83% for Task 2 (Figure 2, Table S2).

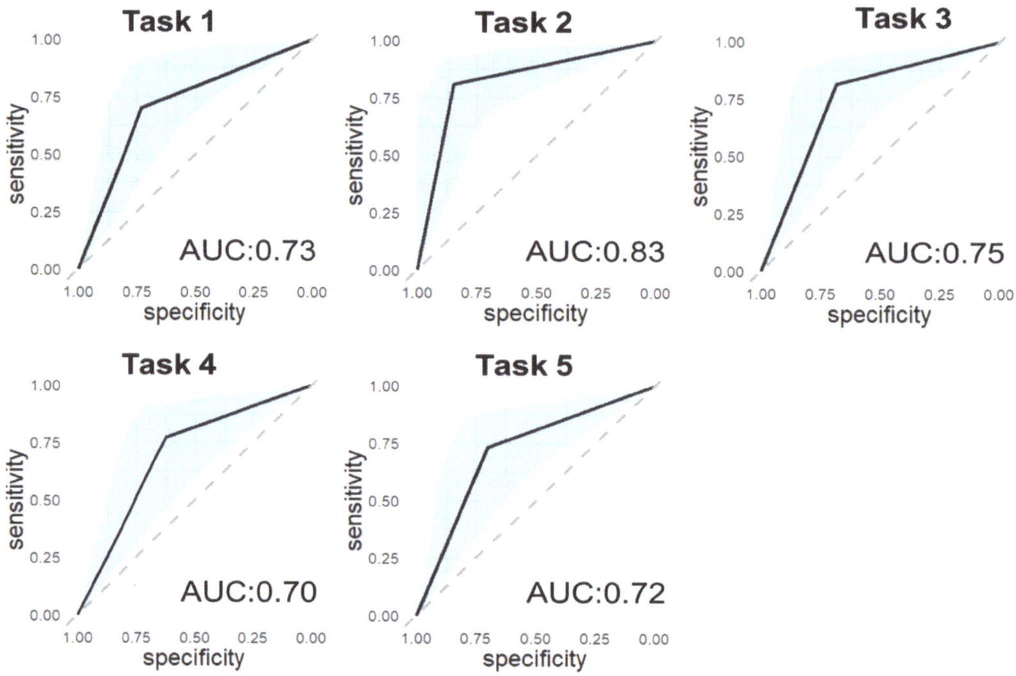

Figure 2. Receiver Operating Characteristic (ROC) curve (solid black line), empirical 95% confidence interval (grey shade), and area under the ROC curve (AUC) of the image-derived Luria-Nebraska Neuropsychological Battery (LNNB) tasks, classifying participants with lower mobility (LNNB score < population mean levels). Dashed grey line indicates the line of no-discrimination. Task 1: the dominant hand clench, Task2: the finger–thumb touching with the dominant hand, Task 3: the non-dominant hand clench, Task 4: the finger–thumb touching with non-dominant hand, Task 5 alternative hand clench.

4. Discussion

This is the first study to develop automatic and image-derived LNNB task scores in adolescents using a novel mathematical approach combining optic flow fields from image sequences on a frame-by-frame basis. This novel image-derived LNNB task-scoring approach accurately estimated an alternative for conventional LNNB measurements and provided good discrimination of participants with lower scores, indicating motility below the mean population values. These findings supported the hypothesis that imaging data can lead to novel automatic tools able to provide a timely sensorimotor screening and discriminate individuals at higher risk of developing sensorimotor impairments and delays.

Our findings were consistent with prior literature showing that the dominant hand is faster [20], more accurate [21], and less variable [20] in movements than the non-dominant hand. This supports the consensus regarding the specialization of the non-dominant system for utilizing proprioceptive feedback [22–24], defined as the central motor ability to sense the position and the movement of a limb in space along with muscular effort and tension [25]. The proprioceptive feedback is also strongly connected with peripheral or central nervous structures [25]. Our findings also showed good accuracy (≥70%) in the estimation of LNNB task score alternatives using image-derived data, especially in evaluating the non-dominant hand clench tasks. This can be explained by the fact that the non-dominant hand provides more accurate information on proprioception [25]. Indeed, this information cannot be compensated for by the advantages of the dominant hand, such as increased training of the hand muscles [26], the enlarged excitability of the dominant motor cortex [27], and the increased excitability of motor-neuronal pool at the level of spinal circuitry [28]. The conventional and image-derived LNNB scores also showed strong correlation coefficients (≥0.70).

To determine the image-derived LNNB task scores, we employed a novel mathematical approach leveraging the image sequence on a frame-by-frame basis and the local motion energy as a function of time. The contribution of this approach was two-fold. First, we used a normalized, robust, and computationally efficient method, which facilitates the detection of movements that otherwise may be discarded or overestimated by the visual count. Second, our strategy may limit time-consuming duties for the trained administrator and may improve the timeliness of the assessment of sensory-motor impairments or delays. Indeed, this novel LNNB scoring approach did not require any administrator and it may be considered as sensorimotor screening during telemedicine services. We then compared our image-derived LNNB task scores with conventional scores, which are highly subjective. In this study, we were able to limit the subjectivity of conventional LNNB task scores with the assessment by multiple trained administrators. This allowed us to properly characterize the adolescents' movement and to correctly estimate the number of actions performed. We finally included the code to facilitate reproducibility and transparency of our scientific results.

Our study population consisted of healthy adolescents, thus limiting the generalizability of our findings to populations of other ages and different sensorimotor abilities. However, these preliminary results showed the potential of recorded videos for telemedicine applications that aim to provide early diagnostics of any sensorimotor impairments and delays, including Parkinsonians-like symptoms. An additional limitation was the relatively small sample size, which led us to internally validate our results and to potentially overfit our findings. To overcome these limitations and improve the performance of these novel image-derived LNNB task-scoring approach, further studies should include larger populations of different ages and sensorimotor abilities.

5. Conclusions

We trained and validated an automatic image-derived LNNB task-scoring approach able to assess an individual's movements. This novel image-derived LNNB task-scoring approach may mitigate administrator's subjectivity, limit time-consuming duties, and provide the groundwork for early diagnostics for any sensorimotor impairments and delays.

Supplementary Materials: The following supporting information can be downloaded at: https://www.mdpi.com/article/10.3390/brainsci12060757/s1, Table S1: Pearson correlation coefficient between the conventional and the image-derived Luria-Nebraska Neuropsychological Battery (LNNB) task scores among 46 PHIME participants. Table S2: Area under the Receiver Operating Characteristic (ROC) curve (AUC) and 95% Confidence Interval (95%CI) of each Luria-Nebraska Neuropsychological Battery (LNNB) task, classifying participants with lower mobility (LNNB score < population mean levels). Figure S1: Boxplots of the Luria-Nebraska Neuropsychological Battery (LNNB) task scores: (a) conventional LNNB measurements; (b) the image-derived LNNB measurements.

Author Contributions: Conceptualization, D.C. and E.C.; Data curation, D.C.; Formal analysis, D.C. and E.C.; Methodology, D.C. and E.C.; Software, D.C.; Supervision, D.P., R.G., R.W., D.R.S., R.G.L. and M.K.H.; Writing—original draft, D.C. and E.C.; Writing—review & editing, D.C., R.G.L., M.K.H. and E.C. All authors have read and agreed to the published version of the manuscript.

Funding: EC was supported by the National Institute of Environmental Health Science (NIEHS): R01ES032242, 5U2CES026555-03, and P30ES023515.

Informed Consent Statement: Informed consent was obtained from all subjects involved in the study.

Conflicts of Interest: The authors declare no conflict of interest.

References

1. Deng, S.; Li, W.G.; Ding, J.; Wu, J.; Zhang, Y.; Li, F.; Shen, X. Understanding the mechanisms of cognitive impairments in developmental coordination disorder. *Pediatr. Res.* **2014**, *75*, 210–216. [CrossRef]
2. Zwicker, J.G.; Missiuna, C.; Harris, S.R.; Boyd, L.A. Developmental coordination disorder: A review and update. *Eur. J. Paediatr. Neurol.* **2012**, *16*, 573–581. [CrossRef] [PubMed]
3. Geuze, R.H. Postural control in children with developmental coordination disorder. *Neural. Plast.* **2005**, *12*, 183–196, discussion 263–272. [CrossRef] [PubMed]
4. Abdelkarim, O.; Ammar, A.; Chtourou, H.; Wagner, M.; Knisel, E.; Hökelmann, A.; Bös, K. Relationship between motor and cognitive learning abilities among primary school-aged children. *Alex. J. Med.* **2017**, *53*, 325–331. [CrossRef]
5. Levit-Binnun, N.; Davidovitch, M.; Golland, Y. Sensory and motor secondary symptoms as indicators of brain vulnerability. *J. Neurodev. Disord.* **2013**, *5*, 26. [CrossRef]
6. Van der Fels, I.M.; Te Wierike, S.C.; Hartman, E.; Elferink-Gemser, M.T.; Smith, J.; Visscher, C. The relationship between motor skills and cognitive skills in 4–16 year old typically developing children: A systematic review. *J. Sci. Med. Sport.* **2015**, *18*, 697–703. [CrossRef]
7. Van Dyck, D.; Deconinck, N.; Aeby, A.; Baijot, S.; Coquelet, N.; Trotta, N.; Rovai, A.; Goldman, S.; Urbain, C.; Wens, V.; et al. Resting-state functional brain connectivity is related to subsequent procedural learning skills in school-aged children. *Neuroimage* **2021**, *240*, 118368. [CrossRef]
8. Bonacina, S.; Huang, S.; White-Schwoch, T.; Krizman, J.; Nicol, T.; Kraus, N. Rhythm, reading, and sound processing in the brain in preschool children. *NPJ Sci. Learn.* **2021**, *6*, 20. [CrossRef]
9. Fortin, C.; Barlaam, F.; Vaugoyeau, M.; Assaiante, C. Neurodevelopment of Posture-movement Coordination from Late Childhood to Adulthood as Assessed from Bimanual Load-lifting Task: An Event-related Potential Study. *Neuroscience* **2021**, *457*, 125–138. [CrossRef]
10. Golden, C.J. Identification of specific neurological disorders using double discrimination scales derived from the standardized Luria neuropsychological battery. *Int. J. Neurosci.* **1979**, *10*, 51–56. [CrossRef]
11. Lucchini, R.G.; Guazzetti, S.; Zoni, S.; Donna, F.; Peter, S.; Zacco, A.; Salmistraro, M.; Bontempi, E.; Zimmerman, N.J.; Smith, D.R. Tremor, olfactory and motor changes in Italian adolescents exposed to historical ferro-manganese emission. *Neurotoxicology* **2012**, *33*, 687–696. [CrossRef]
12. Zoni, S.; Albini, E.; Lucchini, R. Neuropsychological testing for the assessment of manganese neurotoxicity: A review and a proposal. *Am. J. Ind. Med.* **2007**, *50*, 812–830. [CrossRef]
13. Golden, C.J.; Hammeke, T.A.; Purisch, A.D. *The Luria-Nebraska Neuropsychological Battery: Manual*; Western Psychological Services; John Wiley & Sons: Los Angeles, CA, USA, 1980.
14. Pauwels, K.; Van Hulle, M.M. Optic flow from unstable sequences through local velocity constancy maximization. *Image Vis. Comput.* **2009**, *27*, 579–587. [CrossRef]
15. Duric, Z.; Rosenfeld, A. Shooting a smooth video with a shaky camera. *Mach. Vis. Appl.* **2003**, *13*, 303–313. [CrossRef]
16. Adiv, G. Determining three-dimensional motion and structure from optical flow generated by several moving objects. *IEEE Trans. Pattern Anal. Mach. Intell.* **1985**, *7*, 384–401. [CrossRef]
17. Gautama, T.; van Hulle, M. A phase-based approach to the estimation of the optical flow field using spatial filtering. *IEEE Trans. Neural Netw.* **2002**, *13*, 1127–1136. [CrossRef]
18. Barron, J.; Fleet, D.; Beauchemin, S. Performance of optical flow techniques. *Int. J. Comput. Vis.* **1994**, *12*, 43–77. [CrossRef]
19. Corbo, D.; Orban, G.A. Observing Others Speak or Sing Activates Spt and Neighboring Parietal Cortex. *J. Cogn. Neurosci.* **2017**, *29*, 1002–1021. [CrossRef]
20. Elliott, D.; Heath, M.; Binsted, G.; Ricker, K.L.; Roy, E.A.; Chua, R. Goal-Directed Aiming: Correcting a Force-Specification Error with the Right and Left Hands. *J. Mot. Behav.* **1999**, *31*, 309–324. [CrossRef]
21. Carson, R.G.; Goodman, D.; Chua, R.; Elliott, D. Asymmetries in the regulation of visually guided aiming. *J. Mot. Behav.* **1993**, *25*, 21–32. [CrossRef]
22. Goble, D.J.; Lewis, C.A.; Brown, S.H. Upper limb asymmetries in the utilization of proprioceptive feedback. *Exp. Brain. Res.* **2006**, *168*, 307–311. [CrossRef]
23. Goble, D.J.; Brown, S.H. Task-dependent asymmetries in the utilization of proprioceptive feedback for goal-directed movement. *Exp. Brain Res.* **2007**, *180*, 693–704. [CrossRef]
24. Goble, D.J.; Brown, S.H. Upper limb asymmetries in the perception of proprioceptively determined dynamic position sense. *J. Exp. Psychol. Hum. Percept. Perform.* **2010**, *36*, 768–775. [CrossRef]

25. Bravi, R.; Cohen, E.J.; Martinelli, A.; Gottard, A.; Minciacchi, D. When Non-Dominant Is Better than Dominant: Kinesiotape Modulates Asymmetries in Timed Performance during a Synchronization-Continuation Task. *Front. Integr. Neurosci.* **2017**, *11*, 21. [CrossRef]
26. Ozcan, A.; Tulum, Z.; Pinar, L.; Başkurt, F. Comparison of pressure pain threshold, grip strength, dexterity and touch pressure of dominant and non-dominant hands within and between right-and left-handed subjects. *J. Korean Med. Sci.* **2004**, *19*, 874–878. [CrossRef]
27. De Gennaro, L.; Cristiani, R.; Bertini, M.; Curcio, G.; Ferrara, M.; Fratello, F.; Romei, V.; Rossini, P.M. Handedness is mainly associated with an asymmetry of corticospinal excitability and not of transcallosal inhibition. *Clin. Neurophysiol.* **2004**, *115*, 1305–1312. [CrossRef]
28. Adam, A.; De Luca, C.J.; Erim, Z. Hand dominance and motor unit firing behavior. *J. Neurophysiol.* **1998**, *80*, 1373–1382. [CrossRef]

MDPI AG
Grosspeteranlage 5
4052 Basel
Switzerland
Tel.: +41 61 683 77 34

Brain Sciences Editorial Office
E-mail: brainsci@mdpi.com
www.mdpi.com/journal/brainsci

www.ingramcontent.com/pod-product-compliance
Lightning Source LLC
Chambersburg PA
CBHW041407210926
51850CB00013B/3205